Practical Control System Design

Practical Control System Design

Real World Designs Implemented on Emulated Industrial Systems

Adrian Medioli
University of Newcastle
Australia

Graham Goodwin
University of Newcastle
Australia

WILEY

This edition first published 2024

Registered Offices
John Wiley & Sons, Inc., 111 River Street, Hoboken, NJ 07030, USA
John Wiley & Sons Ltd, The Atrium, Southern Gate, Chichester, West Sussex, PO19 8SQ, UK

For details of our global editorial offices, customer services, and more information about Wiley products visit us at www.wiley.com.

Wiley also publishes its books in a variety of electronic formats and by print-on-demand. Some content that appears in standard print versions of this book may not be available in other formats.

Library of Congress Cataloging-in-Publication Data Applied for
Hardback ISBN: 9781394168187

Cover Design: Wiley
Cover Images: © Chuyuss/Shutterstock; Oleksandr Fagaulin/Shutterstock; Yasin; Jarmoluk/Pixabay hm/UnSplash; Abdulaziz Alfaleh/UnSplash

Set in 9.5/12.5pt STIXTwoText by Straive, Chennai, India
Printed and bound by CPI Group (UK) Ltd, Croydon, CR0 4YY

C9781394168187_090124

Contents

Preface

This book provides an introduction to control system design. It would be suitable as the basis of a first or second course in control. It has also been written as a refresher course for graduate engineers who wish to 'up-skill' in the area of practical control system design.

The book has been inspired and informed by the authors' industrial experience in control. Our goal has been to package that experience so that those reading this book and using the associated resources will acquire the tools and skills necessary to tackle real-world control engineering design problems.

The book draws upon many industrial projects conducted by the authors and associates. These projects are used as case studies throughout the book. The authors, belief is that the richest learning experience comes from testing ideas via 'hands-on' laboratory studies that closely reflect the real world. The case studies are organised in the form of virtual laboratories. Readers can explore the ideas at their own pace and to their own level of interest.

The case studies include the following:

- electromechanical servo systems
- fluid storage
- continuous steel casting
- rolling mill centre line gauge control
- rocket dynamics and control
- cross-directional control in paper machines
- audio quantisation
- wind power generation – mechanical aspects
- wind power generation – electrical aspects (including three-phase induction generator)
- boiler control.

Wherever possible, the models used in the case studies use physical dimensions and parameters from real-world systems.

The laboratories can be conducted 'on-line' without needing to access a physical laboratory. Thus, the laboratories would be well suited to remote education. Nonetheless, the laboratories have been designed so as to be as realistic as possible. Indeed sitting in front of the laboratory interface is essentially the same as sitting in an industrial control room with the following two caveats:

(i) It would be completely unrealistic for any one person to gain access to the range of physical systems studied here.
(ii) One can learn by one's mistakes on a virtual laboratory whereas, if such mistakes were ever made in practice, the consequences could be catastrophic.

For ease of use, the architectures of the various control laws have been provided. Readers can use these architectures as a starting point. In many cases, it is also possible to link the process simulations to a control law whose architecture is specified by the reader or by an instructor.[1] In either case, it is suggested that the reader begin with the given architectures to establish a starting point for further study.

The topics covered in the course include:

- modelling
- PID design
- dealing with delays
- impact of actuator limitations
- dealing with sensor limitations
- feedforward
- soft sensing
- dealing with nasty disturbances
- observers and Kalman filtering
- Bode sensitivity constraints
- dealing with periodic disturbances
- multivariable interactions
- sampled data systems
- dealing with resonant behaviour
- controlling systems which are open-loop unstable.

This book contains 30 chapters. Sixteen of these chapters provide a review of the necessary background theory, whilst the remaining 14 are devoted to testing the ideas on real-world (emulated) systems. The treatment of theoretical topics will be relatively brief since the key aim of the book is to study real-world control system design.

The authors' belief is that anybody who has understood the core concepts explained in the book will be well-equipped to tackle real-world control design problems.

Difficult choices need to be made when writing a book of this type. In particular, to present a detailed treatment of any of the topics covered in the theory sections would expand the book enormously. Hence, the driving philosophy has been to limit the presentation to the depth necessary to enable the reader to undertake the real-world designs. In other words, the goal has been to make the presentation 'as simple as possible, but not so simple as to lose important details'.

---

1 Exceptions are the Wind Power Laboratory and the Audio Laboratory which require specific architectures.

About the Authors

Adrian Medioli, BE, PhD, OStJ, MIEEE

Adrian Medioli was born in Newcastle, Australia, and completed a B.E. (Computer) in 1992 at the University of Newcastle. He spent 10 years as a senior systems engineer working on the automation and control of steel manufacturing processes including rolling mills, galvanising lines, batch annealing, gauge control and Sendzimir mills and two years as a director of Omni Automation Pty. Ltd., a control engineering and automation startup company. In 2008, he completed a Ph.D. in Electrical Engineering at the University of Newcastle. From 2008 to 2021, he was employed as a research academic in the ARC/PRC for Complex Dynamic Systems and Control at the University of Newcastle. In 2021, he joined Whitely Corporation where he worked on the implementation of Industry Four protocols. As a professional engineer Adrian has worked on many and varied projects which resulted in extensive experience in the areas of software development, team management, control system design and implementation to steel production and manufacturing. As a researcher, he has developed and implemented new techniques in optimisation and control. Some of his achievements include a new technique for maximal controllability of unstable systems using reduced complexity Model Predictive Control (MPC); development of tools for the classification of coal loader downtime causes; ambulance fluid deployment strategy optimisation; ambulance optimal distribution and rostering to better match demand; design and testing of novel control strategies with application to the development of an artificial pancreas for better blood glucose regulation in diabetes sufferers; and re-design and development of a suite of virtual laboratories for the teaching of control system design. At present, Dr. Medioli's focus is on his research interests, including optimisation-based control strategies such as MPC stabilisation and modelling with particular emphasis on unstable systems; virtual laboratory development; and the application of control concepts and design to medical problems. He is a Member of IEEE and holds a Certificate IV in vocational training and assessment with accreditation under the Australian Quality Training Framework (AQTF).

Graham Goodwin, BSc, BE, PhD, FRS, FIEEE, Hon.FIE.Aust., FTSE, FAA

Graham Goodwin is an Emeritus Laureate Professor of Electrical Engineering at the University of Newcastle. His education includes B.Sc., B.E., and Ph.D. from the University of New South Wales. In 2010, he was awarded the IEEE Control Systems Field Award, in 2011 the ACA Wook Hyuan Kwon Education Award and in 2013 he received the Rufus T. Oldenburger Medal from the American Society of Mechanical Engineers. He was twice awarded the International Federation of Automatic Control triennial Best Engineering Text Book Prize. In 2021, he was awarded the American Control Council John Ragazzini Education Award. He is a Fellow of IEEE; an Honorary Fellow of Institute of Engineers, Australia; a Fellow of the International Federation of Automatic

Control; a Fellow of the Australian Academy of Science; a Fellow of the Australian Academy of Technology, Science and Engineering; a Member of the International Statistical Institute; a Fellow of the Royal Society, London, and a Foreign Member of the Royal Swedish Academy of Sciences. In 2021, he was recognised by the Australian Government by becoming an Officer in the General Division of the Order of Australia. He holds Honorary Doctorates from Lund Institute of Technology, Sweden, and the Technion Israel. He is the co-author of 10 previous books, four edited books, and five hundred papers. He holds 16 International Patents covering rolling mill technology, telecommunications, mine planning, and mineral exploration. His current research interests include power electronics, boiler control systems and management of type 1 diabetes.

Acknowledgements

This book is the culmination of many years of work by many individuals. An abbreviated list of those who worked on developing the laboratories includes Adrian Basitani, Diego Carrasco, Ross Cockwell, Sam Crisafulli, David Ferris, Arie Feuer, Richard Middleton, Galina Mirzaeva, Rob Peirce, Osvaldo Rojas, Frank Sabora, Maria Seron, Eduardo Silva, Bob Skelton, Peter Stepien, James Welsh, Peter Wellstead, and Mark West. Financial support from the Australian Research Council plus the School of Engineering and the College of Engineering, Science and Environment at the University of Newcastle is gratefully acknowledged. The book was typed by Jayne Disney.

Adrian Medioli
Graham Goodwin

About the Companion Website

This book is accompanied by a companion website.

www.wiley.com/go/medioli/practicalcontrolsystemdesign

This website includes:

- Solutions
- Software
- Solutions to revision questions

Part I

Modelling and Analysis of Linear Systems

This part begins the journey through the fascinating world of control system design. Chapter 1 provides a brief historical overview of control. Chapter 2 explains why control is a quintessential example of an inverse problem. Chapter 3 provides an introduction to modelling of physical systems. Chapter 4 describes basic mathematical tools used to analyse linear systems. Chapter 5 contains the first virtual laboratory. A simple door-closing mechanism is used to illustrate basic principles of model building using data collected from a system. Many of the ideas covered in this part of the book would normally be presented in an introductory course on control. Hence, the development will focus on the key concepts that will be used in the real-world designs covered in the sequel.

Practical Control System Design: Real World Designs implemented on Emulated Industrial Systems,
First Edition. Adrian Medioli and Graham Goodwin.

Companion website: www.wiley.com/go/medioli/practicalcontrolsystemdesign

1

Introduction to Control System Design

1.1 Introduction

Control is the hidden yet ubiquitous technology [1]. Essentially, no piece of equipment in the modern world would operate satisfactorily without some form of feedback control. Indeed, beyond physical devices such as automobiles or aircraft, control lies at the core of many other systems, e.g. economic systems and even human psychology.

A fascinating overview of control is contained in reference [2] given in Further Reading for this chapter.

The essence of feedback control is that one wishes to act on a system so that it performs in some desirable fashion. Often the goals are expressed as 'set-points' for the outputs, sometimes called 'Process Variables' (PVs). Action on the system is achieved by changing the inputs, sometimes called 'Manipulated Variables' (MVs). Examples occur in every realm. Just a few examples to think about are:

- making a small company profitable
- ensuring a national economy has a low inflation rate
- delivering electricity to a community or country
- cooling a house
- landing an aircraft
- producing a desired product in a chemical plant
- achieving a personal fitness goal
- designing an adaptive cruise controller for a car
- achieving a certain level of proficiency in a trade or discipline
- taking a drug so as to control a chronic disease.

For each of the above examples, the reader may like to think about what the PVs and MVs could be.

The tasks listed above would be easy if

(i) one knew the exact impact that changing the MVs have on the PVs (i.e. one has an exact model), and
(ii) there were no external disturbances.

Alas, in practice, neither of the above requirements is met. Thus, the art of control system design is to build a 'control' system that achieves the desired goals, as closely as possible, in the face of uncertainty in both the model and external disturbances.

Practical Control System Design: Real World Designs implemented on Emulated Industrial Systems,
First Edition. Adrian Medioli and Graham Goodwin.

Companion website: www.wiley.com/go/medioli/practicalcontrolsystemdesign

1.2 A Brief History of Control

Control underpins the operation of almost every piece of modern equipment, including automobiles, aircraft, chemical plants, national economies and medical devices. It is thus not surprising that people have been thinking about control for hundreds of years. Indeed, Ref.[3] describes a feedback mechanism aimed at regulating the accuracy of a water clock. This device was developed several thousand years ago.

A very well-known application of feedback control was the regulation of the speed of steam engines in the eighteenth-century CE. This was the work of Watt and colleagues and was an enabling technology in the industrial revolution. These speed governors worked amazingly well but were known to sometimes exhibit oscillatory behaviour. James Clarke Maxwell, who is famous for his work on electromagnetic theory, provided a theoretical description in (1868) for this in terms of the properties of ordinary differential equations in the time domain [4].

An important early practical discovery was the use of 'integral action' to achieve zero steady-state error. This was introduced in 1790 in a governor designed by the Perrier brothers [2].

Turning to the process industries, control is an essential technology in all chemical processes. Indeed, modern chemical plants are 'littered' with controllers. The most common controller is a device known as a PID controller. The letters PID stand for Proportional, Integral and Derivative feedback. Though simple, such controllers are incredibly robust and achieve remarkable performance in many cases.

Another remarkable area where control is essential is in powered flight. Indeed, the Wright Brothers, aircraft critically depended on the pilot adjusting the wing surfaces to maintain stability of flight [5].

Moving to an entirely different area, feedback control is a key enabling technology in the telephone. The telephone was invented by Alexander Graham Bell in 1876. However, as the number of repeaters was increased internally generated noise and distortion became intolerable. A major breakthrough was made by Harold Black at Bell Laboratories in 1927 with the development of the negative-feedback amplifier [6]. The essential idea was to use feedback around a high-gain amplifier to reduce the impact of noise on the output signal. In these feedback amplifiers, instability was again sometimes observed (appearing as a 'whistle'). A remarkable engineer, Harry Nyquist, studied the problem of stability at Bell Laboratories [7]. He moved away from the time domain and instead studied how sinusoidal signals propagated around a feedback loop. This was the initial step in using frequency domain analysis of feedback systems.

The frequency domain insights were a huge breakthrough. This was built upon by many people. For example, major contributions were made by Bode in 1940. He recognised that complex variable theory could be used to give deep insights into the frequency domain analysis of feedback systems [8]. In a truly remarkable result, he proved that the integral of log sensitivity with respect to frequency was constant. (Actually zero for an open-loop stable system.) This meant that the action of feedback was simply to shift the impact of disturbances around in the frequency domain. In particular, reducing sensitivity to disturbances at low frequencies would necessarily be accompanied by an increase in sensitivity to disturbances at higher frequencies. This idea underpins many systems used in audio quantisation, industrial electronics and process control.

The above body of work, which emphasises frequency domain concepts, is often classified as part of the 'Classical Control' era. This circle of ideas dominated until the 1960s when a step back to time domain ideas was made. A major contribution was the work of Pontryagin, Boltyanskii, Gamkrelidze and Mishechenko (1962) in the USSR on 'optimal' control [9]. A related stream of research was the development of Dynamic Programming by Richard Bellman in the United States [10]. The linear quadratic version of these ideas gave a particularly elegant solution as shown by Bellman, Kalman, Bucy and many others [11]. A dual problem turns out to be that of estimating the 'state' of a system from output measurements. Algorithms which carry out this process are commonly called 'Observers'. Major contributions to the subject of observers were made by Stratonovich, Luenberger, Kalman, Bucy, Wonham and many others [12–14].

The latter ideas are sometimes described as being part of the 'Modern Control' era. In the late 1980s a return was made again to frequency domain ideas. This included work on robust linear control by Doyle, Glover, Khargonekar and Francis [15] plus many others. It was recognised that these new, time domain, results gave elegant solutions. Not surprisingly, the combination of time domain and frequency domain gives the most powerful view. Thus, the modern view of control involves a mixing of both time and frequency domain ideas.

1.3 Digital Control

Early control laws were implemented in analogue form using hydraulic or pneumatic equipment. An important development that occurred in control was the introduction of digital feedback using computers. Early work emphasised the difference between analogue control and digital control. This led to a new set of challenges in control system design. For example, linear analogue control loops are known to be stable if the roots of the characteristic polynomial lie in the open-loop left-half complex plane, whereas digital control loops are stable if the roots of the characteristic polynomial (expressed in terms of the forward shift operator) lie in the open unit disc. An entirely new machinery was developed to study such problems including shift operators and Z-transforms [16].

As with the time-frequency domain divide, the greatest insights arise when both continuous and discrete viewpoints are combined. In support of this perspective, it has been shown that analogue and digital control loops can be studied under a common framework using the divided difference operator [17]. Using this operator, analogue control can be viewed as a special case of digital control in which the sample period is made arbitrarily small. This line of work has a long history in mathematics, but its relevance to control engineering was brought to the notice of control engineers in a series of papers by one of the authors of the current book (Goodwin) together with co-workers Middleton, Poor, Feuer, Yuz and many others [18].

1.4 Our Selection

The brief history given above gives just a preliminary 'taste of' the huge range of ideas that exist in modern control engineering.

Of course, there are many other ideas and concepts in modern control science. For example, the interested reader is referred to the excellent survey by Åström and Kumar [2]. Other topics of great importance include Adaptive Control, Nonlinear Control, Control of Discrete Event Systems, Model Predictive Control and many others.

It would be impossible in a single book to do justice to even a small fraction of this huge spectrum of ideas. Thus, the focus of the book will be a relatively small subset. The choice is driven by the author's experience in using control to solve real-world control problems in industry. This leads to a particularly simple path.

1.5 Thinking Outside the Box

A recurring theme in the real-world control problems described in this book is that one often needs to 'think outside the box'. Indeed, it is a belief of the authors that, on occasions, excessive emphasis is devoted to the tuning of control law parameters within a given control architecture rather than focusing on the architecture itself.

Undoubtedly, getting the right tuning is important and thus significant space will be devoted to this topic in this book. However, the biggest improvements in control performance often arise from changing the underlying architecture of the system. This can be achieved either by introducing new sensors, by changing the physical structure of the system or by introducing alternative or secondary actuators. As a simple example, pure delays in measured outputs fundamentally limit the achievable performance. (Imagine trying to steer a car with your eyes closed and somebody telling you where the car sits on the road but with a 10-seconds delay!) In some cases, the problem can be remedied by introducing a new sensor. A very nice example of this will be given in Chapter 13. There, a device known as a BISRA gauge will be studied. This device is used for centre line thickness control in rolling mills. Other examples of the benefits of using alternative sensors or actuators will arise in other real-world problems.

1.6 How the Book Is Organised

Chapters 2–4 contain basic modelling and mathematical background necessary to provide a common 'language' that can be used to discuss feedback control problems. It is not surprising that control has its own 'language'. After all, one would not want to keep having to explain what 'PID' means. Instead, it is better to simply view this as part of the control lexicon. From Chapter 5 onwards, the presentation will 'shift gears' and control ideas and language will be utilised in the context of real-world problems of the type illustrated in Figure 1.1. Extensive use will be made of the Virtual Laboratories to motivate and illustrate the evolving circle of ideas. The sequence of laboratories has been organised so that the ideas evolve in a 'structured' fashion proceeding from relatively simple ideas to more complex concepts. New concepts and methods are introduced, if and only if, they are needed for the real-world problems. The reader may feel a little frustrated that Chapters 1–4 are largely of a theoretical nature. However, some basic background is necessary before one can jump into a practical problem. Readers who have background in basic control concepts could go straight to Chapter 5.

The book has been divided into a number of parts. The reason for doing this is to mark the transition in ideas from simple to more complex.

1.7 Testing the Reader's Understanding

Throughout the book, the reader is asked to perform various experiments and to answer some revision questions. At times, these questions are deliberately 'open-ended' so as to challenge the reader to think outside the bounds of the material presented in the book.

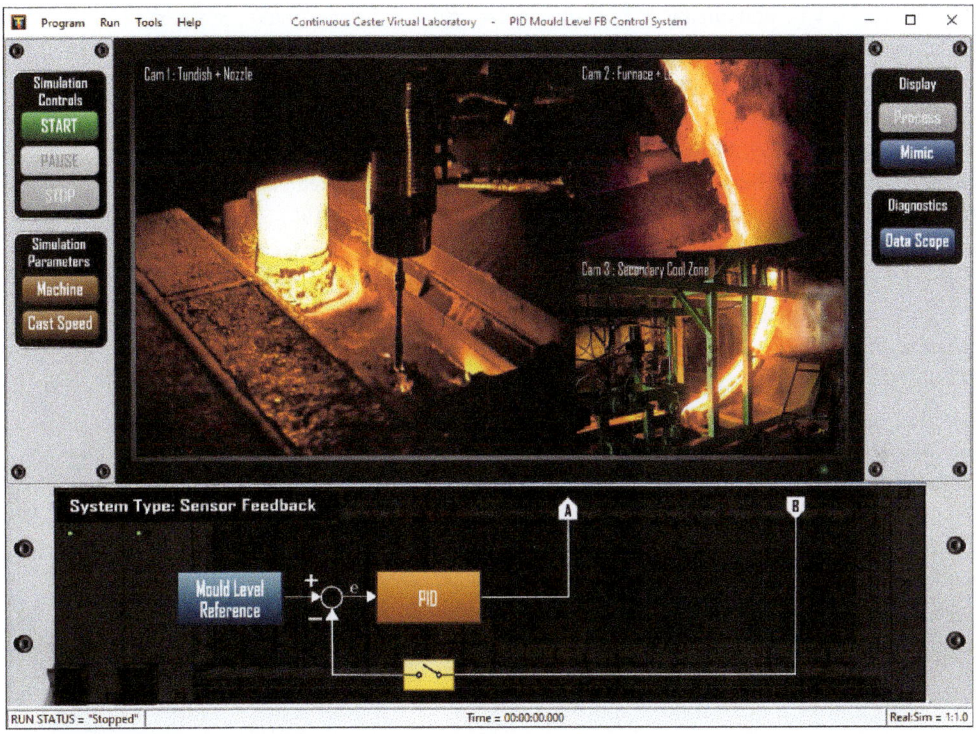

Figure 1.1 Screenshot of Continuous Caster Laboratory.

1.8 Revision Questions

1. List five devices in your possession which depend upon feedback for their operation.
2. The term 'feedback' is often used when somebody provides a commentary on somebody's work. What are the measurements and manipulated variables?
3. Use your answer to Question 2 to illustrate why not all feedback is helpful.

Further Reading

1 K.J. Åström, "Automatic control - the hidden technology", in P. Frend (ed.) *Advances in Control - Highlights of ECC'99* (pp. 1–29), European Control Council, Springer–Verlag.
2 K.J. Åström and P.R. Kumar, "Control a perspective", *Automatica*, 50:3–43, 2014.
3 O. Mayr, *The origins of feedback control*, MIT Press, 1970.
4 J.C. Maxwell, "On governors", *Proceedings of the Royal Society of London*, 16:270–283, 1868.
5 C.S. Draper, "Flight control", *Journal Royal Aeronautical Society*, 59:449–477, 1955.
6 H.S. Black, "Stabilizing feedback amplifies", *Bell System Technical Journal*, 13:1–18, 1934.
7 H. Nyquist, "Regeneration theory", *Bell System Technical Journal*, 11:126–147, 1932.
8 H. Bode, *Network analysis and feedback amplifier design*, Van Nostrand, New York, 1945.

9 L.S. Pontryagin, V.G. Boltyanskii, R.V. Gamkrelidze and E.F. Mishechenko, *Mathematical theory of optimal processes*, John Wiley & Sons, Inc., New York, 1962.

10 R.E. Bellman, *Dynamic programming*, Princeton University Press, Princeton, New Jersey, 1957.

11 R.E. Kalman,"Contributions to the theory of optimal control", *Boletin de la Sociedad Matematica Mexicana*, 5:102–119, 1956.

12 R.E. Kalman and R.S. Bucy, "New results in linear filtering and prediction theory", *Transactions of the ASME, Series D, Journal of Basic Engineering*, 83(1):95–108, 1961.

13 D.G. Luenberger,"Observing the state of a linear system", *IEEE Transactions on Military Electronics*, 8(2):74–80, 1964.

14 W. Wonham,"Dynamic observers - geometric theory", *IEEE Transactions on Automatic Control*, 15(2):258–259, 1970.

15 J. Doyle, K. Glover, P. Khargonedor and B. Francis, "State- space solutions to standard H_2 and H_∞ control problems", *IEEE Transactions on Automatic Control*, 34(8):831–847, 1989.

16 J.R. Ragazzini and L.A. Zadeh, "The analysis of sampled-data systems", *Transactions of the American Institute of Electrical Engineers, Part II: Applications and Industry*, 71(5):225–234, 1952.

17 G.C. Goodwin, R.H. Middleton and H.V. Poor, "High-speed digital signal processing and control, *Proceedings of the IEEE*, 80:240–259, 1992.

18 G.C. Goodwin, J.C. Agüero, M.E.C. Garridos, M.E. Salgado and J.I. Yuz, "Sampling and sampled-data models. The interface between the continuous world and digital algorithms", *IEEE Control Systems Magazine*, 33(5):34–53, 2013.

2

Control as an Inverse Problem

2.1 Introduction

This chapter takes a first look at control system design. The presentation begins by viewing control as a simple example of an inverse problem. Indeed, inversion lies at the core of all control system design procedures and is thus a unifying theme.

2.2 The Elements

The most elementary control problem is shown in Figure 2.1.

With reference to Figure 2.1, say that one wishes to act on the process, P, via the input u, so as to cause the output, y, to reach (as closely as possible) a given 'set-point', y^*, in the presence of disturbances, d.

At a conceptual level, the system of Figure 2.1 can be described as follows:

$$y = Pu + d. \tag{2.1}$$

For the moment, the reader can assume that y, P, u and d are all simple scalar real numbers.

Say that it is desired that, y be equal to y^*, then it is clear from (2.1) that the following choice for the input u will 'do the trick'.

$$u = P^{-1}(y^* - d). \tag{2.2}$$

From Eq. (2.2) we note that it contains an inverse and it is evident that the following information is needed:

- knowledge of P,
- measurement of d,
- and the ability to sensibly invert P.

The astute reader will immediately raise serious objections to the model (2.1) and point out that the real-world problems described in the introduction are never so simple that they could possibly be described by a simple algebraic gain. This is, of course, true. However, remarkably, it will be shown later that a model consisting of a simple algebraic gain covers a huge range of real cases provided one uses a simple trick of thinking about the model in a particular way. For those readers familiar with eigenvalues and eigenvectors from linear algebra, the trick is to use this idea. Other readers may be familiar with frequency response which, again, uses the same concept.

Practical Control System Design: Real World Designs implemented on Emulated Industrial Systems,
First Edition. Adrian Medioli and Graham Goodwin.

Companion website: www.wiley.com/go/medioli/practicalcontrolsystemdesign

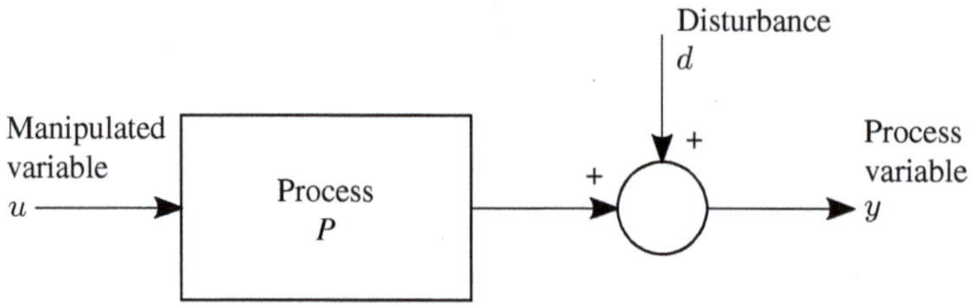

Figure 2.1 Schematic of a simple single-input single-output system.

2.3 Using Eigenvalue Analysis

To illustrate the idea of how complicated systems can be represented by simpler (scalar) systems, a brief diversion is taken to use the idea of eigenvalues to simplify a multi-input, multi-output system. Readers who are not familiar with eigenvalues need not read the remainder of this section but can immediately jump to Section 2.4.

Consider a system described by a linear (non-dynamic) matrix operation

$$Y = MU, \tag{2.3}$$

where Y is an $n \times 1$ output vector and U is an $n \times 1$ input vector, M is an $n \times n$ matrix of reals.

Say that one can diagonalise the Matrix operator M by a transformation of the form:

$$SMS^{-1} = \Lambda, \tag{2.4}$$

where S is a matrix 'diagonaliser' and

$$\Lambda = diag(\lambda_1, \ldots, \lambda_n), \tag{2.5}$$

where $\lambda_1, \ldots, \lambda_n$ are the eigenvalues. Substituting (2.4) into (2.3) yields

$$Y = S^{-1}\Lambda SU \tag{2.6}$$

or

$$SY = \Lambda SU. \tag{2.7}$$

Hence, instead of using the original input vector U and the original output vector Y, one might use instead

$$\overline{Y} = SY, \tag{2.8}$$

$$\overline{U} = SU. \tag{2.9}$$

Then, the model becomes

$$\overline{Y} = \Lambda\overline{U}. \tag{2.10}$$

However, this is nothing more than a set of scalar gain systems, i.e.

$$\overline{Y}_i = \lambda_i\overline{U}_i; \quad i = 1, \ldots, n. \tag{2.11}$$

The book will return to this idea frequently. Indeed, this is the core principle underlying Laplace transforms and frequency domain analysis. Specifically, the Laplace operator turns out to be the eigenvalue of systems described by linear differential equations. For the moment, it suffices to consider the simple scalar algebraic model of Eq. (2.1). However, the reader will be encouraged to know that these simple algebraic ideas will extend rather easily to systems described by high-order linear differential equations, as shown later in Chapter 4. Later in the book, specifically in Section 21.6, an industrial application of the above idea will be described that directly uses the core idea given in Eq. (2.11).

2.4 The Effect of Process and Disturbance Errors

Another criticism of the solution (2.2) is that it ignores errors in the knowledge available about both the process P and of the disturbance d.

To illustrate this difficulty, say that it is believed that the model is P_0 and the disturbance is d_0, however, in reality, the real plant and real disturbance satisfy

$$P = P_0 + \Delta P, \tag{2.12}$$

$$d = d_0 + \Delta d, \tag{2.13}$$

where ΔP and Δd are errors.

Equation (2.1) then becomes

$$y = [P_0 + \Delta P]\, u + [d_0 + \Delta d]\,. \tag{2.14}$$

Alas, the control law (2.2) can only utilise what is believed to be true, i.e. it is only feasible to set

$$u = P_0^{-1}\, [y^* - d_0]\,. \tag{2.15}$$

Substituting (2.15) into (2.14) yields

$$y = [P_0 + \Delta P]\, [P_0^{-1}\, (y^* - d_0)] + (d_0 + \Delta d)\,. \tag{2.16}$$

It can then be seen that what is actually achieved is

$$y = y^* + \left[\frac{\Delta P}{P_0}\right] y^* - \left[\frac{\Delta P}{P_0}\right] d_0 + \Delta d. \tag{2.17}$$

Then

$$y - y^* = \left[\frac{\Delta P}{P_0}\right] y^* - \left[\frac{\Delta P}{P_0}\right] d_0 + \Delta d \tag{2.18}$$

or

$$\left[\frac{y - y^*}{y^*}\right] = \left[\frac{\Delta P}{P_0}\right] - \left[\frac{\Delta P}{P_0}\right] \frac{d_0}{y^*} + \frac{\Delta d}{y^*}. \tag{2.19}$$

The term $[(y - y^*)/y^*]$ is termed the 'relative tracking error'. It follows from (2.19) that the relative tracking error is proportional to the relative error in one's knowledge of the model, i.e. $[\Delta P/P_0]$, and the relative size of the error in the disturbance, i.e. $\Delta d/y^*$.

2.5 Feedback Control

The state of affairs described in Section 2.4 is rather unsatisfactory since relative errors in P and d are reflected, undiminished, in the output response. Fortunately, there is a rather simple way that one can reduce the impact of the relative errors $\Delta P/P_0$ and $\Delta d/y^*$ on the output. The key idea is to

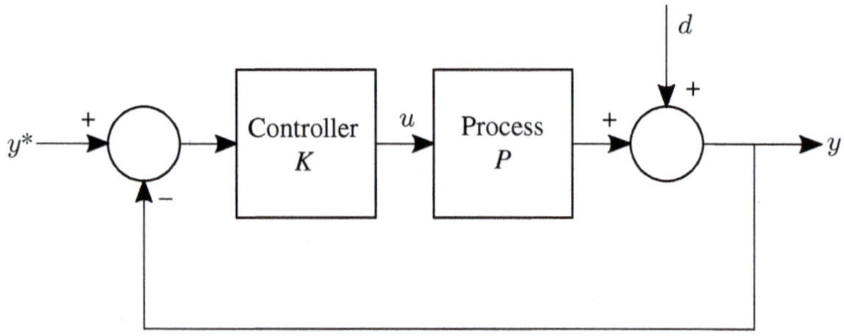

Figure 2.2 Schematic of simple feedback system.

use the 'magic of feedback', i.e. to replace the 'open-loop' control system shown in Figure 2.1 by a 'closed-loop' control system.

The basic idea inherent in feedback is to change the problem formulation via the use of feedback. The core idea is to compare the output, i.e. y, to what is desired, i.e. y^* and then to use the error $e = y^* - y$ to adjust u via a feedback 'gain' K.

This leads to a feedback (or closed-loop) structure of the type conceptually illustrated in Figure 2.2.

From Figure 2.2, the following algebraic relationships can be seen to hold. (Here it is again assumed that all quantities are simply scalar real numbers:)

$$y = Pu + d, \tag{2.20}$$

$$u = K\left[y^* - y\right]. \tag{2.21}$$

Hence,

$$y = PK\left[y^* - y\right] + d \tag{2.22}$$

or

$$y = \left[\frac{PK}{1+PK}\right]y^* + \left[\frac{1}{1+PK}\right]d. \tag{2.23}$$

Examining the two terms on the right-hand side of (2.23) leads to a rather magical conclusion, namely, if one could only make the gain K very large then, independent of P or d, the following result would be achieved.

$$y = y^*. \tag{2.24}$$

It thus follows that if one uses high gain feedback (i.e. K large), then this will cause y to equal y^* independent of knowledge of P or d.

In case the reader should feel that this is of esoteric theoretical interest only, it will be shown in Section 4.10 that almost every practical control system uses (a special form of) infinite gain to achieve perfect tracking and disturbance rejection (for constant signals) without requiring knowledge of the true model of the system. This goes under the heading of 'integral action'.

2.6 The Effect of Measurement Noise

Another type of error that is frequently encountered in real systems is measurement error. Indeed, the nature of sensors is such that the information they collect and communicate is usually corrupted by measurement errors and noise. This means that the measurements will have 'noise' superimposed on the information that is measured. Also, sensors can deliver inaccurate results

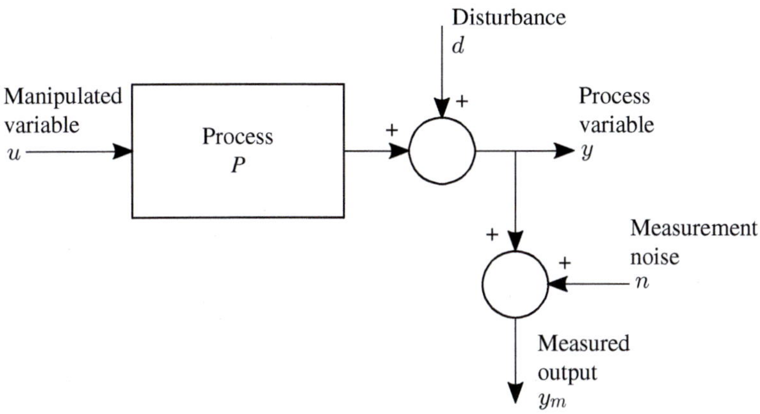

Figure 2.3 Schematic of a simple single-input single-output system with measurement noise.

for many reasons, including the inherent difficulty of measuring certain variables. These errors can be lumped into a measurement 'noise' term, *n*.

Figure 2.3 shows a simple process schematic, but now with measurement noise, *n*, shown explicitly. Note that the output that is measured is described by the following equation:

$$y_m = Pu + d + n, \tag{2.25}$$

where n is the measurement noise and y_m is the measured noisy output of the system.

The effect of noise on the simple feedback system can now be analysed.

A feedback system with both a disturbance, d, and measurement noise, n, is shown in Figure 2.4. From Figure 2.4, the following algebraic relationships follow:

$$y = Pu + d, \tag{2.26}$$

$$y_m = Pu + d + n, \tag{2.27}$$

$$u = K\left[y^* - y_m\right]. \tag{2.28}$$

Hence,

$$y = PK\left[y^* - y - n\right] + d \tag{2.29}$$

or

$$y = \left[\frac{PK}{1+PK}\right]y^* + \left[\frac{1}{1+PK}\right]d - \left[\frac{PK}{1+PK}\right]n. \tag{2.30}$$

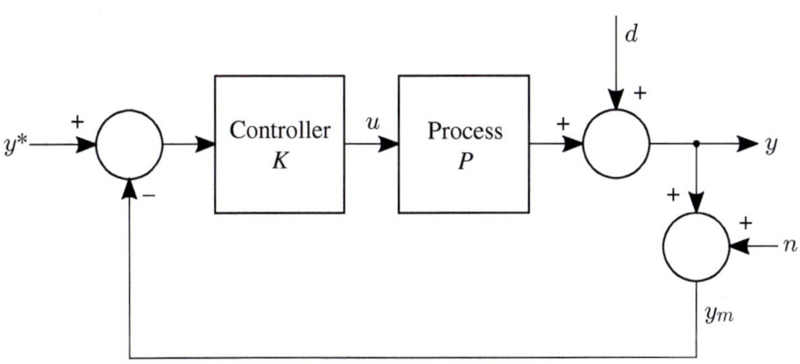

Figure 2.4 Schematic applying feedback control to a simple single-input single-output system with measurement noise.

Note that now, instead of the conclusion from Section 2.5, namely, that high gain feedback is sufficient to cause $y \simeq y^*$, now it is not true, in general, that $y \simeq y^*$ due to the presence of measurement noise. Specifically making K large 'impresses' the measurement noise, n, on the output response.

2.7 Sensitivity Functions

Examining Eq. (2.30) shows that the right-hand side depends on three key functions of the underlying quantities P and K. These functions play a key role in feedback control and they are given special names and symbols as follows:

$$\text{Loop Gain} : L = PK, \tag{2.31}$$

$$\text{Sensitivity Function} : S = \frac{1}{1+PK} = \frac{1}{1+L}, \tag{2.32}$$

$$\text{Complementary Sensitivity Function} : T = \frac{PK}{1+PK} = \frac{L}{1+L}. \tag{2.33}$$

Using these functions, then (2.30) can be rewritten as

$$y = Ty^* + Sd - Tn. \tag{2.34}$$

Equation (2.34) suggests that S and T play a core role in feedback control problems. Indeed, understanding S and T will turn out to be one of the central themes as the ideas in the book evolve.

2.8 Reducing the Impact of Disturbances and Model Error

It can be seen from (2.32) and (2.33), that simply making K large, causes $T \to 1$ and $S \to 0$. This is great news since we see from the first two terms on the right-hand side of (2.34) that, in the absence of noise, n, the result $y = y^*$ follows.

2.9 Impact of Measurement Noise

The third term in (2.34) shows a fundamental difficulty, namely, making K large, then $T \to 1$. This leads to the conclusion that the measurement noise is 'impressed' on the output, i.e. y, becomes 100% sensitive to measurement noise.

This leads to a fundamental design trade-off in feedback design, namely,

'One cannot *simultaneously* make the output of a closed-loop system insensitive to both disturbances and measurement noise.'

This idea will also be a recurring theme as the book evolves.

2.10 Other Useful Sensitivity Functions

In addition to the sensitivity functions S and T, one can also define

$$S_i = \frac{P}{1+PK}, \tag{2.35}$$

$$S_u = \frac{K}{1+PK}. \tag{2.36}$$

It is easily checked that S_i and S_u are, respectively, the functions relating an input disturbance to the output and the set point to the input. The four sensitivity functions S, T, S_i, S_u are sometimes referred to as the 'gang of four'. They capture all possible combinations of P and K in a closedloop.

Whilst all four sensitivity functions have a role to play in control system design, it often suffices to focus on S and T. Thus, emphasis will be given to these two sensitivity functions in the sequel.

2.11 Stability (A First Look)

Based on the arguments presented above, one might be tempted to say, provided measurement noise is small, then one can obtain a near-perfect feedback system provided one simply makes the feedback controller gain, K, very large. This is a reasonable hypothesis. Indeed, much of the detail of control system design is aimed at making K large whilst avoiding other difficulties. This begs the question, 'What are the, so called, other difficulties?' One of these difficulties is that high-loop gain will lead to large input signals which may not be feasible. Another issue is that of closed-loop stability.

To explain the idea of closed-loop stability, the reader may have witnessed the high-pitched whistling noise when a musical band has its speaker system too close to the microphone. What happens is that a certain note enters the microphone. It gets amplified and exits the speakers. It then returns to the microphone larger than it entered in the first place. This continues with the note increasing in volume until it saturates the amplifier or speakers. This phenomenon is commonly called, in popular culture, 'feedback'. It results from having a loop gain greater than one plus some other conditions related to phase. The latter issues will be central to much of the subsequent development. For the moment, it suffices to say that having a loop gain greater than one gives insensitivity to the model and disturbances but comes at the risk of having an excessively large input and/or a sustained oscillation. One remedy that readers may have noticed is that band members face the speakers in a different orientation so that the loop gain is reduced below one. This usually fixes the problem.

The moral of the above short anecdote is that high gain in a feedback loop is desirable from the point of view of achieving insensitivity to disturbances and model errors but comes at the risk of producing an excessively large input and/or self-sustained oscillations. The latter phenomenon is termed 'instability'. Much more will be said on this topic, as the book proceeds.

2.12 Sum of Sensitivity and Complementary Sensitivity

Examination of Sections 2.8 and 2.9 leads to an important conclusion, namely, that the sum of the Sensitivity Function, S, and Complementary Sensitivity Function, T is 1, i.e.

$$S + T = 1. \tag{2.37}$$

This fact has important consequences in feedback control.

Indeed, control system design can be thought of, at least in a simple context, of obtaining an appropriate trade-off between making S and T small whilst retaining closed-loop stability.

2.13 Revision Questions

1. Consider a simple non-dynamic system governed by the model:

$$y = u^{\frac{1}{3}}, \tag{2.38}$$

   where u is the input, y is the output.
   (a) What is an ideal (open-loop) control law that yields $y = y^*$?
   (b) Why do you think this control law may be a bad idea in practice?
2. (a) Calculate the eigenvalues of

$$M = \begin{bmatrix} 2 & 1 \\ 1 & 2 \end{bmatrix}. \tag{2.39}$$

   (b) What Transformation Matrix diagonalises M?
3. Consider the following non-dynamic system

$$y = (u + d)^2, \tag{2.40}$$

   where u is the input, y is the output and d is an input disturbance.
   (a) What is the ideal u that brings y to 9 assuming $d = 0$?
   (b) Say that $d = 1$. What impact does this have on the result in part (a)?
4. Consider again the system in Question 3. Rather than using inversion, as in Question 3, consider a feedback policy

$$u = -K(y - y^*). \tag{2.41}$$

   (a) Obtain an expression for u as a function of K, y^* and d for the closed-loop system. (Hint: Use the block diagram in 2.2 to obtain an expression for u.)
   (b) What value of u does this feedback solution give for y^* with K large?
   (c) Substitute your value for u with K large into the expression for y. What is the result?
5. Repeat Question 4 when there is measurement noise, i.e.

$$y = (u + d)^2 + n, \tag{2.42}$$

   where n is measurement noise.
6. In digital control, it is often necessary to convert an analogue signal, u^*, into a set of quantised values. A suitable quantiser for a scalar signal is given by

$$u = \arg\min_{r \in R} ||r - u^*||, \tag{2.43}$$

   where $||\ ||$ denotes a suitable norm. Say that R consists of the finite set $\{-1, 0, 1\}$. Describe the quantiser via 'if statements'.
7. Consider the nonlinear element described in Question 6. What difficulties do you think you might encounter trying to use an inverse function for the control law?

Further Reading

The following book adopts the view that control is at heart an inverse problem:

1 G.C. Goodwin, S.F. Graebe and M.E. Salgado, *Control system design*, Prentice Hall, 2000.

3

Introduction to Modelling

3.1 Introduction

Chapter 2 began our journey along the road to understanding control system design by describing a system as a simple gain, P. Indeed, based on understanding the gain P, it was possible to say many things about control and the action of feedback. Of course, nobody really believes that such a simple model describes real systems. This provides motivation to delve more deeply into the issue of building models for systems.

This chapter takes a first look at the problem of modelling real-world systems. Several simple examples will be used to illustrate the core ideas. More sophisticated examples will be presented throughout the book when the real-world case studies are described. In particular, Chapter 11 will explore the mechanical aspects of wind power generator, Chapter 28 will derive a nonlinear model for a rocket and Chapter 29 will derive a nonlinear model for a steam generator. These topics will be delayed since they do not impact the initial ideas presented in the book.

3.2 Physical Modelling

A common approach to developing a model for a process or system is to apply the fundamental laws of physics. These laws include conservation of mass, energy and momentum. These ideas are illustrated below with some simple examples. More complex examples will appear when some of the physical systems used for the laboratories are describeds (See, for example, Chapters 11, 28 and 29.)

3.2.1 Radio Telescope Positioning

Say one wishes to design a control system for positioning the azimuth of a radio telescope dish. A photograph of a real radio telescope is given in Figure 3.1. A schematic diagram of some of the components is shown in Figure 3.2. A natural starting point to build a model is to use the most basic physical law (known to most high school students), namely, Newton's Second Law ($F = ma$) which states that the rate of change of momentum (mv) is equal to the applied force. If mass is constant,

Practical Control System Design: Real World Designs implemented on Emulated Industrial Systems,
First Edition. Adrian Medioli and Graham Goodwin.

Companion website: www.wiley.com/go/medioli/practicalcontrolsystemdesign

Figure 3.1 Radio telescope – Parkes, NSW, Australia.

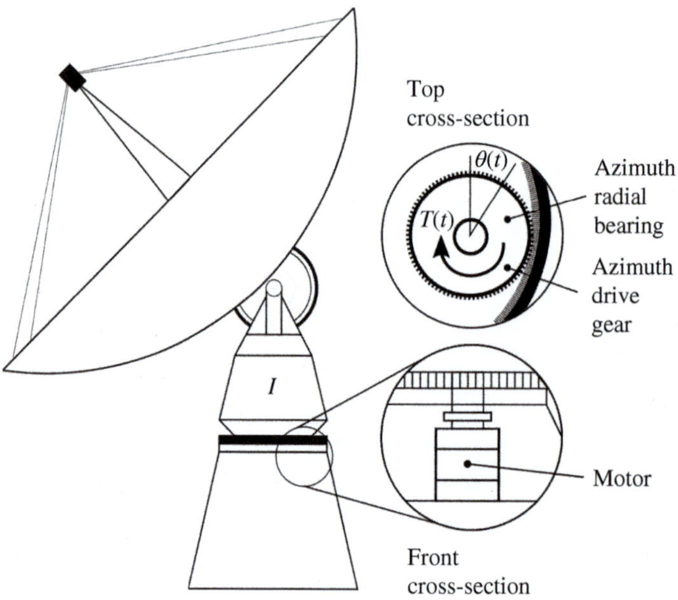

Figure 3.2 Radio telescope positioning.

then force is the product of mass and acceleration. In rotational dynamics, this is equivalent to torque being the product of inertia and angular acceleration, i.e.

$$T(t) = I\frac{d^2\theta(t)}{dt^2}, \tag{3.1}$$

where I is the inertia, $T(t)$ is the torque and $\theta(t)$ is the shaft angle. In this example, the input (or manipulated variable) is the torque and the output (or process variable) is the dish positioning shaft angle.

As a simple embellishment, one could assume the presence of some frictional torque that increases (linearly) with angular velocity. Then the model (3.1) becomes

$$T(t) = I\frac{d^2\theta(t)}{dt^2} + C\frac{d\theta(t)}{dt}, \tag{3.2}$$

with C denoting the friction coefficient.

Dividing by I and rearranging gives the following second-order differential equation model:

$$\frac{d^2\theta(t)}{dt^2} + \frac{C}{I}\frac{d\theta(t)}{dt} = \frac{1}{I}T(t). \tag{3.3}$$

3.2.2 Band-Pass Filter

As another illustration of the use of physical modelling, a model for a band-pass filter is developed. Such a system has many applications in signal processing including electrocardiogram (ECG) for extracting accurate heart rhythms. The corresponding electrical circuit is shown in Figure 3.3.

Kirchhoff's laws lead to the following model:

$$v_0(t) = L\frac{di(t)}{dt}. \tag{3.4}$$

$$\frac{v_i(t) - v_0(t)}{R_1} = i(t) + C\frac{dv_0(t)}{dt} + \frac{v_0(t)}{R_2}. \tag{3.5}$$

Also, the model for the inductance is

$$L\frac{di(t)}{dt} = v_0(t). \tag{3.6}$$

3.2.3 Inverted Pendulum

Some readers may be familiar with the idea of balancing a broom on the end of one's finger. A physical system that duplicates this idea, albeit in one dimension, is to balance a rod on a cart having wheels. This is often called an 'inverted pendulum'.

A sketch of an inverted pendulum is shown in Figure 3.4.

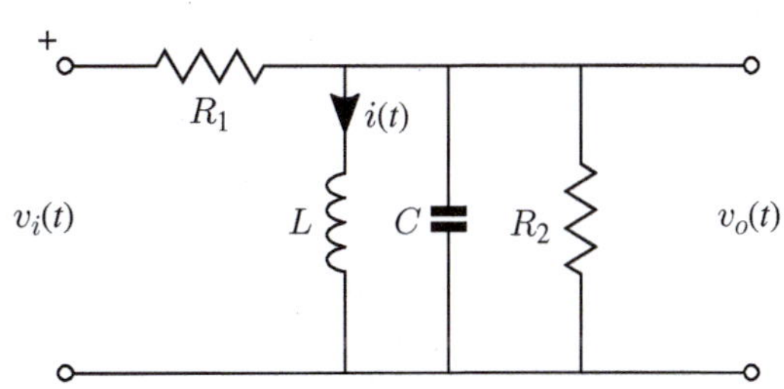

Figure 3.3 Band-pass filter electrical circuit.

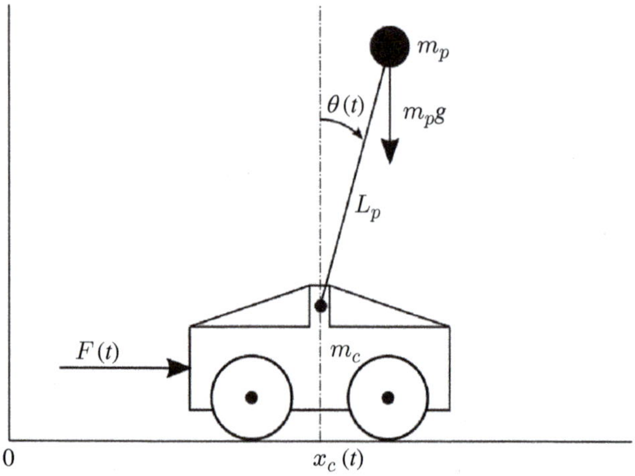

Figure 3.4 Inverted pendulum schematic.

In Figure 3.4, the following notation is used:

$x_c(t)$	distance of cart from some reference point
$\theta(t)$	angle of pendulum
m_c	mass of cart
m_p	mass of pendulum (assumed concentrated at tip)
L_p	length of pendulum
$F(t)$	forces applied to the cart

Application of Newtonian physics to this system leads to the following model:

$$\ddot{x}_c(t) = \frac{1}{\lambda_m + \sin^2\theta(t)}\left[\frac{F(t)}{m_p} + \dot{\theta}^2(t)L_p \sin\theta(t) - g\cos\theta(t)\sin\theta(t)\right], \tag{3.7}$$

$$\ddot{\theta}(t) = \frac{1}{L_p[\lambda_m + \sin^2\theta(t)]}\left[-\frac{F(t)}{m_p}\cos\theta(t) + \dot{\theta}^2(t)L_p \sin\theta(t)\cos\theta(t) + (1-\lambda_m)g\sin\theta(t)\right], \tag{3.8}$$

where $\lambda_m = (m_c/m_p)$ and g is the gravitational constant.

Those readers who are proficient in dynamics may like to derive the above equations. For others, it suffices to take the equations as a given fact. The resultant model is part of the model for a rocket (see Chapter 28.)

3.2.4 Flow of Liquid out of a Tank

Consider a tank of liquid where the height of the liquid is H. Liquid flows out of the tank through a hole of cross-sectional area A, see Figure 3.5.

Therefore the force acting on the liquid above the hole is ρHAg, where ρ is the liquid density and g is the gravitational constant.

Rate of change of momentum out of tank is equal to the applied force. Thus,

$$F = \rho HAg = \frac{d}{dt}[mv], \tag{3.9}$$

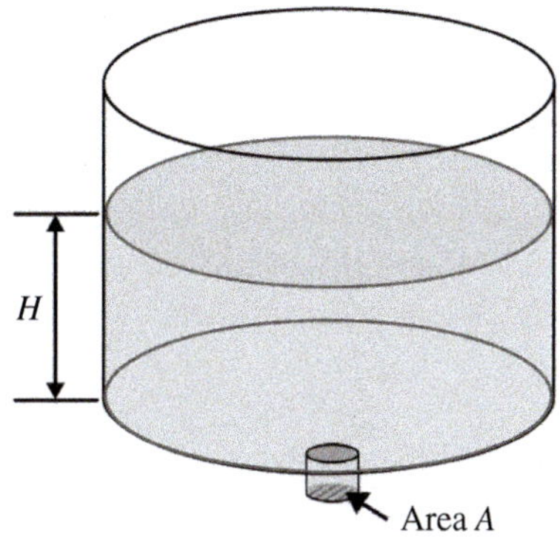

Figure 3.5 Flow out of tank.

where

m mass of liquid exiting the tank
v linear velocity out of the tank.

Assuming steady flow, then

$$v = \frac{f}{\rho A}, \tag{3.10}$$

where f is the mass flow out of the tank in kg/s.

By definition, we have

$$f = \frac{d}{dt}m, \tag{3.11}$$

and substituting (3.11) into (3.9) gives

$$\rho HAg = f\left(\frac{f}{\rho A}\right). \tag{3.12}$$

In steady state, one can solve the above equation for f. The result is

$$f = \rho A\sqrt{Hg}. \tag{3.13}$$

This model will be revisited in Chapter 12.

3.3 State-Space Model Representation

In many cases, it is helpful to express models, especially for higher-order systems, as a set of first-order differential equations. Such a model is usually called a 'state-space model'. When the model is linear, then one obtains a linear-state-space model. The advantage of using a state-space model is that, instead of needing to study high-order differential equations, one need only deal with a (set of coupled) first-order systems.

Chapters 22 and 23 will present results for linear-state-space models. These ideas are so compelling that the reader may feel that state-space ideas are the panacea to control system design. Whilst it is certainly true that state-space ideas give beautifully insightful ideas on feedback systems, they are not without their limitations. An illustration of these limitations will be given in Chapter 16 where Bode Sensitivity Integrals will be introduced. These ideas will be illustrated in the context of audio quantisation in Chapter 18.

3.3.1 Systems Without Zeros

The following steps can be used to transform a simple physical model described by one or more ordinary differential equations into state-space form.

1. Rearrange each physical equation so that the highest-order derivative is the only variable on the left-hand side of each equation.
   For example, consider
$$\frac{d^2h(t)}{dt^2} + G\frac{dh(t)}{dt} + h(t) = Km(t), \tag{3.14}$$
$$N\frac{d^3j(t)}{dt^3} - G\frac{dj(t)}{dt} + h(t) = 0, \tag{3.15}$$
   rearranged to
$$\frac{d^2h(t)}{dt^2} = -G\frac{dh(t)}{dt} - h(t) + Km(t), \tag{3.16}$$
$$\frac{d^3j(t)}{dt^3} = \frac{G}{N}\frac{dj(t)}{dt} - \frac{h(t)}{N}. \tag{3.17}$$
2. For each rearranged physical equation, define as many state variables as the order of each equation making them unique. Continuing with the example, the states allocated to (3.16) are x_1 and x_2 and those allocated to (3.17) will be x_3, x_4 and x_5.
3. Starting with each rearranged equation from step 1, one can define the first state variable for each equation to be the variable that is being differentiated, e.g. $x_1(t) = h(t)$ and each subsequent state to be the derivative of the previous one and repeat this for each equation. For example:
$$x_1(t) = h(t), \tag{3.18}$$
$$x_2(t) = \frac{dh(t)}{dt} = \frac{dx_1(t)}{dt}, \tag{3.19}$$
$$x_3(t) = j(t), \tag{3.20}$$
$$x_4(t) = \frac{dj(t)}{dt} = \frac{dx_3(t)}{dt}, \tag{3.21}$$
$$x_5(t) = \frac{d^2j(t)}{dt^2} = \frac{dx_4(t)}{dt}. \tag{3.22}$$
4. Substitute all the state variables into the equations resulting from step 1
$$\frac{dx_2(t)}{dt} = -Gx_2(t) - x_1(t) + Km(t), \tag{3.23}$$
$$\frac{dx_5(t)}{dt} = \frac{G}{N}x_4(t) - \frac{x_1(t)}{N}. \tag{3.24}$$
5. Take the first derivative of all state variables that do not already have a first-order differential equation as determined in the previous step and collate them in a matrix as follows:
$$\frac{d}{dt}\begin{bmatrix} x_1(t) \\ x_2(t) \\ x_3(t) \\ x_4(t) \\ x_5(t) \end{bmatrix} = \begin{bmatrix} 0 & 1 & 0 & 0 & 0 \\ -1 & -G & 0 & 0 & 0 \\ 0 & 0 & 0 & 1 & 0 \\ 0 & 0 & 0 & 0 & 1 \\ \frac{1}{N} & 0 & 0 & \frac{G}{N} & 0 \end{bmatrix} \begin{bmatrix} x_1(t) \\ x_2(t) \\ x_3(t) \\ x_4(t) \\ x_5(t) \end{bmatrix} + \begin{bmatrix} 0 \\ K \\ 0 \\ 0 \\ 0 \end{bmatrix} m(t). \tag{3.25}$$

6. Identify the output of the system or the PV. In this example, the state, $x_3(t)$, is selected as the single output. The output $y(t)$ can then be expressed as:

$$y(t) = \begin{bmatrix} 0 & 0 & 1 & 0 & 0 \end{bmatrix} \begin{bmatrix} x_1(t) \\ x_2(t) \\ x_3(t) \\ x_4(t) \\ x_5(t) \end{bmatrix}. \tag{3.26}$$

These can be expressed in a more compact form as

$$\dot{x} = Ax + Bu, \tag{3.27}$$

$$y = Cx + Du, \tag{3.28}$$

where x denotes the vector of states, u denotes the inputs or MVs, y denotes the outputs or PVs and A, B, C, and D are matrices of appropriate dimensions.

Hence, for this example, using $u = m(t)$, $y = j(t)$, then

$$A = \begin{bmatrix} 0 & 1 & 0 & 0 & 0 \\ -1 & -G & 0 & 0 & 0 \\ 0 & 0 & 0 & 1 & 0 \\ 0 & 0 & 0 & 0 & 1 \\ \frac{1}{N} & 0 & 0 & \frac{G}{N} & 0 \end{bmatrix} \quad B = \begin{bmatrix} 0 \\ K \\ 0 \\ 0 \\ 0 \end{bmatrix} \quad C = \begin{bmatrix} 0 & 0 & 1 & 0 & 0 \end{bmatrix} \quad D = 0. \tag{3.29}$$

Note that here, and in the sequel, for brevity, the time dependence of states, inputs and outputs is assumed but not explicitly denoted.

3.3.2 Systems Which Depend on Derivatives of the Input

In some cases, the model can depend upon differentials of the input as well as the output. Thus, consider the following linear model linking $y(t)$ to $u(t)$:

$$\frac{d^n}{dt^n}y + a_{n-1}\frac{d^{n-1}}{dt^{n-1}}y + \cdots + a_0 y = b_{n-1}\frac{d^{n-1}}{dt^{n-1}}u + \cdots + b_0 u. \tag{3.30}$$

One can then define the following state-space equations:

$$\frac{d}{dt}x_1 = -a_0 x_n + b_0 u, \tag{3.31}$$

$$\frac{d}{dt}x_2 = x_1 - a_1 x_n + b_1 u, \tag{3.32}$$

$$\frac{d}{dt}x_3 = x_2 - a_2 x_n + b_2 u, \tag{3.33}$$

$$\vdots$$

$$\frac{d}{dt}x_n = x_{n-1} - a_{n-1} x_n + b_{n-1} u, \tag{3.34}$$

$$y = x_n. \tag{3.35}$$

To show how this works, one can work down from the top by successive differentiation. Using (3.35), this leads to

$$\frac{d}{dt}x_1 = -a_0 y + b_0 u, \tag{3.36}$$

$$\frac{d}{dt}x_2 = x_1 - a_1 y + b_1 u. \tag{3.37}$$

Hence,

$$\begin{aligned}\frac{d^2}{dt^2}x_2 &= \frac{d}{dt}x_1 - a_1\frac{dy}{dt} + b_1\frac{du}{dt}\\ &= -a_0 y + b_0 u - a_1\frac{dy}{dt} + b_1\frac{du}{dt}.\end{aligned} \tag{3.38}$$

Similarly,

$$\frac{dx_3}{dt} = x_2 - a_2 x_n + b_2 u. \tag{3.39}$$

Hence ,

$$\frac{d^3x_3}{dt^3} = \frac{d^2x_2}{dt^2} - a_2\frac{d^2y}{dt^2} + b_2\frac{d^2u}{dt^2} \tag{3.40}$$

$$= -a_0 y + b_0 u - a_1\frac{dy}{dt} + b_1\frac{du}{dt} - a_2\frac{d^2y}{dt^2} + b_2\frac{d^2u}{dt^2} \tag{3.41}$$

$$\vdots$$

$$\frac{d^n x_n}{dt^n} = -a_0 y + b_0 u - a_1\frac{dy}{dt} + b_1\frac{du}{dt} \dots - a_{n-1}\frac{d^{n-1}y}{dt^{n-1}} + b_{n-1}\frac{d^{n-1}}{dt^{n-1}}u. \tag{3.42}$$

Finally, using (3.35) again, (3.42) yields (3.30).

The state-space model (3.31)–(3.35) is given a special name. It is called the 'Observer Canonical Form'. The above development emphasises that state-space models and high-order differential equation models are simply alternative ways of expressing the model equations.

3.3.3 Example: State-Space Representation

To complete this section, the example of the radio telescope positioning physical model from Section 3.2 will be used to produce the equivalent state-space representation.

Recall that the physical model for the motor positioning of a radio telescope dish is given by

$$\frac{d^2\theta(t)}{dt^2} + \frac{C}{I}\frac{d\theta(t)}{dt} = \frac{1}{I}T(t). \tag{3.43}$$

Following the procedure described above,

Step 1 gives

$$\frac{d^2\theta(t)}{dt^2} = -\frac{C}{I}\frac{d\theta(t)}{dt} + \frac{1}{I}T(t). \tag{3.44}$$

In Steps 2 and 3, the states are defined as

$$x_1(t) = \theta(t), \tag{3.45}$$

$$x_2(t) = \frac{d\theta(t)}{dt}, \tag{3.46}$$

Step 4 gives

$$\frac{dx_1(t)}{dt} = x_2(t), \tag{3.47}$$

$$\frac{dx_2(t)}{dt} = -\frac{C}{I}x_2(t) + \frac{1}{I}T(t). \tag{3.48}$$

Finally, using steps 5 and 6, the Eqs. (3.47) and (3.48) can be represented in state-space form as

$$\dot{x} = Ax + Bu,$$

$$y = Cx + Du,$$

where input is $u = T(t)$ and the output is $y = \theta(t)$, and where

$$A = \begin{bmatrix} 0 & 1 \\ 0 & -\frac{C}{I} \end{bmatrix} \quad B = \begin{bmatrix} 0 \\ \frac{1}{I} \end{bmatrix} \quad C = \begin{bmatrix} 1 & 0 \end{bmatrix} \quad D = 0. \tag{3.49}$$

The conversion of the other physical models from Section 3.2, into their respective linear-state-space representation is left as an exercise for the reader. The inverted pendulum will be discussed in more detail in Section 3.4.1.

3.4 Linearisation and Approximation

Many insights can be gained through the use of models. This statement is particularly true for linear models. Once non-linearities are introduced, the analysis and the extraction of meaningful insights become more difficult. Hence, the first 'port of call' when presented with a non-linear control problem is usually to develop a linear approximation to the non-linear model. There are many approaches to obtaining a linear model but the most common is referred to as 'small signal analysis' or linearisation.

In small signal linearisation, it is assumed that the process operates around some given 'operating point' and only deviates from that operating point by small amounts.

To describe the idea, consider a general non-linear state-space model written in vector form as

$$\dot{x} = f(x, u), \quad y = h(x), \tag{3.50}$$

where x, u and y denote the state, input and output, respectively.

Say that the current value x_0, u_0 describe an equilibrium or operating point i.e.

$$f(x_0, u_0) = 0. \tag{3.51}$$

The deviations from x_0, u_0 can be expressed as

$$x = x_0 + \tilde{x}; \quad u = u_0 + \tilde{u}. \tag{3.52}$$

Then a first-order Taylor's series expansion can be used to find a model describing the perturbations. In particular,

$$\begin{aligned} \frac{d}{dt}(x_0 + \tilde{x}) = \dot{\tilde{x}} &\approx f(x_0, u_0) + \left.\frac{\partial f}{\partial x}\right|_{x_0,u_0} \tilde{x} + \left.\frac{\partial f}{\partial u}\right|_{x_0,u_0} \tilde{u}, \\ &\approx \left.\frac{\partial f}{\partial x}\right|_{x_0,u_0} \tilde{x} + \left.\frac{\partial f}{\partial u}\right|_{x_0,u_0} \tilde{u}, \end{aligned} \tag{3.53}$$

since it is assumed that (x_0, u_0) is an equilibrium point, i.e. $f(x_0, u_0) = 0$.

3.4.1 Linearisation of Inverted Pendulum Model

The inverted pendulum model is recalled from Section 3.2 which is non-linear when derived from physical laws and has the following form:

$$\ddot{x}_c(t) = \frac{1}{\lambda_m + \sin^2\theta(t)}\left[\frac{F(t)}{m_p} + \dot{\theta}^2(t)L_p\sin\theta(t) - g\cos\theta(t)\sin\theta(t)\right], \tag{3.54}$$

$$\ddot{\theta}(t) = \frac{1}{L_p[\lambda_m + \sin^2\theta(t)]} \times \left[-\frac{F(t)}{m_p}\cos\theta(t) + \dot{\theta}^2(t)L_p\sin\theta(t)\cos\theta(t) + (1-\lambda_m)g\sin\theta(t)\right], \tag{3.55}$$

where $\lambda_m = (m_c/m_p)$ and g is the gravitational constant.

For the inverted pendulum, the operating point is selected to be when the pendulum is vertical and not moving. This point is denoted by the couple $(\theta_0, \dot{\theta}_0)$ which has the values $\theta_0 = 0$ and $\dot{\theta}_0 = 0$. Also, this point can be achieved with $F(t) = u_0 = 0$. Substituting these values into (3.54) and (3.55) leads to the approximate linear model:

$$\ddot{\tilde{x}}_c(t) = \frac{1}{\lambda_m}\left[\frac{\tilde{F}(t)}{m_p} - g\tilde{\theta}(t)\right], \tag{3.56}$$

$$\ddot{\tilde{\theta}}(t) = \frac{1}{L_p\lambda_m}\left[-\frac{\tilde{F}(t)}{m_p} + (1-\lambda_m)g\tilde{\theta}(t)\right]. \tag{3.57}$$

The model centre is converted to linear state-space form, with input $\tilde{u}(t) = \tilde{F}(t)$ and output $y(t) = \tilde{x}_c(t)$, by introducing

$$\begin{aligned} x_1(t) &= \tilde{x}_c(t), \\ x_2(t) &= \dot{\tilde{x}}_c(t), \\ x_3(t) &= \tilde{\theta}(t), \\ x_4(t) &= \dot{\tilde{\theta}}(t), \end{aligned}$$

This leads to a linear state-space model

$$\dot{x} = Ax + Bu, \tag{3.58}$$

$$y = Cx + Du, \tag{3.59}$$

where

$$A = \begin{bmatrix} 0 & 1 & 0 & 0 \\ 0 & 0 & \frac{-m_p g}{m_c} & 0 \\ 0 & 0 & 0 & 1 \\ 0 & 0 & \frac{(m_c+m_p)g}{m_c L} & 0 \end{bmatrix}; \quad B = \begin{bmatrix} 0 \\ \frac{1}{m_c} \\ 0 \\ -\frac{1}{m_c L} \end{bmatrix}; \quad C = \begin{bmatrix} 1 & 0 & 0 & 0 \end{bmatrix}; \quad D = 0. \tag{3.60}$$

Remark 3.1 Modern computational packages include special commands to compute linearised models around a user-defined (precomputed) operating point. In the case of MATLAB®, the appropriate commands are **linmod** (for continuous-time systems) and **dlinmod** (for discrete-time and hybrid systems).

Remark 3.2 Obviously, linearised models are only approximate models. Thus, these models should be used with appropriate caution (as indeed all models should). In the case of linearised models, the next term in the Taylor series expansion can often be employed to say something about the size of the associated modelling error. In the context of control system design, it is important to recognise that all real-world models involve some level of approximation. In the context of control system design, the issue of the impact of model error is commonly treated under the heading of 'robustness'. Much more will be said on this issue in the sequel.

Remark 3.3 The model given in (3.60) is very closely related to rocket dynamics. This is easily seen since a rocket sitting on its thruster is much like an inverted pendulum.

Control of a simple version of (3.60) will be studied in Chapter 15.

3.5 Revision Questions

1. Consider the radio telescope described in Section 3.2.1.
   (a) What happens to the moment of inertia for rotation about the vertical axis when the telescope is leant over?
   (b) Say that the dish is modelled as a flat disc of mass m with radius R. Assume the mass is uniformly distributed on the disc. Obtain an expression for the inertia about the vertical axis when the telescope is pointing upwards.
   (c) Repeat if the disc is tilted at an angle θ.
2. Consider a mass m located at the end of a rod of length ℓ. What is the moment of inertia?
3. Consider a PI control law:

$$\frac{du}{dt} = K\left(\tau\frac{de}{dt} + e\right).$$

   Write this control law in state-space form.
4. Consider the model:

$$\dot{x} = u^2 - 3u - x.$$

$$y = x.$$

   (a) What values of u would bring y to 4 in steady state?
   (b) For the larger of the two values found in (a), linearise the model about the equilibrium point.
5. Figure 3.6 shows a second-order $\Sigma\Delta$ modulator. Such a modulator is often used in signal processing. The block Q denotes a nearest neighbour quantiser. The action of Q is often modelled as a noise source, q. Derive a model linking u to u^* and q.
6. Revisit your answer to Question 3 in Chapter 1. Give a more nuanced explanation of the fact that overreacting to feedback can lead to instability.

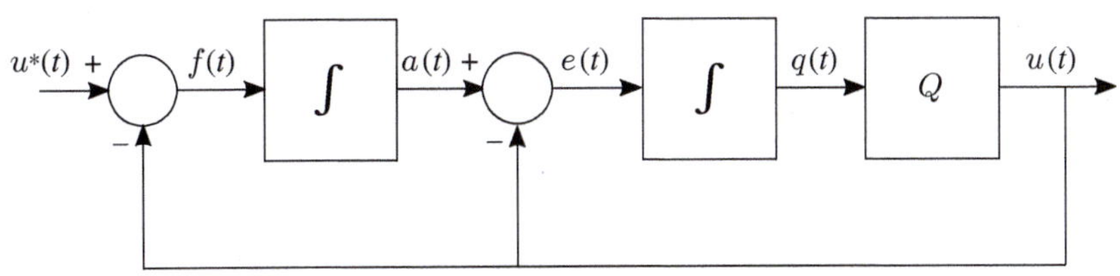

Figure 3.6 Second-order $\Sigma\Delta$ modulator.

7. Consider again the tank in Section 3.2.4. Say that there is a flow of water into the tank of f_{in} kg/s. Modify the model to include this flow. (Note the height, H, now becomes a state variable.)
8. Say that a heat flow of Q joules/s is applied to the liquid in the tank. Also, assume that the liquid flowing into the tank has temperature T_{in}. Derive a model for the temperature of liquid exiting the tank. (Assume internal energy and enthalpy are proportional to temperature.)
9. Rearrange your model found in Question 8 to state-space form using T_{out} and H as the state variables.

Further Reading

1 C.M. Close, D.K. Frederick and J.C. Newell, *Modelling and analysis of dynamic systems*, John Wiley & Sons, Inc., 2002.

2 S. Heinz, *Mathematical modelling*, Springer-Verlag, 2011.

3 L. Ljung and T. Glad, *Modelling of dynamic system*, Prentice Hall, 1994.

4 J.A. Mandles, "Modelling for control analysis and design in complex industrial separation and liquefaction processes", *Journal of Process Control*, 10(2–3):167–175, 2006.

4

Continuous-Time Signals and Systems

4.1 Introduction

Chapter 3 gave a brief introduction to modelling. Models for other real-world systems will be explored as the book proceeds. The next step is to study some simple tools and techniques that can be used to manipulate and analyse models.

Many of the models introduced in Chapter 3 were quite simple. They are special cases of a broad class of systems known, collectively, as linear dynamical systems.

The advantage of being able to describe a modelling problem, at least approximately, via a linear dynamical system is that subsequent analysis can draw on the wealth of information that is available about the operation of such systems. Indeed, this theory is so powerful that it is almost always beneficial to use a linear (approximate) model even if later it is embellished by adding some nonlinear features. This chapter covers the fundamentals of the theory of linear models. Specific topics to be covered include the following:

- linear high-order differential equation models;
- Laplace transforms, which convert linear differential equations into algebraic equations;
- general linear state-space models;
- methods for assessing the stability of linear dynamic systems; and
- frequency response.

Readers who are familiar with the above concepts could progress to Chapter 5 where the ideas will be applied to a real-world system. Others may like to revise the contents of this chapter quickly and return to it later when the concepts appear in the virtual laboratory problems.

4.2 Linear Continuous-Time Models

The brief overview of the analysis of linear systems begins by considering the type of models previously considered in Eq. (3.31), Section 3.3. Thus, consider an nth-order differential equation model of the form:

$$\frac{d^n y(t)}{dt^n} + a_{n-1}\frac{d^{n-1}y(t)}{dt^{n-1}} + \cdots + a_0 y(t) = b_{n-1}\frac{d^{n-1}}{dt^{n-1}}u(t) + \cdots + b_0 u(t), \tag{4.1}$$

where $a_0, \ldots, a_{n-1}, b_0, \ldots, b_{n-1}$ are (constant) coefficients.

Practical Control System Design: Real World Designs implemented on Emulated Industrial Systems,
First Edition. Adrian Medioli and Graham Goodwin.

Companion website: www.wiley.com/go/medioli/practicalcontrolsystemdesign

To simplify the notation, it is convenient to use operator notation to denote the operation of differentiation. Consequently, the Heaviside, or differential, operator, $\rho\langle\circ\rangle$, is introduced by defining:

$$\rho\langle f(t)\rangle = \rho f(t) \triangleq \frac{df(t)}{dt} \tag{4.2}$$

$$\rho^n\langle f(t)\rangle = \rho^n f(t) = \rho\langle \rho^{n-1}\langle f(t)\rangle\rangle = \frac{df^n(t)}{dt^n} \tag{4.3}$$

In terms of this operator, the model (4.1) can be written more simply as

$$\rho^n y(t) + a_{n-1}\rho^{n-1}y(t) + \cdots + a_0 y(t) = b_{n-1}\rho^{n-1}u(t) + \cdots + b_0 u(t). \tag{4.4}$$

Of major importance for linear systems is that the *Principle of Superposition* holds. This implies that, if two inputs are applied simultaneously, then the response due to these inputs is simply the sum of the responses resulting from applying the inputs separately. This has huge implications. For example, one can understand the response to complex inputs by decomposing them into elementary subcomponents which are considered one at a time. This idea will be used in Section 4.12, leading to new, and important, insights associated with the frequency response of linear systems.

4.3 Laplace Transforms

A relatively formal introduction to Laplace transforms is given here. However, when using the ideas, it is often sufficient to adopt a mechanistic point of view. For those readers who feel more comfortable with a more heuristic style, they can jump to Section 4.5.

Of all the methods available for studying linear differential equations, one particularly useful tool is provided by Laplace transforms. A powerful feature of this transform is that it converts linear differential equations into algebraic equations. The reader is referred to the comments made in Chapter 2 about the surprising generality of algebraic models. It will be shown below that such algebraic models are actually rather general.

Definition 4.1 ***(Transform)*** Consider a continuous-time signal $y(t), 0 \leq t < \infty$. The Laplace-transform pair associated with $y(t)$ is defined as follows:

$$\mathscr{L}\,[y(t)] = Y(s) = \int_{0^-}^{\infty} e^{-st} y(t)dt, \tag{4.5}$$

$$\mathscr{L}^{-1}\,[Y(s)] = y(t) = \frac{1}{2\pi j}\int_{\sigma-j\infty}^{\sigma+j\infty} e^{st} Y(s)ds, \tag{4.6}$$

$Y(s)$ is referred to as the Laplace transform of $y(t)$. A full treatment of Laplace transforms will not be given. This is not necessary to understand their application to control system design. The interested reader is referred to the more in-depth treatment given in the books cited in Section Further Reading. Note, in passing, that the transform pair is well-defined if there exist $\sigma \in \mathbb{R}$ and a positive constant $k < \infty$ such that

$$|y(t)| < ke^{\sigma t}\ ;\ \forall t \geq 0. \tag{4.7}$$

The region $\mathscr{R}\{s\} \geq \sigma$ is known as the *region of convergence* of the transform.

This concept is necessary if one wishes to use Laplace transforms in sophisticated situations, e.g. fancy signal processing. However, it is generally not necessary in the context of control system design.

Table 4.1 Laplace transform table.

$f(t)$ $(t \geq 0)$	$\mathscr{L}[f(t)]$	Region of convergence
1	$\frac{1}{s}$	$\sigma > 0$
$\delta_D(t)$	1	$\lvert\sigma\rvert < \infty$
t	$\frac{1}{s^2}$	$\sigma > 0$
t^n $\quad n \in \mathbb{Z}^+$	$\frac{n!}{s^{n+1}}$	$\sigma > 0$
$e^{\alpha t}$ $\quad \alpha \in \mathbb{C}$	$\frac{1}{s-\alpha}$	$\sigma > \mathscr{R}\{\alpha\}$
$te^{\alpha t}$ $\quad \alpha \in \mathbb{C}$	$\frac{1}{(s-\alpha)^2}$	$\sigma > \mathscr{R}\{\alpha\}$
$\cos(\omega_o t)$	$\frac{s}{s^2+\omega_o^2}$	$\sigma > 0$
$\sin(\omega_o t)$	$\frac{\omega_o}{s^2+\omega_o^2}$	$\sigma > 0$
$e^{\alpha t}\sin(\omega_o t+\beta)$	$\frac{(\sin\beta)s+\omega_o\cos\beta-\alpha\sin\beta}{(s-\alpha)^2+\omega_o^2}$	$\sigma > \mathscr{R}\{\alpha\}$
$t\sin(\omega_o t)$	$\frac{2\omega_o s}{(s^2+\omega_o^2)^2}$	$\sigma > 0$
$t\cos(\omega_o t)$	$\frac{s^2-\omega_o^2}{(s^2+\omega_o^2)^2}$	$\sigma > 0$
$\mu(t)-\mu(t-\tau)$	$\frac{1-e^{-s\tau}}{s}$	$\lvert\sigma\rvert < \infty$

Equations (4.5) and (4.6) provide a way to take signals and system descriptions to the Laplace or s-domain (where simple algebra can be used). Equation (4.6) is rarely used to obtain the inverse Laplace transform, because the Laplace transforms of most signals of interest are rational fractions in s. Thus, a partial-fraction approach is normally used to evaluate the inverse transform by comparing the transform with standard results.

The preceding transform pair can be used to derive a table of Laplace transforms. Examples of transforms frequently used in control applications are given in Table 4.1. The reader is encouraged to derive some of the results from first principles using (4.5).

There are many interesting properties that flow from the definition of the transforms. Some of these are listed in Table 4.2. Again, the reader is encouraged to derive some of these properties from first principles.

4.4 Application of Laplace Transforms to Linear Differential Equations

Laplace transforms play an important role in studying linear differential equations. The associated ideas are briefly described in the sequel.

The reader may care to verify the following key result (see Table 4.2) regarding the transform of the derivative of a function

$$\mathscr{L}\left[\frac{dy(t)}{dt}\right] = sY(s) - y(0^-), \tag{4.8}$$

Table 4.2 Properties of Laplace transforms.

Time-Domain	S-Domain	Names
$af(t) + bg(t)$	$aF(s) + bG(s)$	Linearity
$\frac{df}{dt}$	$sF(s) - f(0^-)$	Derivative
$\frac{d^n f}{dt^n}$	$s^n F(s) - \sum_{k=1}^{n} s^{(n-t)} f^{(k-1)}(0^-)$	General derivative
$\frac{1}{t} f(t)$	$\int_s^\infty F(\sigma)d\sigma$	Frequency domain integration
$\int_0^t F(\tau)d\tau$	$\frac{1}{s}F(s)$	Time domain integration
$e^{at}f(t)$	$F(s-a)$	Frequency shifting
$f(t)g(t)$	$\frac{1}{2\pi j}\int_{\sigma-j\infty}^{\sigma+j\infty} F_1(\xi)F_2(s-\xi)d\xi$	Time domain multiplication
$\int_0^t f(\tau)g(t-\tau)dc$	$F(s)G(s)$	Time domain convolution
$f(0^+)$	$\lim_{s\to\infty} sF(s)$	Initial value
$f(\infty)$	$\lim_{s\to\infty} sF(s)$[All poles of $f(s)$ in LHP]	Final value

where $\mathscr{L}\,[y(t)] = Y(s)$. This result can be derived directly from (4.5). This result can be used to convert differential equations into algebraic equations in the variable s. Indeed, ignoring the initial condition term, $y(0^-)$ in (4.8), it can be seen that one simply replaces derivatives by ' s'.

Two simple examples are used below to illustrate the key idea.

4.4.1 Example: Angle of Radio Telescope

Consider again the radio telescope problem described in Chapter 3. This model is a second-order version of (4.1). By way of illustration, let $a_1 = 2$, $b_0 = 1$. (Note that, in this particular example, $a_0 = 0$). Then, applying (4.8), and taking Laplace transforms, one obtains

$$s^2\Theta(s) + 2s\Theta(s) - (s+2)\theta(0^-) - \dot{\theta}(0^-) = V_a(s), \tag{4.9}$$

where $\Theta(s)$ and $V_a(s)$ are the Laplace transforms of the variables $\theta(t)$ and $V_a(t)$, respectively. Also, $\theta(0^-)$ and $\dot{\theta}(0^-)$ denote initial conditions for θ and $\dot{\theta}$.

Say that the initial conditions are specified as $\theta(0^-) = 0, \dot{\theta}(0^-) = 0$, and say that the input is a unit step applied at $t = 0$. Then from (4.9)

$$\begin{aligned}\Theta(s) &= \left(\frac{1}{s^2+2s}\right)V_a(s)\\ &= \left(\frac{1}{s^2+2s}\right)\frac{1}{s}\end{aligned} \tag{4.10}$$

Expanding in partial fractions [1] gives

$$\Theta(s) = \frac{1}{4(s+2)} + \frac{1}{2s^2} - \frac{1}{4s}. \tag{4.11}$$

1 You can use the MATLAB® command **residue**.

Then, applying the results from Table 4.1, the output response for $t \geq 0$ is

$$\theta(t) = \frac{1}{4}e^{-2t} + \frac{1}{2}t - \frac{1}{4}. \tag{4.12}$$

Thus, Laplace transforms have been used to find the time response of the second-order differential equation for the given input and initial conditions.

4.4.2 Example: Modelling the Angular Velocity of Radio Telescope

In the above example, the response from the input voltage to the output shaft was examined. An even simpler problem arises if one examines the response from the input voltage to the output angular velocity. To examine the latter, it is assumed that the initial condition is zero. Then the Laplace transform of the output satisfies $\omega(s) = \left(\frac{1}{s+2}\right) V_a(s)$.

Next, say that a step input is applied, i.e. $V_a(s) = \frac{1}{s}$.

Expanding in partial fractions yields:

$$\omega(s) = \frac{\frac{1}{2}}{s} - \frac{\frac{1}{2}}{s+2}. \tag{4.13}$$

Hence, the output angular velocity response can be evaluated using Table 4.1. The result is

$$\omega(t) = \frac{1}{2} - \frac{1}{2}e^{-2t}. \tag{4.14}$$

4.5 A Heuristic Introduction to Laplace Transforms

A simple (heuristic) explanation of Laplace transforms is given below. Consider a special input signal comprising a single exponential of the form:

$$u(t) = ke^{\beta t}. \tag{4.15}$$

Interestingly, if this input is applied to a linear system, then the output has the same form. To verify that this is true, consider a linear nth-order differential equation as in (4.4). It will be shown that the output (for zero initial conditions) is also of the form:

$$y(t) = k'e^{\beta t}. \tag{4.16}$$

The key concept needed to show this result is the fact that $\frac{d}{dt}(e^{\beta t}) = \beta e^{\beta t}$ and hence $\frac{d^n}{dt^n}(e^{\beta t}) = \beta^n e^{\beta t}$. Then, to test that (4.16) actually satisfies (4.4) when the input is as in (4.15), one simply substitutes (4.15) and (4.16) into (4.4) leading to

$$\left[\beta^n + a_{n-1}\beta^{n-1} + \cdots + a_0\right] k'e^{\beta t} = \left[b_{n-1}\beta^{n-1} + \cdots + b_0\right] ke^{\beta t}. \tag{4.17}$$

It then follows that $y(t) = k'e^{\beta t}$ does indeed satisfy (4.4) (i.e. is a solution of the model) when the input is (4.15) provided the 'gain', i.e. $\frac{k'}{k}$, satisfies:

$$\frac{k'}{k} = \frac{b_{n-1}\beta^{n-1} + \cdots + b_0}{\beta^n + a_{n-1}\beta^{n-1} + \cdots + a_0}. \tag{4.18}$$

Note that the gain, $\frac{k'}{k}$, changes with the choice of β. Hence, one can express $\frac{k'}{k}$ as a function of β. The gain function $G(\beta)$ is then

$$G(\beta) = \frac{b_{n-1}\beta^{n-1} + \cdots + b_0}{\beta^n + a_{n-1}\beta^{n-1} + \cdots + a_0}. \tag{4.19}$$

The β-dependent gain function, $G(\beta)$, is called the system 'transfer function' for input $ke^{\beta t}$. The function $G(\beta)$ can be manipulated using the nominal rules of algebra. A slightly more formal derivation of the function $G(\beta)$ will be given in Section 4.6.1.

4.6 Transfer Functions

4.6.1 High-Order Differential Equation Models

Consider again the general linear (high-order) differential equation model (4.1). Taking Laplace transforms, converts this differential equation into the following algebraic equation:

$$s^n Y(s) + a_{n-1}s^{n-1}Y(s) + \cdots + a_0 Y(s) = b_{n-1}s^{n-1}U(s) + \cdots + b_0 U(s) + f(s, x_o) \tag{4.20}$$

where $f(s, x_o)$ stands for a function of the initial conditions. In the case of *zero initial conditions*, it follows that

$$Y(s) = G(s)U(s), \tag{4.21}$$

where

$$G(s) = \frac{B(s)}{A(s)}, \tag{4.22}$$

and

$$A(s) = s^n + a_{n-1}s^{n-1} + \cdots + a_0 \tag{4.23}$$

$$B(s) = b_{n-1}s^{n-1} + b_{n-2}s^{n-2} + \cdots + b_0 \tag{4.24}$$

To place Section 4.5 into the context of Laplace transforms, it is to be noted that the Laplace transform of the input (4.15) is

$$U(s) = \frac{k}{s - \beta}. \tag{4.25}$$

Utilising (4.21), it can be seen that the output (for zero initial conditions) is

$$Y(s) = G(s)\frac{k}{s - \beta}. \tag{4.26}$$

Carrying out a partial fraction expansion and assuming that β is not a zero of the denominator of $G(s)$ yields

$$Y(s) = \frac{kG(\beta)}{s - \beta} + \text{terms that depend upon the zeros of the denominator of } G(s). \tag{4.27}$$

Hence, assuming zero initial conditions, the extra terms referred to above are zero and hence

$$Y(s) = \frac{kG(\beta)}{s - \beta}, \tag{4.28}$$

or, finding the inverse transform:

$$y(t) = kG(\beta)e^{\beta t}. \tag{4.29}$$

Those readers who have read both Sections 4.4 and 4.5 will see that (4.29) is the same quantity, namely, $G(\beta)$ as the β dependent gain.

The reader will see, in the above developments that, for zero initial conditions, one simply replaces the derivative operator by the Laplace transform variable. Thus, the power of using Laplace transforms is that one moves from needing to study the impact of derivatives on signals to being able to use simple algebra. These algebraic ideas will be used in the sequel to give deep insights into the response of models to various inputs. Of particular interest in this context will be the quantity $G(s)$ defined in Eq. (4.22).

The quality $G(s)$ will turn out to be of central importance. It is thus given a special name, the *transfer function*. The decomposition (4.22) is also a very useful one. It can be used to gain deep insights into various control-design questions. In particular, it is now clear that the simple algebraic model described in Section 2.2 is directly applicable to high-order differential equations provided one allows the system gain to be a function of s. (The Laplace transform variable.)

4.6.2 Example: Transfer Function for Radio Telescope

The transfer function for the system in Example 4.4.1 is

$$G(s) = \frac{1}{s^2 + 2s}. \tag{4.30}$$

4.6.3 Transfer Functions for Continuous-Time State-Space Models

Laplace transform can also be used to *turn* a state-space model into a set of algebraic equations. For example, taking the Laplace transform in (3.62) and (3.63) yields

$$sX(s) - x(0) = \mathrm{A}X(s) + \mathrm{B}U(s) \tag{4.31}$$

$$Y(s) = \mathrm{C}X(s) + \mathrm{D}U(s) \tag{4.32}$$

where $X(s)$, $U(s)$ and $Y(s)$ are the Laplace transforms of the state, input and output-time responses, respectively, and where $x(0)$ is the initial value of x.

From (4.31) and (4.32), it follows that

$$X(s) = (s\mathrm{I} - \mathrm{A})^{-1}x(0) + (s\mathrm{I} - \mathrm{A})^{-1}\mathrm{B}U(s) \tag{4.33}$$

$$Y(s) = [\mathrm{C}(s\mathrm{I} - \mathrm{A})^{-1}\mathrm{B} + \mathrm{D}]U(s) + \mathrm{C}(s\mathrm{I} - \mathrm{A})^{-1}x(0) \tag{4.34}$$

These equations can be used to derive the formula for the solution to the state-space model. In particular, note that a simple matrix generalisation of row 5 of table 4.1 leads to

$$\mathscr{L}\left[e^{\mathrm{A}t}\right] = (s\mathrm{I} - \mathrm{A})^{-1}, \tag{4.35}$$

where $e^{\mathrm{A}t} = I + \mathrm{A}t + \frac{1}{2}\mathrm{A}^2t^2 + \cdots$.

Thus, with zero initial conditions, the Laplace transform of the output, $Y(s)$, is given by:

$$Y(s) = \mathrm{G}(s)U(s) \tag{4.36}$$

$$\mathrm{G}(s) = \mathrm{C}(s\mathrm{I} - \mathrm{A})^{-1}\mathrm{B} + \mathrm{D} \tag{4.37}$$

As before G(s) relates the input to output. It is thus the 'system transfer function'.

An interesting fact arising from (4.33) is that knowledge of $x(0)$ (the initial 'state') and the input from time 0 onwards uniquely determines the system response. This is an important property of the 'state' of a system. Specifically, knowledge of the current state compactly summarises everything that has happened to a system up to this point in time. Also, it uniquely defines the future response to given inputs. Hence, if one has a static (or time independent) objective, then it follows that the control can only be a function (possibly nonlinear) of the current state of the system plus the desired future trajectory of the output. This idea will be further explored in Chapter 22.

4.6.4 Example: Inverted Pendulum

Consider again the linearised form of the model for the inverted pendulum from Chapter 3. The transfer function from U to Y for this system is

$$\begin{aligned} \mathrm{G}(s) &= \mathrm{C}(s\mathrm{I} - \mathrm{A})^{-1}\mathrm{B}, \\ &= \frac{1}{M}\frac{(s^2 - b^2)}{s^2(s^2 - a^2)}, \\ &= \frac{B(s)}{A(s)} \end{aligned} \tag{4.38}$$

where $a = \sqrt{\frac{(M+m)g}{M\ell}}$; $b = \sqrt{\frac{g}{\ell}}$.

4.6.5 Poles, Zeros and Other Properties of Transfer Functions

Further terms used in connection with transfer functions are defined below.

Consider the transfer function given in Eqs. (4.22)–(4.23).

For simplicity, it is assumed, at least for the moment, that $B(s)$ and $A(s)$ are not simultaneously zero for the same value of s. (In this case, they are said to be 'relatively prime polynomials'.) Also, let the degree of $A(s)$ and $B(s)$ be n and m, respectively. The following terms are defined:

(i) The roots of $B(s) = 0$ are called the system *zeros*.
(ii) The roots of $A(s) = 0$ are called the system *poles*.
(iii) If $A(s) = 0$ has n_k roots at $s = \lambda_k$, the pole λ_k is said to have *multiplicity* n_k.
(iv) The difference in degree between $A(s)$ and $B(s)$ is called the *relative degree*.
(v) If $m < n$, then the model is said to be *strictly proper*. This means positive relative degree.
(vi) If $m = n$, then the model is said to be *biproper*. This means zero relative degree.
(vii) If $m \leq n$, then the model is said to be *proper*, but it may not be strictly proper.
(viii) If $m > n$, then the model is said to be *improper* (or has *negative relative degree*).

Remark 4.1 Real systems are almost always strictly proper. However, some controller-design methods lead to transfer functions for the control law that are biproper.

4.6.6 Time Delays

Often practical systems have a time delay between inpu and output. This is usually associated with the transport of material from one point to another. For example, if there is a conveyor belt or pipe connecting different parts of a plant, then these will invariably introduce a delay.

The transfer function of a pure delay of time length T_d is of the following form (see time shifting in table 4.2):

$$H(s) = e^{-sT_d}, \tag{4.39}$$

where T_d is the delay (in seconds). T_d will typically vary with the transportation speed.

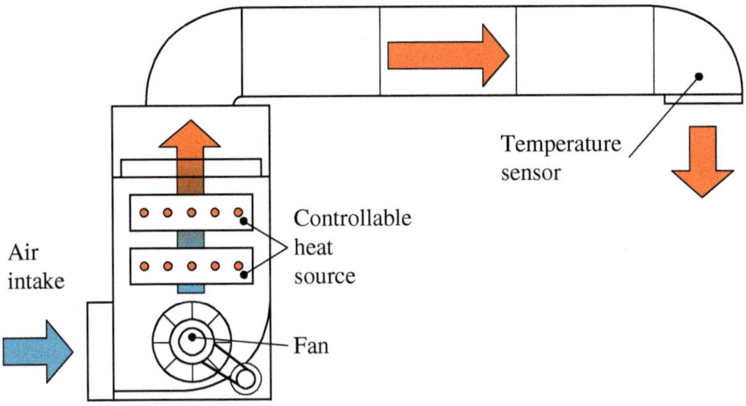

Figure 4.1 Electric forced air-heating system.

4.6.7 Heuristic Development of Transfer Function of Delay

Equation (4.39) can also be appreciated using the heuristic ideas introduced in Section 4.5. In particular, if the input to a system is as in Eq. (4.15) and the system consists of a simple delay of T_d, then the output satisfies

$$y(t) = ke^{s(t-T_d)} = e^{-sT_d}[ke^{st}] = e^{-sT_d}u(t). \tag{4.40}$$

Hence, the (s-dependent) gain of this system can be seen to be

$$G(s) = e^{-sT_d}, \tag{4.41}$$

exactly as in (4.39).

4.6.8 Example: Heating System

As a simple example of a system having a pure time delay, consider the electric forced air-heating system shown in Figure 4.1.

The transfer function from the input (the voltage applied to the heating element) to the output (the temperature as seen by the thermocouple) is approximately of the form:

$$H(s) = \frac{Ke^{-sT_d}}{(\tau s + 1)}. \tag{4.42}$$

Note that $K,\ T_d,\ and\ \tau$ all depend on the speed of the fan, which changes the transport lag from the heater to the measured output as well as various heat-transfer coefficients.

Although this is a very simple example, the kind of model given in (4.42) is very frequently used in process-control applications. An industrial problem where time delay plays a central role will be studied in Chapter 13.

The core conclusions regarding transfer functions can be summarised as follows:

Transfer functions describe the input–output properties of linear systems in algebraic form.

4.7 Stability of Transfer Functions

It has been shown that the response of a system having transfer function $G(s)$ is of the form:

$$Y(s) = G(s)U(s) + \sum_{k=1}^{p}\sum_{i=1}^{n_k} \frac{\beta_{ki}}{(s-\lambda_k)^i}, \tag{4.43}$$

where every β_{ki} is a function of the initial conditions and where, for completeness, it has been assumed that the pole, at $s = \lambda_k$, has multiplicity n_k. This latter assumption implies that $n_1 + n_2 + \cdots + n_p = n$.

The system is said to be *stable* if any bounded input produces a bounded output for all bounded initial conditions. To gain insight into what stability means, a partial-fraction expansion can be used to decompose the total response of the system into the response of each pole taken separately. For continuous-time systems, it can be seen that the response comprises a sum of exponential terms. These exponential terms decay if the real part of the exponential is negative. Moreover, an exponent having a negative real part coincides with the system poles having negative real part. Hence, a linear time-invariant system is stable when the system poles all have negative real parts.

Summarising

For stability, the system poles need to be in the open left-half plane (OLHP) of the complex plane [s].

4.7.1 Example: Poles of the Radio Telescope Model

Consider the system of example 4.4.1. The poles of this transfer function are −2 and 0. These do not lie in the open left-half plane since 0 is in the closed left-half plane (CLHP). Thus, the system is not stable. Indeed, the reader should verify that a non-zero constant input will produce an output that increases without bound.

4.8 Impulse Response of Continuous-Time Linear Systems

4.8.1 Impulse Response

Of special interest in the study of linear systems is the response to a Dirac-delta function. This can be considered as the limit as $\Delta \to 0$ of the pulse shown in Figure 4.2. In practice, if one wants to apply a Dirac-delta function to a system, then one would use an input similar to the one in Figure 4.2 with Δ chosen sufficiently small. The Laplace transform of the Dirac Delta is 1 (see Table 4.1). Hence, if one were to apply such an input to a system with zero initial conditions, then the output response would simply be $Y(s) = G(s)U(s) = G(s)$. This observation can be summarised as follows:

The transfer function of a continuous-time system is the Laplace transform of its response to a unit impulse (Dirac's delta) with zero initial conditions.

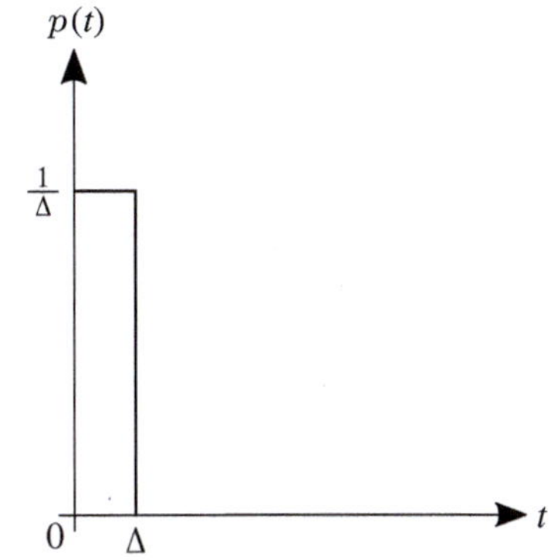

Figure 4.2 Discrete pulse.

4.8.2 Convolution and Transfer Functions

An alternative interpretation can now be given to the development of Chapter 2. In particular, for zero initial conditions, it follows that the input and output Laplace transforms satisfy

$$Y(s) = G(s)U(s), \tag{4.44}$$

then using the convolution property in Table 4.2, it must be true that

$$y(t) = \int_0^\infty g(\tau)u(t-\tau)d\tau, \tag{4.45}$$

where $g(\tau)$ is the system impulse response, i.e. the function whose Laplace transform is $G(s)$. Hence, the system output, $y(t)$, is the convolution of the system impulse response, $g(\tau)$, with the input, $u(t)$. As a special case, say that the system is a simple gain having magnitude p, then the response to a unit impulse, $\delta(t)$, will be

$$y(t) = p\delta(t). \tag{4.46}$$

Thus, the system impulse response, $g(t)$, in this case is $p\delta(t)$.

Substituting into (4.45) gives, for this special case, the following time domain result:

$$y(t) = pu(t). \tag{4.47}$$

Alternatively, using Table 4.1,

$$\mathscr{L}[g(t)] = \mathscr{L}[p\delta(t)] = p = G(s). \tag{4.48}$$

Hence, (4.44) yields

$$Y(s) = pU(s). \tag{4.49}$$

Thus, the simple scalar gain results in Chapter 2 have both a time domain and Laplace domain interpretation since the associated transfer function, in this case, is simply $G(s) = p$.

4.9 Step Response

A closely related property of the impulse response is a system's step response, the response to $U(s) = 1/s$. For zero initial conditions, this leads to

$$Y(s) = G(s)\frac{1}{s}. \tag{4.50}$$

4.10 Steady-State Response and Integral Action

Application of the final-value theorem (see Table 4.2) to (4.50) shows that the steady-state output response (provided that it exists) for a unit step is given by

$$\lim_{t\to\infty} y(t) = y_{ss} = \lim_{s\to 0} sG(s)\frac{1}{s} = G(0). \tag{4.51}$$

If the system is stable, then the transient part of the step response will decay exponentially to zero, and hence the steady-state value, y_{ss}, will exist. Note that, if $G(s)$ has one or more zeros at $s = 0$, then it follows that $y_{ss} = 0$. This simple observation will be extensively used in the sequel to design control systems which ensure zero steady-state tracking error to constant inputs by including an integrator in the control law. (The reader may recall that the term PID was introduced in Chapter 1. The ' *I*' in this phrase refers to the inclusion of this integral term. Also, the reader may recall the comment made in Section 2.5 of Chapter 2 that almost all real-world control systems use (a special form of) infinite gain to achieve perfect tracking and disturbance rejection without depending on knowledge of the true model of the system.)

There are three simple ways that one can see why this works. Three possible explanations are given below:

(i) Putting 's' in the denominator of the control law means that the controller has infinite gain for constant signals. Hence, the simple argument given in Section 2.5 of Chapter 2 establishes that one will achieve perfect tracking of constant signals.

(ii) An alternative, transfer function, argument is as follows: if the control transfer function, $C(s)$, contains an integrator, then $C(s) = C'(s)/s$. The sensitivity function $S(s)$ then becomes

$$S(s) = \frac{1}{1 + G(s)C(s)} = \frac{s}{s + G(s)C'(s)}. \tag{4.52}$$

Since $Y(s) = S(s)D(s)$, then for a step output disturbance, i.e. $D(s) = 1/s$, $Y(s) = S(s)1/s$. Application of the Final-Value Theorem (see Table 4.2) then yields $\lim_{t\to\infty} y(t) = \lim_{s\to\infty} sS(s)/s = 0$.

Hence, constant output disturbances result in zero steady-state output responses for constant disturbances.

(iii) Using the heuristic ideas of Section 4.5, it is possible to think of a step as a very slow settling exponential, i.e. a unit step can be thought of as e^{-st} for s small.

The corresponding output response is then $y(t) = G(s)e^{-st}$ for s small.

Hence, the appropriate transfer function can be thought of as $lim_{s\to 0} G(s) = G(0)$.

4.11 Terms Used to Describe Step Responses

It is useful to define a set of parameters based on a system's step response that succinctly describes certain time-domain properties of a system's dynamics. To introduce these definitions, consider a stable transfer function having the step response of the generic form shown in Figure 4.3.

Then the following properties are defined to describe a system's step response:

Steady-state value, y_{ss}: the final value of the step response.

Rise time, t_r: the time elapsed up to the instant at which the step response reaches, for the first time, the value $k_r y_{ss}$. The constant k_r is often chosen as 0.9.

Overshoot, m_p: the maximum instantaneous amount by which the step response exceeds its final value. (It is usually expressed as a percentage of y_{ss}.)

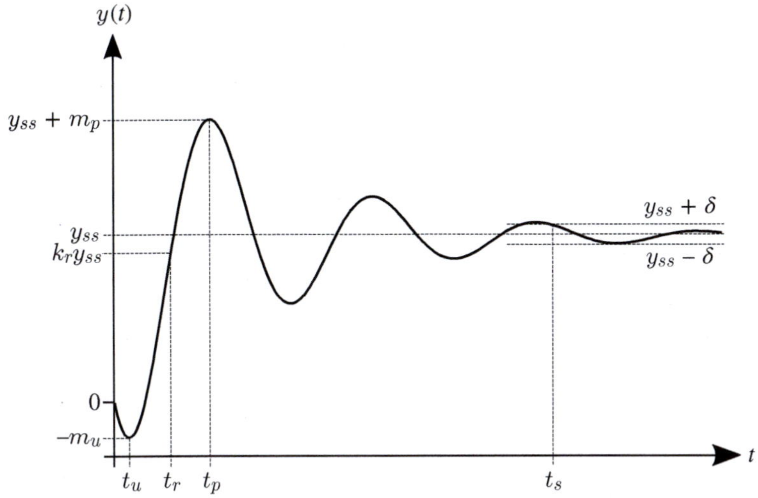

Figure 4.3 Step-response indicators.

Undershoot, m_u: the (absolute value of the) maximum instantaneous amount by which the step response falls below zero.

Settling time, t_s: the time lapsed until the step response enters (without leaving it afterwards) a specified deviation band, $\pm\delta$, around the final value. The deviation, δ, is usually defined as a percentage of y_{ss}, say 2% or 5%.

4.12 Frequency Response

This section explores the response of a linear system to a very special input, namely, a sine wave. The reader may recall the comments made about Harry Nyquist's work at Bell Laboratories – see Section 1.2 of Chapter 1. The reason for considering sinusoidal inputs is that the response to sine waves also contains rich information about the response of linear systems to *all* other signals. This can be appreciated from Fourier analysis, which tells us that any signal defined on an interval $[t_o, t_f]$ can be represented as a linear combination of sine waves of frequencies $0, \omega_{of}, 2\omega_{of}, 3\omega_{of}, \ldots$, where $\omega_{of} = 2\pi/(t_f - t_o)$ is the fundamental frequency. The Principle of Superposition introduced in Section 4.2, then allows one to combine the response to the individual sine waves to determine the response to the composite waveform. Thus, sine wave analysis essentially tells us all that needs to be known about the response of linear systems to arbitrary input signals.

The remainder of this section uses some simple properties of complex exponential functions.[2] Those readers who are unfamiliar with these ideas can jump to the boxed summary given after Eq. (4.58).

To proceed, consider a linear stable system described in transfer-function form by,

$$H(s) = K\frac{\sum_{i=0}^{m} b_i s^i}{s^n + \sum_{k=1}^{n-1} a_k s^k}. \tag{4.53}$$

2 Many readers will be familiar with the identity $e^{j\pi} + 1 = 0$. This is an amazing result that brings together three quantities, namely, $j = \sqrt{-1}$, π and e. This equation was called 'the Most Beautiful Equation in Mathematics' by Richard Feynman in his Lectures on Physics.

Also, let the system input be an exponential, $e^{s_o t}$, with Laplace transform $\frac{1}{s-s_o}$. For simplicity, it is assumed that the system poles are distinct and that none of them is equal to s_o. Then the Laplace transform of the system response can be computed by using a partial-fraction expansion to yield the decomposition:

$$Y(s) = \frac{H(s_o)}{s - s_o} + Y_h(s), \tag{4.54}$$

where the first term is the Laplace transform of the response resulting from the input (the, so-called, 'forced response') and the second term is the decaying response due to initial conditions (the 'natural response'). The corresponding time-domain response is

$$y(t) = H(s_o)e^{s_o t} + \sum_{k=1}^{n} C_k e^{\lambda_k t}, \tag{4.55}$$

where $\lambda_k, k = 1, 2, \ldots, n$ are the system natural frequencies – that is, the poles of $H(s)$, and the $C_k' s$ depend on the initial conditions.

Note that Eq. (4.55) is exactly as described earlier in Section 4.6.1.

The second term on the right-hand side of (4.55) will decay to zero if the system is stable. Thus, the focus is shifted, at least for the moment, to the first term, i.e. $H(s_o)e^{s_o t}$.

Note that a sine wave can be written as the combination of two complex exponentials, i.e.

$$\sin(\overline{\omega}t) = \frac{1}{2j}(e^{j\overline{\omega}t} - e^{-j\overline{\omega}t}). \tag{4.56}$$

Hence, the system response to a sine wave input of frequency $\overline{\omega}$ can be computed by combining the complex response to $e^{s_o t}$ with $s_o = j\overline{\omega}$ and that to $e^{s_o t}$ with $s_o = -j\overline{\omega}$.

Note that evaluating $H(s)$ at $s = j\omega$ yields a complex number, which can be represented conveniently by its amplitude, $|H(j\omega)|$, and phase, $\varphi(\omega)$, in polar coordinates, as

$$H(j\omega) = |H(j\omega)|e^{j\varphi(\omega)}. \tag{4.57}$$

The function $H(j\omega)$, where ω takes any values from 0 to ∞, is called the system 'frequency response function'. The steady-state response to a sine wave input of frequency $\overline{\omega}$ can then be obtained directly from (4.55) to (4.57), via superposition, i.e. the output response is the sum of the responses to $e^{j\overline{\omega}t}$ and $e^{-j\overline{\omega}t}$. Using (4.56) and (4.57)

$$\begin{aligned} y(t) &= \frac{1}{2j}\left[H(j\overline{\omega})e^{j\overline{\omega}t} - H(-j\overline{\omega})e^{-j\overline{\omega}t}\right] \\ &= \frac{1}{2j}\left[|H(j\overline{\omega})|e^{j(\overline{\omega}t+\varphi(\overline{\omega}))} - |H(j\overline{\omega})|e^{-j(\overline{\omega}t+\varphi(\overline{\omega})}\right] \\ &= |H(j\overline{\omega})|\sin(\overline{\omega}t + \varphi(\overline{\omega})) \end{aligned} \tag{4.58}$$

where the fact that, for a transfer function having real coefficients, it is true $|H(j\overline{\omega})| = |H(-j\overline{\omega})|$ and $\varphi(\overline{\omega}) = -\varphi(-\overline{\omega})$ has been used.

Summary

In steady state, for a stable linear system, a sine wave input of frequency $\overline{\omega}$ leads to a sine wave output at the same frequency. Moreover, the amplitude of the output sine wave is the input amplitude modified by multiplication by the magnitude of $H(j\omega)$ at $\omega = \overline{\omega}$, and the phase is shifted by a quantity equal to the phase of $H(j\omega)$ at $\omega = \overline{\omega}$.

An interesting observation is that, if $H(j\omega)$ is known at least for q non-zero different frequencies, where $q = 1 +$ integer part of $[(m+n)/2]$, then $H(s)$ is uniquely determined at all other frequencies as well. This idea can, in principle, be utilised to build a model by injecting a finite number of sinusoidal components at different frequencies. This idea will be used in Section 5.3.5 of Chapter 5 and in a real-world setting in Section 30.2 of Chapter 30.

Remark 4.2 If the system contains a time delay, that is, if the transfer function (4.53) is changed to

$$H(s) = e^{-s\tau} \frac{\sum_{i=0}^{m} b_i s^i}{s^n + \sum_{k=1}^{n-1} a_k s^k}, \tag{4.59}$$

then it can be seen that equation (4.58) becomes

$$y(t) = \begin{cases} 0 & \text{if } t < \tau \\ |H(j\overline{\omega})| \sin(\overline{\omega}t + \varphi(\overline{\omega})) + \sum_{k=1}^{n} C_k e^{\lambda_k (t-\tau)} & \text{if } t \geq \tau, \end{cases} \tag{4.60}$$

where $\varphi(\overline{\omega})$ now includes an additional term $-\tau\overline{\omega}$. Thus, a pure time delay leads to additional phase lag in the output response. The last term in (4.60) is formed by the natural modes, and the constants C_k depend on the initial conditions. It will become clear later that phase lag inhibits the design of 'tight' feedback control and is thus an important impediment to achieving 'good' control.

□□□

Frequency responses are a very useful tool for all aspects of analysis, synthesis and design of controllers and filters. Because of their importance, special plots are used to graph them.

4.12.1 Nyquist Diagrams

One way to illustrate the frequency responses of a system is to draw a polar diagram where the magnitude and phase of $H(j\omega)$ are plotted as ω varies from 0 to ∞. This kind of plot is called a Nyquist plot. Much more will be said about Nyquist diagrams in Chapter 6.

4.12.2 Bode Diagrams

In addition to the work of Hendrik Bode on sensitivity integrals described in Section 1.2 of Chapter 1, Bode also introduced a convenient way of plotting the frequency response of the system. This representation is now simply called a 'Bode Diagram'.

Bode diagrams consist of a pair of plots. One of these plots displays the magnitude of the frequency response as a function of the angular frequency, ω; the other displays the phase of the frequency response, also as a function of the angular frequency, ω.

Usually, Bode diagrams are drawn with special axes:

- The abscissa axis is linear in $\log_{10}(\omega)$, where the $\log_{10}$ is to the base 10. This allows a compact representation of the frequency response over a wide range of frequencies. The unit on the horizontal axis is the *decade*, where a decade is the distance between ω_1 and $10\omega_1$ for any value of ω_1.
- The magnitude of the frequency response is measured in *decibels* [dB], that is, in units of $20\log_{10}|H(j\omega)|$. This has several advantages, including the ability to cover a wide range of values of $|H(j\omega)|$, enabling simple approximations for $20\log_{10}|H(j\omega)|$, and the fact that the frequency response of cascade systems can be obtained by adding the individual frequency responses.
- The angle is measured on a linear scale in radians or degrees.

Software packages such as MATLAB® and GNU Octave provide special commands to compute the frequency response and to plot Bode diagrams. However, some simple rules allow one to sketch an approximation for the magnitude and phase diagrams. Consider a transfer function given by

$$H(s) = K\frac{\prod_{i=1}^{m}(\beta_i s + 1)}{s^k \prod_{i=1}^{n}(\alpha_i s + 1)}. \tag{4.61}$$

Then,

$$20\log_{10}|H(j\omega)| = 20\log_{10}(|K|) - 20k\log_{10}|\omega| + \sum_{i=1}^{m} 20\log_{10}|\beta_i j\omega + 1| - \sum_{i=1}^{n} 20\log_{10}|\alpha_i j\omega + 1|, \tag{4.62}$$

$$\angle(H(j\omega)) = \angle(K) - k\frac{\pi}{2} + \sum_{i=1}^{m}\angle(\beta_i j\omega + 1) - \sum_{i=1}^{n}\angle(\alpha_i j\omega + 1). \tag{4.63}$$

It, thus, can be seen that the Bode diagram of any transfer function can be obtained by either adding or subtracting magnitudes (in [dB]) and phases of simple factors. The following observations can be made:

- A simple gain K has constant magnitude and phase Bode diagram. The magnitude diagram is a horizontal line at $20\log_{10}|K|$ (dB), and the phase diagram is a horizontal line either at 0 (rad) (when K is positive) or at π (rad) (when K is negative).
- The factor s^k has a magnitude diagram that is a straight line with slope equal to 20 k (dB/decade) and has constant phase equal to $k\pi/2$. This line crosses the horizontal axis (0 (dB)) at $\omega = 1$.
- The factor $as + 1$ has a magnitude Bode diagram that can be asymptotically approximated as follows:
  - For $|a\omega| \ll 1$, $20\log_{10}|aj\omega + 1| \approx 20\log_{10}(1) = 0[dB]$–that is to say, for low frequencies, this magnitude is a horizontal line. This is known as the *low-frequency asymptote.*
  - For $|a\omega| \gg 1$, $20\log_{10}|aj\omega + 1| \approx 20\log_{10}(|a\omega|)$–that is to say, for high frequencies, this magnitude is a straight line with a slope of 20 (dB/decade) which crosses the horizontal axis (0 (dB)) at $\omega = |a|^{-1}$. This is known as the *high-frequency asymptote.*
  - The phase response is more complex. It roughly changes over two decades. One decade below $|a|^{-1}$ the phase is approximately zero. One decade above $|a|^{-1}$ the phase is approximately sign $(a)0.5\pi$ (rad). Connecting the points $(0.1|a|^{-1}, 0)$ and $(10|a|^{-1}, 0)$ by a straight line gives sign $(a)0.25\pi$ for the phase at $\omega = |a|^{-1}$. This is a very rough approximation.
- For $a = a_1 + ja_2$, the phase Bode diagram of the factor $as + 1$ corresponds to the angle of the complex number with real part $1 - \omega a_2$ and imaginary part $a_1\omega$.

4.12.3 Example: Simple Transfer Function

Consider a transfer function given by

$$H(s) = 640\frac{(s+1)}{(s+4)(s+8)(s+10)}. \tag{4.64}$$

To draw the asymptotic behaviour of the gain diagram, the function $H(s)$ is first rearranged into the form shown in Eq. (4.61); this leads to

$$H(s) = 2\frac{(s+1)}{(0.25s+1)(0.125s+1)(0.1s+1)}. \tag{4.65}$$

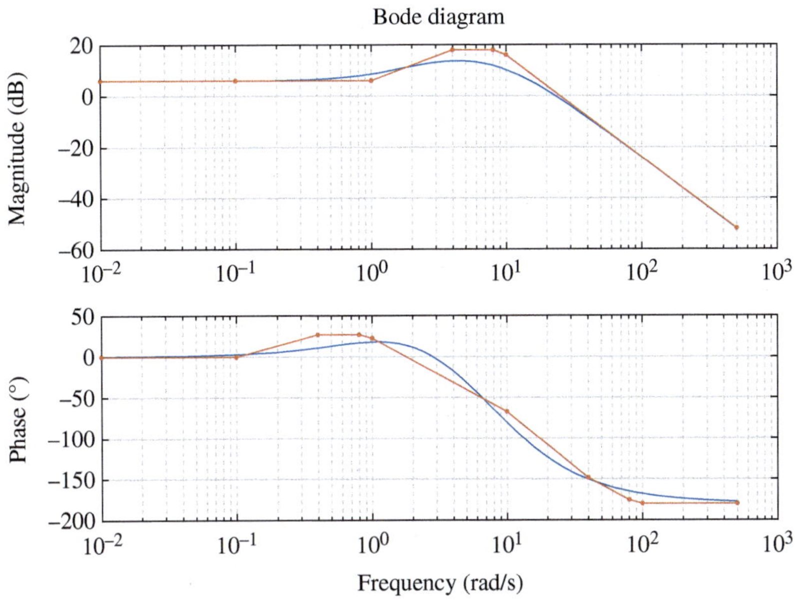

Figure 4.4 Exact (blue) and asymptotic (red) Bode plots.

The Bode diagram, thus, has a constant factor ($K = 2$) and four factors of the form $as + 1$. Using the rules outlined above, the asymptotic magnitude diagram and the asymptotic phase diagram can be obtained. Both of these are shown in Figure 4.4, together with the exact Bode plots.

Remark 4.3 Asymptotic approximations of the type outlined above are useful to gain insight. However, exact Bode diagrams can be easily obtained using commonly available packages such as MATLAB® or GNU Octave.

4.13 Revision Questions

1. What is the time response, $y(t)$, if $\mathscr{L}\,[y(t)] = Y(s) = \frac{1}{s(s+1)}$?
2. Consider the following state-space model:

$$\dot{x}_1 = x_2 + u \tag{4.66}$$

$$\dot{x}_2 = -x_2 + u \tag{4.67}$$

$$y = x_1 \tag{4.68}$$

   What is the transfer function from u to y?
3. What are the poles and zeros of the system in Question 2?
4. What the impulse response and step response of the system found in Question 2?
5. What are the poles and zeros of the system in Eq. (4.38) in the text?
6. Consider a system having transfer function $T(s) = \frac{1}{s+3}$.
   (a) Is the system stable?
   (b) What is the steady-state output if the input is $u(t) = 6\sin(4t)$?
7. If $y(0) = 1$; and an input $u(t) = \cos(4t)$ is applied at $t = 0$ to the system in Question 6, what is the response from time 0 onwards?

8. Plot the Nyquist plot of $T(s) = \frac{1}{s+1}$.
9. Plot the straight-line approximation to the Bode diagram for $T(s) = \frac{1}{s+10}$.
10. For the system in Question 9, what is the true magnitude and phase at $\omega = 10$?
11. Consider again the system described in Question 5 of Chapter 3.
    Find the transfer functions linking u^* and q to u.
12. For the system in Question 11,
    (a) What are the poles of the system?
    (b) What are the zeros of the transfer function linking u^* to u?
    (c) What are the zeros of the transfer function linking q to u?

Further Reading

There exists a vast literature on linear system theory. The following sources offer further insights beyond those described in this chapter.

1 J.J. DiStefano, A.R. Stubberud and I.J. Williams, *Schaum's outline of theory and problems of feedback and control systems 2/e*, McGraw-Hill, 1995.
2 E. Euler, "De constructive aquatorium", *Opera Omnia*, 22:150–161, 1744.
3 M.F. Gardner, F. Murray and J.L. Barnes, *Transients in linear systems, studied by the Laplace transformation*, John Wiley & Sons, Inc., New York, 1942.
4 G.C. Goodwin, S.F. Graebe and M.E. Salgado, *Control system design*", Prentice Hall, 2000.
5 A.V. Oppenheim, A.S. Willsky and N.S. Hanid, *Signals and systems*, Prentice Hall, Englewood Cliffs, New Jersey, 1997.
6 L.R. Rabiner and B. Gold, *Theory and application of digital signal processing*, Prentice Hall, Englewood Cliffs, New Jersey, 1975.

5

Laboratory 1: Modelling of an Electromechanical Servomechanism

5.1 Introduction

Chapters 2–4 introduced a number of important 'tools' that allow one to manipulate and understand linear feedback systems. These tools all depend on having a model for the system of interest. Thus, modelling is a core concept in control system design. This chapter will expand the reader's capacity to understand control concepts by building a model for a simple real system.

In particular, the virtual laboratory system will be used to experimentally explore a simple servo system.

5.2 The Physical Apparatus

Electromechanical servomechanisms are frequently encountered in practice. They can be used as positioning systems or drive systems. At the core of these systems, one often finds a DC (direct current) motor. The DC motor is an actuator that converts electrical energy into rotational mechanical energy. The latter is then applied to the load typically via a gearbox. DC motors are used in many control applications including robotics, machine tools and valve actuators, as well as, disk and CD drives. A real-world application to yaw control of a wind power generator will be studied in Section 11.2 of Chapter 11. In a sense, such systems represent the simplest possible dynamics and are thus a worthy place to begin examining control system design. Surprisingly, one can learn a great deal by exploring the control of such systems.

The goal of this emulated laboratory is to study the dynamics of a DC electromechanical servo system. The virtual laboratory is based on an automatic door-opening system for an elevator. The physical equipment providing motivation for this virtual laboratory is shown in Figure 5.1. The test setup consists of the door opening mechanics, drive belt, the servo motor (including a gearbox and load equivalent in weight to the doors), an embedded microcontroller to implement the control strategy and a data logger to collect and analyse the I/O signals of the door opener.

Screenshots of the emulated virtual laboratory setup are shown in Figures 5.2 and 5.3.

Note that several non-ideal features of the physical setup are included in this virtual laboratory including power amplifier saturation limits, position travel limits and signal noise. Thus, the laboratory is more than a theoretical exercise since the emulation captures many of the real-world phenomena met in practice.

Practical Control System Design: Real World Designs implemented on Emulated Industrial Systems,
First Edition. Adrian Medioli and Graham Goodwin.

Companion website: www.wiley.com/go/medioli/practicalcontrolsystemdesign

Figure 5.1 The physical equipment.

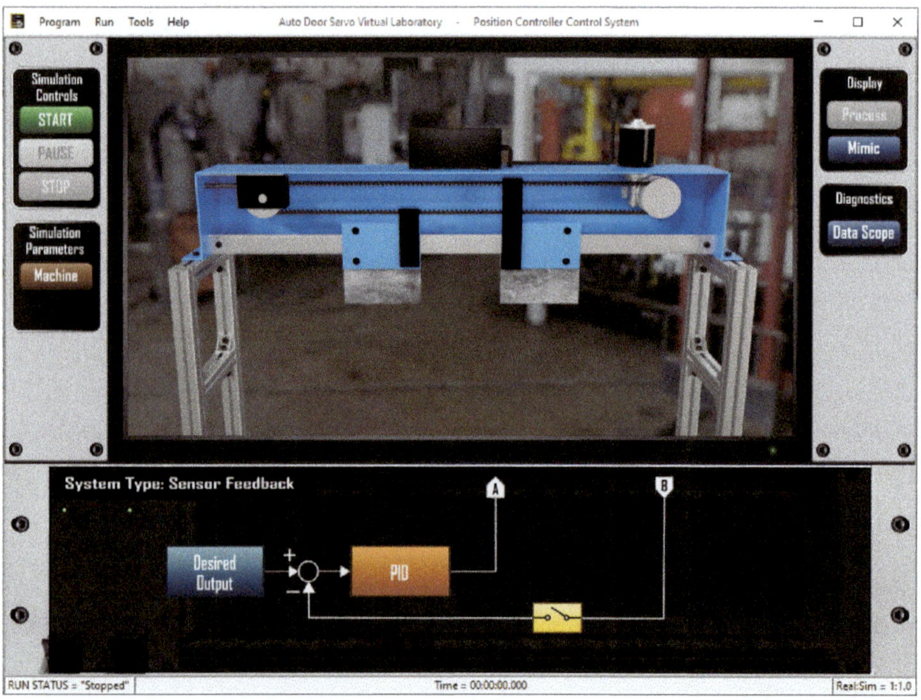

Figure 5.2 Screen shot of program.

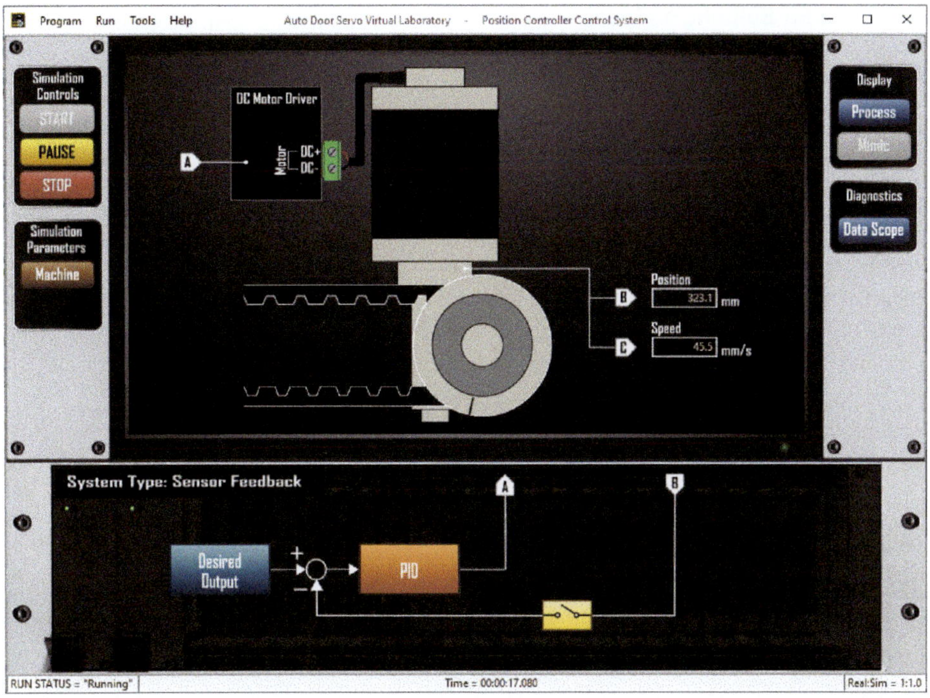

Figure 5.3 The servomechanism.

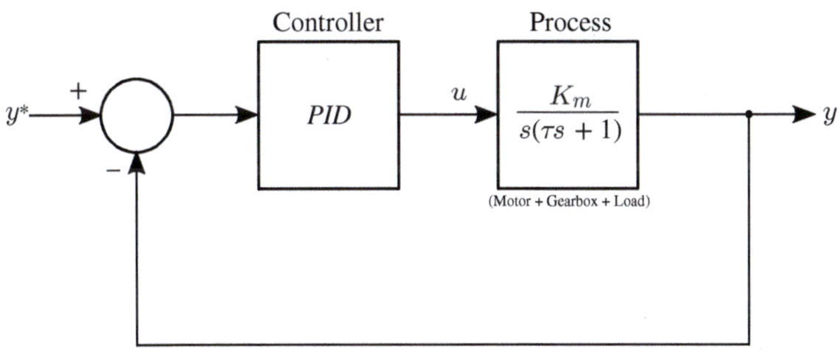

Figure 5.4 Block diagram of the servo system.

A simplified representation of the system is given in Figure 5.4. In describing the system, it is not necessary to keep track of the preamplifier and power amplifier gains or the gear box reduction ratio. Instead it suffices to refer to the combined gain, K_m, as the 'DC gain of the motor'. It combines all of the above gain terms including the DC gain of the motor itself. Likewise, the output, y, will be taken to be the position of the door edge which is linearly related to the angular displacement of the gearbox output shaft.

5.3 Estimation of Motor Parameters

The purpose of the following experiments is to determine parameters K_m and τ in the process model as shown in Figure 5.4. To obtain these parameters, two alternative tests will be used, namely a step response and the frequency response.

5.3.1 Motivation for Building a Model

The reader may well feel that this is such a simple system that building a model is an unnecessary step. To motivate the step of building a model, it is suggested that the reader carry out the following experiment:

5.3.2 Experiment: Why Build a Model?

1. Start the virtual laboratory.
2. Close the loop by ensuring that the switch in the feedback path is in the closed position.
3. Choose the 'Desired Output' as a sine wave with off-set 100, frequency 0.05 Hz, duty cycle 50%. Use a proportional controller with gain 1.
4. Observe the output response. Do you feel it is satisfactory?

The reader probably found that the output response was bizarre! understanding why this has occurred requires that a simple model be built.

5.3.3 Step Response Testing

Before doing this test, it is suggested that the reader revise some of the material presented in Chapter 4.

The model linking the input voltage, u to, the output position, y, is of the form:

$$\tau \frac{d\dot{y}}{dt} = -\dot{y} + K_m u \tag{5.1}$$

where $\dot{y}$ denotes 'rate of change' or velocity.

Using Laplace Transforms, the solution to this equation for a step input of size M is

$$\dot{y}(t) = \dot{y}(0) + K_m M \left[1 - e^{-\frac{t}{\tau}}\right] \tag{5.2}$$

where τ is the time constant.

Observe that, as $t \to \infty$, then the change in the steady state velocity becomes

$$\dot{y}_{ss} - \dot{y}(0) = K_m M. \tag{5.3}$$

Thus, the gain can be determined by measuring the steady-state change in velocity resulting from a step input. In the physical equipment, it is not sensible to apply a constant voltage for a long period of time since the door opener has travel limits, i.e. fully closed and fully open. Instead a sequence of (sufficiently long) plus and minus steps will be used to sequentially open and close the door. The step size is taken to be M and the steps are held for a sufficiently long time such that the velocity (almost) reaches a constant value but not long enough for the door to reach its limits. Note that the input goes from a value $+M$ to $-M$. Hence,

$$K_m = \frac{\text{Change in Velocity}}{2M}. \tag{5.4}$$

Next, to measure the time constant, an estimate is made of the time at which the velocity has reached 63.2% of its final value. To see how this works, note that at $t = \tau$, for zero initial conditions,

$$\begin{aligned} \dot{y}(t = \tau) &= K_m M \left[1 - e^{-1}\right] \\ &= 0.632\, \dot{y}_{ss}. \end{aligned} \tag{5.5}$$

Thus, the time constant τ is equal to the time required for the output to increase to 63.2% of its ultimate steady state value in response to a step increase in input.

In case, the reader should feel that step-response testing is too simplistic to be used in practice, the authors have frequently used this tool in practice to study complex real-world problems. For example, they recently used this tool to obtain a model of a two-input-two output system for a sugar mill furnace. The main practical issue was that the system needed to be operated in open-loop which may not always be feasible.

5.3.4 Experiment: Measuring the Open-Loop Gain and Time Constant

1. Start the virtual laboratory program.
2. Before proceeding it may be helpful for the reader to explore the operation of the signal generator and data scope. For example, the reader should explore the travel limits of the device. (Hint one end is at zero and the other is at about 200 units of position travel.)
3. This experiment will be performed 'in open-loop', i.e. without feedback between the system outputs and inputs. To ensure this is the case, confirm that the switch, in the position feedback path, is in the open state.
4. Next, drive the system using the direct output of the reference signal generator. To achieve this click on the PID block and set the proportional term K_p to a unity gain whilst ensuring that K_i and K_d are zero.
5. Use the reference signal generator to apply a suitable square wave.[1] Note that using a square wave input causes the output to run backwards and forwards. Increasing the voltage increases the speed and a negative voltage results in the shaft rotating in the opposite direction.
6. Using the Data Logger, observe the position reference $y*$ (tag 'y*') and the output speed and estimate the gain and time constant.
7. Note that the door might hit one of the end-stops (i.e. it is fully open or fully closed). The reader will appreciate that this issue is one that is consistent with the operation of the real system. Thus, finding a way around this problem is a genuine practical aspect of this particular experiment.

The output velocity is directly measured. Actually, this is an estimate of the velocity obtained by differencing two successive position measurements and dividing by the time-step.

5.3.5 Frequency Response

An alternative method is next used to find the system parameters based on sinusoidal signals. An immediate problem is encountered, namely, the system is not easy to operate in open loop. This is easily remedied by utilising a very simple control law to give acceptable closed-loop operation. In particular, it is suggested that the reader close the feedback loop and set the proportional gain of the PID controller to a value of 2.0.

Next the resulting closed-loop system should be driven by a sinusoidal reference of amplitude 1.0 and different frequencies and the relative magnitude and phase of the input and output measurement.

The input to the servo and the velocity of the output will be measured. Using ideas from Chapter 4, it follows that the corresponding frequency response is the ratio of the output to input and takes the specific form:

$$G(j\omega) = \frac{K_m}{(\tau j\omega + 1)} \tag{5.6}$$

1 Try: Offset = 100, Peak amplitude = 10, Frequency $\simeq$ 0.03 Hz, Duty cycle = 50% and Noise SD = 0.

Thus, it can be seen that at the particular frequency $\omega = 1/\tau$, the response between input voltage and output velocity is

$$G(j/\tau) = \frac{K_m}{(j+1)}, \\ = \frac{K_m}{\sqrt{2}} e^{-j\frac{\pi}{4}}. \quad (5.7)$$

Thus, the frequency at which the phase shift is 45° is $\omega = 1/\tau$. Also the gain is $K_m = \sqrt{2}|G(j/\tau)|$.

An industrial application of the idea of measuring the ratio between output and input when in closed loop will be described in Section 30.2 of Chapter 30.

5.3.6 Experiment: Measuring Frequency Response

1. Start the virtual laboratory program.
2. This experiment will be performed 'in closed-loop' i.e. with feedback between the system outputs and inputs. To ensure this is the case, confirm that the switch, in the position feedback path, is in the closed state.
3. Next, drive the system using the reference signal generator. To achieve this click on the PID block and set the proportional term K_p to a gain of 1.0 and the derivative term $K_d = 0.5$ whilst ensuring that K_i and T_s are zero.
4. Use the reference signal generator to apply a suitable sinusoidal signal.[2]
5. Using the Data Logger, observe the motor input u (tag 'u') and the motor speed (tag 'omega').
6. Adjust the input frequency of the reference signal, until the phase difference is $-\pi/4$ (45° lag).[3] Measure the corresponding frequency and amplitude ratio and estimate the gain and time constant.
7. Again be careful of saturation at the two ends of travel. This will require that the reference signal be offset.

5.3.7 Experiment: Alternative Measurement of Frequency Response

The same idea can be used by comparing the input with the output position save that now it is necessary to look for the frequency that gives 135° phase lag since 90° phase shift is introduced by the integrals between $\dot{y}$ and y. Also, the gain is now $\sqrt{2}\omega^*|G(j\omega^*)|$ where ω^* is the frequency at which the phase shift is 135°.

1. Start the virtual laboratory program.
2. This experiment will be performed 'in closed-loop', i.e. with feedback between the system outputs and inputs. To ensure this is the case, confirm that the switch, in the position feedback path, is in the closed state.
3. Next drive the system using the reference signal generator. To achieve this click on the PID block and set the proportional term K_p to a gain of 2.0 whilst ensuring that K_i and K_d are zero.
4. Use the reference signal generator to apply a suitable sinusoidal signal.[4]
5. Using the Data Scope, observe the motor input u (tag 'u') and the motor position (tag 'y_m').
6. Adjust the input frequency of the reference signal, until the phase difference is 135°. Measure the corresponding frequency and amplitude ratio and estimate the gain and time constant.

---

2 Hint: Offset = 100, Peak amplitude = 10, Frequency = Try 0.03 Hz and Noise SD = 0.

3 This is easily done by comparing the times between zero crossings as a fraction of the sine wave period.

4 Hint: Offset = 100, Peak amplitude = 10, Frequency ≃ 0.03 Hz and Noise SD = 0.

5.4 Revision Questions

1. Consider the door opening system with transfer function

$$G(s) = \frac{K}{s(\tau s + 1)}$$

   Derive an expression for the magnitude and phase of $G(s)$ as a function of frequency ω.
2. What is the magnitude and phase as $\omega \to 0$?
3. What is the magnitude and phase as $\omega \to \infty$?
4. What is the magnitude and phase at $\omega = \frac{1}{\tau}$?
5. What is the magnitude and phase at $\omega = \frac{2}{\tau}$?
6. What is the magnitude and phase at $\omega = \frac{1}{2\tau}$?
7. Compare the exact magnitude and phase at $\omega = \frac{1}{\tau}$ with the values given by the straight line approximation to the Bode diagram.

Further Reading

Servo mechanisms are treated in many introductory books on control. The following two books provide useful further information.

1 J. DiStefano, A. Stubberud and I. Williams, *Feedback and control systems*, McGraw-Hill, New York, 1976.
2 G.C. Goodwin, S.F. Graebe and M.E. Salgado, *Control system design*, Prentice Hall, 2000.

Part II

Control System Design Techniques for Linear Single-Input Single-Output Systems

So far, the book has described methods that can be used to analyse the response of linear systems. The next step is how to design a feedback controller for a linear system. This part of the book will present several cornerstone design methods for linear single-input single-output systems. Chapter 6 develops some necessary background theories. Chapter 7 describes some simple design methods for linear single-input single-output systems. Chapter 8 uses the virtual laboratory for the door-opening system to explore the basic principles of linear feedback design. Chapter 9 studies a continuous-casting machine and explores simple linear control system design techniques for this problem. Chapter 10 uses the virtual laboratory system to study level control in single and two-tank systems. Chapter 11 uses the virtual laboratory to study the mechanical aspects of wind power generation. Chapter 12 extends the theory to cover systems having pure time delays. Chapter 13 uses the virtual laboratory to study control system design for a rolling mill when time delays are included. Chapter 14 extends the background theory to cover open-loop unstable systems. Chapter 15 uses the virtual laboratory to study the control of a rocket in two-dimensional flight. This is a real-world example of a system that is open-loop unstable.

Practical Control System Design: Real World Designs implemented on Emulated Industrial Systems,
First Edition. Adrian Medioli and Graham Goodwin.

Companion website: www.wiley.com/go/medioli/practicalcontrolsystemdesign

6

Analysis of Linear Feedback Systems

6.1 Introduction

Chapter 5 was a first excursion into the real world of control system analysis. In that chapter, simple modelling ideas were applied to a real system. (The door-opening servo.) Here a step further is taken. In particular, for a given controller and plant connected in feedback, an answer is provided to the following questions:

- Is the loop stable?
- What are the sensitivities to various disturbances?
- What is the impact of modelling errors?

Several analysis tools are also introduced, e.g. root locus (RL) and Nyquist stability.

These ideas will further enrich the 'language' that can be used to talk about control problems.

The contents of this chapter are an abridged version of the theoretical development explained in more detail in the references provided at the end of the chapter.

6.2 Feedback Structures

Recall the preliminary introduction to feedback control given in Chapter 2. The topic will now be specifically focused on linear single-input single-output (SISO) control systems. As an extension of the algebraic insights introduced in Chapter 2, feedback will be shown to have many desirable properties, such as the capacity to reduce the impact of disturbances and to decrease sensitivity to model errors. However, it will also be shown that ill-applied feedback can make a previously stable system unstable, or result in high sensitivity to measurement noise. All of these ideas were seen, in a simple algebraic setting, in Chapter 2. Here, these concepts will be extended to models expressed in the Laplace and Frequency domains. The key conceptual change is that, whereas in Chapter 2, the system 'gain' was a simple real scalar, the 'gain' will now be viewed as a transfer function, $G(s)$, i.e. a function of the Laplace Transform variable s or equivalently, when $s = j\omega$, of the frequency domain variable, ω.

The analysis of feedback systems will commence with the linear SISO configuration shown in Figure 6.1. Initial attention is given to the so-called *nominal loop*, that is, the effect of the controller interacting with the *nominal plant model* in feedback; later, in Section 6.10, attention will turn to the impact of modelling errors that arise when the controller is applied to the true system, which will invariably differ from the nominal model.

Practical Control System Design: Real World Designs implemented on Emulated Industrial Systems,
First Edition. Adrian Medioli and Graham Goodwin.

Companion website: www.wiley.com/go/medioli/practicalcontrolsystemdesign

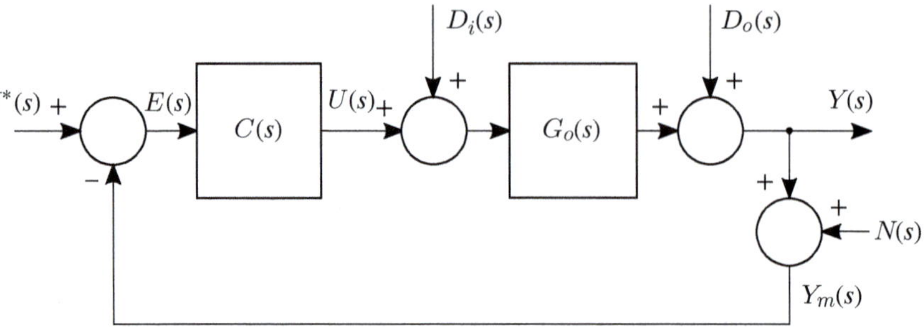

Figure 6.1 Simple feedback control system.

In the loop shown in Figure 6.1, transfer functions and Laplace transforms have been utilised to describe the relationships between signals in the loop. In particular, $C(s)$ and $G_o(s)$ denote the transfer functions of the controller and the nominal plant model, respectively. These can be represented as rational transfer functions in the Laplace Transform variable, s, as follows:

$$C(s) = \frac{P(s)}{L(s)}, \tag{6.1}$$

$$G_o(s) = \frac{B_o(s)}{A_o(s)}, \tag{6.2}$$

where $P(s)$, $L(s)$, $B_o(s)$, and $A_o(s)$ are polynomials in s. $Y^*(s)$, $U(s)$ and $Y(s)$ will denote the Laplace transforms of set-point, control signal and plant output, respectively, $D_i(s)$, $D_o(s)$, and $N(s)$ will denote the Laplace transforms of input disturbance, output disturbance and measurement noise, respectively.

Using the Laplace transform ideas explained in Chapter 4, the following relationships are seen to hold between the variables in Figure 6.1. (Here, and in the sequel, it will be assumed that the system has zero initial conditions.)

$$Y(s) = G_o(s)U(s) + D_o(s) + G_o(s)D_i(s), \tag{6.3}$$

$$U(s) = C(s)Y^*(s) - C(s)Y(s) - C(s)N(s), \tag{6.4}$$

$$= C(s)\left(Y^*(s) - N(s) - G_o(s)U(s) - D_o(s) - G_o(s)D_i(s)\right). \tag{6.5}$$

The above equations can be rearranged to yield:

$$U(s) = \frac{C(s)}{1 + G_o(s)C(s)}\left(Y^*(s) - N(s) - D_o(s) - G_o(s)D_i(s)\right) \tag{6.6}$$

and

$$Y(s) = \frac{1}{1 + G_o(s)C(s)}\left[G_o(s)C(s)(Y^*(s) - N(s)) + D_o(s) + G_o(s)D_i(s)\right]. \tag{6.7}$$

The closed-loop control configuration shown in Figure 6.1, is called a *one-degree-of-freedom* (one-d.o.f.) architecture. This term reflects the fact that there is only one-d.o.f. available, namely, $C(s)$, to shape the transfer functions from $Y^*(s)$, $N(s)$, $D_o(s)$ and $D_i(s)$ to $Y(s)$. Hence, if the controller transfer function $C(s)$ is designed to give a particular reference-signal response, for example,

$$\frac{Y(s)}{Y^*(s)} = \frac{G_o(s)C(s)}{1 + G_o(s)C(s)} = T_o(s), \tag{6.8}$$

then this induces a unique output-disturbance response,

$$\frac{Y(s)}{D_o(s)} = \frac{1}{1 + G_o(s)C(s)} = S_o(s), \tag{6.9}$$

without any further design freedom.

Analogously to the functions previously defined in Chapter 2, the functions $L(s) = G_o(s)C(s)$, $T_o(s) = \frac{G_o(s)C(s)}{1+G_o(s)C(s)}$ and $S_o(s) = \frac{1}{1+G_o(s)C(s)}$ are called the loop transfer function, the nominal complementary sensitivity and nominal sensitivity, respectively.

6.3 Nominal Sensitivity Functions

It can be seen from Eqs. (6.6) and (6.7) that the closed-loop response is governed by four transfer functions, which are collectively known as the *sensitivity functions*. These functions were previously described in Section 2.7 of Chapter 2 in a purely algebraic setting.

Using Eqs. (6.1) and (6.2), these sensitivity functions are seen to be as given below in (6.10) and (6.13).

$$T_o(s) \triangleq \frac{G_o(s)C(s)}{1 + G_o(s)C(s)} = \frac{B_o(s)P(s)}{A_o(s)L(s) + B_o(s)P(s)}. \tag{6.10}$$

$$S_o(s) \triangleq \frac{1}{1 + G_o(s)C(s)} = \frac{A_o(s)L(s)}{A_o(s)L(s) + B_o(s)P(s)}. \tag{6.11}$$

$$S_{io}(s) \triangleq \frac{G_o(s)}{1 + G_o(s)C(s)} = \frac{B_o(s)L(s)}{A_o(s)L(s) + B_o(s)P(s)}. \tag{6.12}$$

$$S_{uo}(s) \triangleq \frac{C(s)}{1 + G_o(s)C(s)} = \frac{A_o(s)P(s)}{A_o(s)L(s) + B_o(s)P(s)}. \tag{6.13}$$

$T_o(s)$: Nominal complementary sensitivity.

$S_o(s)$: Nominal sensitivity.

$S_{io}(s)$: Nominal input-disturbance sensitivity.

$S_{uo}(s)$: Nominal control sensitivity.

All of the sensitivity functions share a common denominator of $A_{cl} = A_o(s)L(s) + B_o(s)P(s)$.

The polynomial $A_{cl} \triangleq A_o(s)L(s) + B_o(s)P(s)$ is called the nominal closed-loop characteristic polynomial.

The sensitivity functions are algebraically related. These relations are one of the key manifestations of trade-offs inherent in feedback control. These relationships can be easily derived from the definitions (6.10)–(6.13). In particular, the following relationships (shown in (6.14)–(6.16)) hold.

$$S_o(s) + T_o(s) = 1, \tag{6.14}$$

$$S_{io}(s) = S_o(s)G_o(s) = \frac{T_o(s)}{C(s)}, \tag{6.15}$$

$$S_{uo}(s) = S_o(s)C(s) = \frac{T_o(s)}{G_o(s)}. \tag{6.16}$$

We note that (6.14) is the dynamic version of (2.32) in Chapter 2.

With these sensitivities, the closed-loop in Figure 6.1 is described by

$$Y(s) = T_o(s)\,(Y^*(s) - N(s)) + S_o(s)D_o(s) + S_{io}(s)D_i(s), \tag{6.17}$$

$$U(s) = S_{uo}(s)\left(Y^*(s) - N(s) - D_o(s) - G_o(s)D_i(s)\right). \tag{6.18}$$

Equations (6.10)–(6.13) can be expressed in more compact form as

$$\begin{bmatrix} Y_o(s) \\ U_o(s) \end{bmatrix} = \frac{\begin{bmatrix} G_o(s)C(s) & G_o(s) & 1 & -G_o(s)C(s) \\ C(s) & -G_o(s)C(s) & -C(s) & -C(s) \end{bmatrix}}{1 + G_o(s)C(s)} \begin{bmatrix} H(s)Y^*(s) \\ D_i(s) \\ D_o(s) \\ N(s) \end{bmatrix}. \tag{6.19}$$

One can now see from (6.10) to (6.13) that the polynomial A_{cl} plays a central role. It determines both the stability and, together with $L(s)$, $P(s)$, $B_o(s)$ and $A_o(s)$, the transient characteristics of the nominal control loop.

6.4 Analysing Stability Using the Characteristic Polynomial

Various tools are next explored to analyse nominal stability.[1]

The following definition is introduced:

Definition 6.1 ***(Internal stability)*** The nominal loop is said to be internally stable if and only if all four transfer functions seen in Eq. (6.19) are stable.

The above definition is equivalent to demanding that all signals in the loop be bounded for every set of bounded inputs $r(t), d_i(t), d_o(t)$ and $d_m(t)$. Comparison of (6.19) with (6.10)–(6.13) shows that the nominal closed-loop is internally stable if and only if the polynomial $A_{cl}(s) = A_o(s)L(s) + B_o(s)P(s)$ has stable factors. The polynomial $A_{cl}(s)$ is called the Characteristic Polynomial This idea is summarised as follows:

Theorem 6.1 ***(Result – nominal internal stability)*** Consider the nominal closed-loop depicted in Figure 6.1 with the model and the controller defined by the representation given by Eqs. (6.2) and (6.1), respectively. Assume that no unstable roots have been cancelled. Then the nominal closed-loop is internally stable if and only if the zeros of the nominal closed-loop characteristic equation

$$A_o(s)L(s) + B_o(s)P(s) = 0, \tag{6.20}$$

all lie in the open left-half plane.

Proof: Straightforward from Eq. (6.19) and the ideas described in Section 4.7.

Note that the idea of internal stability implies more than just the stability of the reference to output transfer function. It requires that, in addition, there are no cancellations of unstable poles between controller and the plant. This is illustrated by the following example.

---

1 A nominal control loop results from connecting a controller to the nominal model.

6.4.1 Example: Pole-Zero Cancellation

Assume that

$$G_o(s) = \frac{3}{(s+4)(-s+10)} \qquad C(s) = \frac{-s+10}{s}. \tag{6.21}$$

It can be seen that $T_o(s)$ is stable (after cancellation); however, the nominal input-disturbance sensitivity is unstable, because

$$S_{io}(s) = \frac{3s}{(-s+10)(s^2+4s+3)}. \tag{6.22}$$

Thus, this system is not internally stable and does not satisfy the conditions of (6.20), because $A_o(s)L(s) + B_o(s)P(s) = (-s+10)(s^2+4s+3)$.

6.5 Stability and Polynomial Analysis

Further issues that surround whether or not a given polynomial corresponds to a stable polynomial (i.e. has all zeros in the open left-half complex plane) are next explored. In particular, some simple graphical procedures will be developed that can be used to test stability. Beyond stability, these procedures also often give valuable insights into the proximity to instability.

6.5.1 Stability via Evaluation of the Roots

Consider a polynomial $p(s)$ defined by

$$p(s) = s^n + a_{n-1}s^{n-1} + \cdots + a_1 s + a_0, \tag{6.23}$$

where $a_i \in \mathbb{R}$.

Obviously, stability of $p(s)$ could be directly tested by calculating the roots of the equation $p(s) = 0$. This could be applied to the closed-loop characteristic polynomial given in (6.20). However, this is rather 'brute force' and yields little insight into what is causing stability or instability to arise. Some graphical tools which can be a useful way of gaining further insight into control loop performance will be described below.

Before doing that, it is to be noted that a necessary (but not sufficient) condition for closed-loop stability is that all of the coefficients of the closed-loop characteristic polynomial be positive. This can be easily seen by examining the expanded form of

$$P(s) = (s+\alpha_1)(s+\alpha_2)\ldots(s+\alpha_n), \tag{6.24}$$

where $\{\text{real part of}\, \alpha_i\} > 0$.

6.6 Root Locus (RL)

A simple classical tool used to study stability of equations of the type given in (6.20) is *Root Locus* (RL). The RL approach can be used to examine the location of the roots of the characteristic polynomial as one parameter is varied. This can be used to give insight into the loop performance. A brief introduction to the idea is provided below.

Assume that the nominal plant model is given by $G_o(s) = K_p\overline{G}_o(s)$ where $\overline{G}_o(s)$ is the ratio of two monic polynomials. Also, assume that the controller has a transfer function described as $C(s) = K_cC_a(s)$, where $C_a(s)$ is a known quotient of two monic polynomials in s, and where K_c is a positive constant whose value can vary. Then the closed-loop poles are the roots of

$$1 + KC_a(s)\overline{G}_o(s) = 0; \quad K = K_pK_c. \tag{6.25}$$

The set of all points in the complex plane that satisfy (6.25) as the value of K is varied over all positive values is known as the *RL*. This particular problem can be stated more generally as follows:

Consider the following equation

$$1 + \lambda F(s) = 0 \quad \text{where} \quad F(s) = \frac{M(s)}{D(s)}, \tag{6.26}$$

with $\lambda \geq 0$ and

$$M(s) = s^m + b_{m-1}s^{m-1} + \cdots + b_1s + b_0 = \prod_{i=1}^{m}(s - c_i), \tag{6.27}$$

$$D(s) = s^n + a_{n-1}s^{n-1} + \cdots + a_1s + a_0 = \prod_{i=1}^{n}(s - p_i), \tag{6.28}$$

where the coefficients of the polynomials $M(s)$ and $D(s)$ are real numbers. Then the solution to the RL problem requires one to find the set of all points in the complex plane that are solutions for (6.26) for all positive values of λ.

Observe that the solution of Eq. (6.26) is also the solution to the equation

$$D(s) + \lambda M(s) = \prod_{i=1}^{n}(s - p_i) + \lambda\prod_{i=1}^{m}(s - c_i) = 0. \tag{6.29}$$

Prior to modern computers, RL methods were an important technique used by designers to get a 'feel' for the impact that one parameter (usually the controller gain) has on the stability and dynamic behaviour of the closed-loop. The RL is easily found using software, tools. Nevertheless, understanding the basic underlying patterns of RL behaviour can still be useful in understanding broad aspects of a feedback-control loop. The rules for constructing a RL follow relatively easily by thinking about the impact of Eq. (6.26). RL building rules include the following:

Rule 1: The root locus has $\max\{m, n\}$ *branches*.

Rule 2: From Eq. (6.26), we observe that s_0 belongs to the root locus (for $\lambda \geq 0$) if and only if

$$\arg F(s_0) = (2k + 1)\pi \quad \text{for} \quad k \in \mathbb{Z}. \tag{6.30}$$

Rule 3: From Eq. (6.26), it follows that if s_0 belongs to the root locus, the corresponding value of λ is λ_0, where

$$\lambda_0 = \frac{-1}{F(s_0)} \tag{6.31}$$

Rule 4: A point s_0 on the real axis, $s_0 \in \mathbb{R}$, is part of the root locus (for $\lambda \geq 0$) if and only if it is located to the left of an odd number of poles and zeros (so that (6.30) is satisfied).

Rule 5: When λ is close to zero, then n of the roots are located at the poles of $F(s)$, that are, at $p_1, p_2, \ldots, p_n$ and, if $n < m$, the other $m - n$ roots are located at ∞. (This statement will be made more precise below.)

Rule 6: When λ is close to ∞, then m of these roots are located at the zeros of $F(s)$, that is, at $c_1, c_2, \dots, c_m$ and, if $n > m$, the other $n - m$ roots are located at ∞. (This statement will also be made more precise below.)

Rule 7: If $n > m$ and λ tends to ∞, then, $n - m$ roots asymptotically tend to ∞, following asymptotes that intersect at $(\sigma, 0)$, where

$$\sigma = \frac{\sum_{i=1}^{n} p_i - \sum_{i=1}^{m} c_i}{n - m} \tag{6.32}$$

The angles of these asymptotes are $\eta_1, \eta_2, \dots, \eta_{n-m}$, where

$$\eta_k = \frac{(2k-1)\pi}{n-m}; \quad k = 1, 2, \dots, n - m \tag{6.33}$$

Rule 8: If $n < m$ and λ tends to zero, then $m - n$ roots asymptotically tend to ∞, following asymptotes that intersect at $(\sigma, 0)$, where

$$\sigma = \frac{\sum_{i=1}^{n} p_i - \sum_{i=1}^{m} c_i}{m - n} \tag{6.34}$$

The angles of these asymptotes are $\eta_1, \eta_2, \dots, \eta_{m-n}$, where

$$\eta_k = \frac{(2k-1)\pi}{n-m}; \quad k = 1, 2, \dots, m - n \tag{6.35}$$

Examples using root locus for design will be given in Section 7.3.1 of Chapter 7.

6.7 Nominal Stability Using Frequency Response

The stability tests described above are of a 'time-domain' nature since they directly apply to the differential equation model. However, as mentioned in Chapters 2 and 4, the system frequency response gives equivalent information, albeit in an entirely different format. This motivates an exploration of alternative stability tests based purely on frequency domain ideas.

A classical and important tool that can be used to assess the stability of a feedback loop in the frequency domain is Nyquist stability theory. In this approach, stability of the closed-loop is predicted by using the open-loop frequency response of the system. This is a beautifully elegant and simple method for assessing stability of a linear system. A polar diagram of the product $G_o(s)C(s)$ is first plotted as a function of $s = j\omega$ for ω varying from 0 to ∞. Then the number of encirclement of the (−1,0) points are counted. Why this works is explained below.

Consider first an arbitrary transfer function $F(s)$ (not necessarily related to closed-loop control. Nyquist stability theory depends upon mappings between two complex planes:

- The independent variable s
- The dependent variable F

The basic idea of Nyquist stability analysis is as follows:

The case of a simple function $F(s) = s - c$ with c lying *inside* a region enclosed by $\mathscr{C}_s$ is first studied. This is illustrated in Figure 6.2a.

It can be seen that, as s moves clockwise along $\mathscr{C}_s$, the angle of $F(s)$ changes by -2π (rad): the curve $\mathscr{C}_F$ will enclose the origin in $\boxed{F}$ once in the clockwise direction.

In Figure 6.2b, the case when c lies *outside* the region enclosed by $\mathscr{C}_s$ is illustrated. In this case, the angle of $F(s)$ does not undergo a complete rotation as s moves around the curve $\mathscr{C}_s$ and thus, no encirclement of the origin by $\boxed{F}$ occurs.

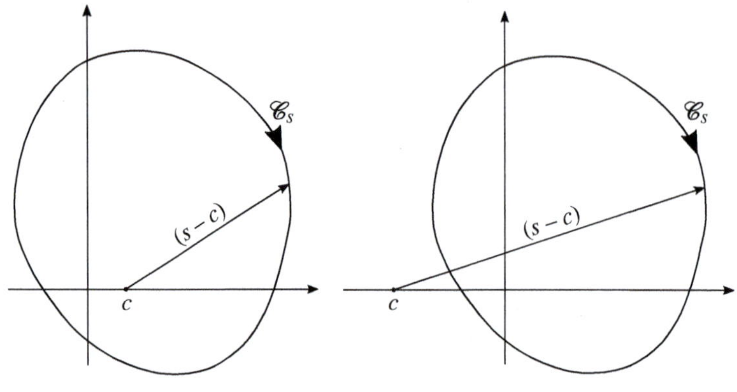

Figure 6.2 Single-zero function and Nyquist analysis. (a) c inside region and (b) c outside region.

By using similar reasoning, it follows that, for the function $F(s) = (s - p)^{-1}$ (where the pole p lies *inside* the region determined by $\mathscr{C}_s$), there is an angle change of $+2\pi$ (rad) when s moves clockwise along $\mathscr{C}_s$. This is equivalent to saying that the curve $\mathscr{C}_F$ encloses the origin in $\boxed{F}$ once in the counter-clockwise direction. The phase change is zero if p is *outside* the region, leading again to no encirclement of the origin in $\boxed{F}$.

Consider now the case in which $F(s)$ takes the following form:

$$F(s) = K\frac{\prod_{i=1}^{m}(s - c_i)}{\prod_{k=1}^{n}(s - p_k)}. \tag{6.36}$$

Then any net change in the angle of $F(s)$ results from the sum of the changes of the angle due to the factors $(s - c_i)$ minus the sum of the changes in the angle due to the factors $(s - p_k)$. This can be summarised as follows:

> Consider a function $F(s)$ as in (6.36) and a closed curve $\mathscr{C}_s$ in $\boxed{S}$. Assume that $F(s)$ has Z zeros and P poles inside the region enclosed by $\mathscr{C}_s$. Then, as s moves clockwise along $\mathscr{C}_s$, the resulting curve $\mathscr{C}_F$ encircles the origin in $\boxed{F}$ $Z-P$ times in a clockwise direction.

This result can be used to determine closed-loop stability for a feedback system. Specifically, to apply this result, a special function, $F(s)$, is considered which is related to the open-loop transfer function of plant, G_o, and a controller $C(s)$ as follows:

$$F(s) = 1 + G_o(s)C(s). \tag{6.37}$$

Note that the zeros of $F(s)$ are the closed-loop poles in a unity feedback control system having plant transfer function $G_o(s)$ and controller $C(s)$. The poles of $F(s)$ are the open-loop poles of plant and controller.

It is assumed that $G_o(s)C(s)$ is strictly proper, so that $F(s)$ in (6.37) satisfies

$$\lim_{|s|\to\infty} F(s) = 1. \tag{6.38}$$

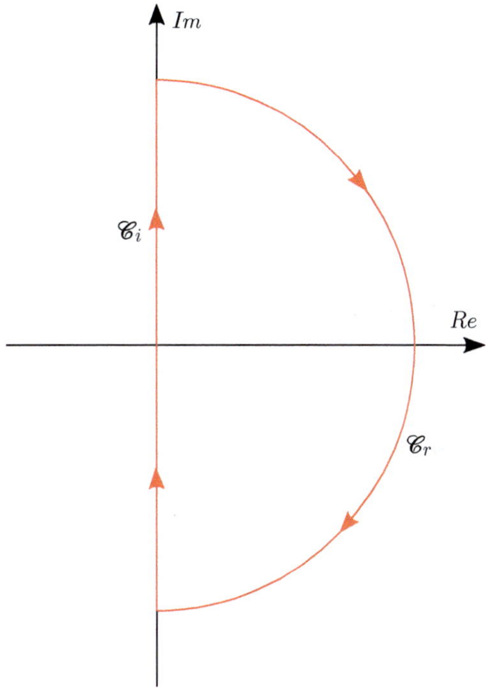

Figure 6.3 Nyquist path.

In the context of stability assessment, particular interest focuses on the number of closed-loop poles (if any) that lie in the right-half plane (RHP). Hence a particular curve $\mathscr{C}_s$ is considered which completely encircles the RHP in $\boxed{s}$ in the *clockwise* direction.

This curve combines the imaginary axis $\mathscr{C}_i$ and the return curve $\mathscr{C}_r$ (a semicircle of 'infinite' radius), as shown in Figure 6.3. This choice of $\mathscr{C}_s$ is known as the Nyquist path.

The closed curve $\mathscr{C}_F$ in the complex plane $\boxed{F}$ that results from evaluating $F(s)$ for every $s \in \mathscr{C}_s$ is next computed. $F(s)$ satisfies Eq. (6.38), so the whole mapping of $\mathscr{C}_r$ collapses to the point (1,0) in $\boxed{F}$. Thus, only the mapping of $\mathscr{C}_i$ has to be computed, i.e. one needs only to plot the *frequency response* $F(j\omega)$, $\forall\omega \in (-\infty, \infty)$ in the complex plane $\boxed{F}$. This is a polar plot of the frequency response, known as the Nyquist plot.

Since $F(s) = 1 + G_0(s)C(s)$, the zeros of $F(s)$ correspond to the closed-loop poles. Moreover, it can be seen that $F(s)$ and $G_0(s)C(s)$ share exactly the same poles (the open-loop poles). It can also be seen that the origin of the complex plane $\boxed{F}$ corresponds to the point (−1,0) in the complex plane $\boxed{G_0C}$. Thus, the Nyquist plot for F can be replaced by that of G_0C by simply counting encirclements about the −1 point. The basic Nyquist theorem, which derives from this line of reasoning, is as follows:

Theorem 6.2 ***(Nyquist stability)*** If a proper open-loop transfer function $G_0(s)C(s)$ has P (open-loop) poles in the open RHP and none on the imaginary axis, then the closed-loop has Z poles in the open RHP if and only if the polar plot $G_0(j\omega)C(j\omega)$ encircles the point (−1,0) clockwise $N = Z - P$ times.

From this theorem, it can be concluded that the following holds:

- If the system is open-loop stable ($P = 0$), then, for the closed-loop to be internally stable, it is necessary and sufficient that no unstable cancellations occur and that the Nyquist plot of $G_o(s)C(s)$ *not encircle the point* $(-1, 0)$.
- If the system is open-loop unstable, with P poles in the open RHP, then, for the closed-loop to be internally stable, it is necessary and sufficient that no unstable cancellations occur and that the Nyquist plot of $G_o(s)C(s)$ *encircle the point* $(-1, 0)$ P *times counter-clockwise.*
- If the Nyquist plot of $G_o(s)C(s)$ passes through the point $(-1, 0)$, there exists an $\omega_o \in \mathbb{R}$ such that $F(j\omega_o) = 0$, i.e. the closed-loop has poles located exactly on the imaginary axis. This situation is known as *marginal stability*.

There is one important remaining issue, namely, how to apply Nyquist's theory when there are open-loop poles exactly on the imaginary axis. The main difficulty in this case can be seen in Figure 6.2a. If c is located exactly on the curve $\mathscr{C}_s$, then the change in the angle of the vector $s - c$ is impossible to determine. To deal with this problem, a *modified Nyquist path* is employed, as shown in Figure 6.4. The modification is illustrated for the simple case when there is one open-loop pole at the origin.

The Nyquist path $\mathscr{C}_s$ is now composed of three curves: $\mathscr{C}_r$, $\mathscr{C}_a$, and $\mathscr{C}_b$. The resulting closed curve $\mathscr{C}_F$ will differ from the previous case only when s moves along $\mathscr{C}_b$. This is a semicircle of radius ϵ, where ϵ is an infinitesimal quantity. (In that way the encircled region is still the whole RHP, except for an infinitesimal area.)

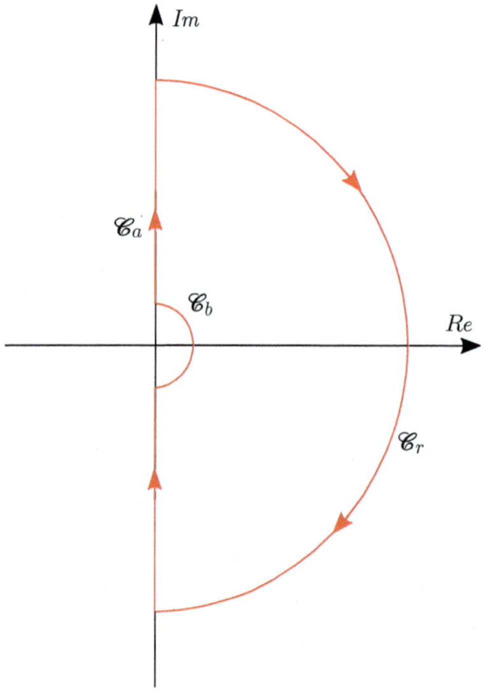

Figure 6.4 Modified Nyquist path.

Using the analysis developed above, it can be seen that one must compute the change in the angle of the vector $(s - p)$ when s travels along $\mathscr{C}_b$ with $p = 0$. This turns out to be $+\pi$ (rad), i.e. the factor s^{-1} maps the curve $\mathscr{C}_b$ into a semicircle with *infinite radius and clockwise direction.*

To compute the number and direction of encirclements, these infinite radius semi-circles must be considered, because they are effectively part of the Nyquist diagram.

The modified form of the Nyquist theorem to accommodate the above changes is shown in Figure 6.4.

The modified form of Theorem 6.2 is as follows:

Theorem 6.3 ***(Nyquist theorem)*** Consider a proper open-loop transfer function $G_o(s)C(s)$ with P poles in the open RHP, then the closed-loop has Z poles in the open RHP if and only if the plot of $G_o(s)C(s)$ encircles the point $(-1,0)$ clockwise $N = Z - P$ times when s travels along the modified Nyquist path.

6.8 Relative Stability: Stability Margins and Sensitivity Peaks

It has been shown above that the Nyquist criterion gives insights into the issue of whether, or not, a closed-loop system is stable. However, in control-system design, one often needs to go beyond the

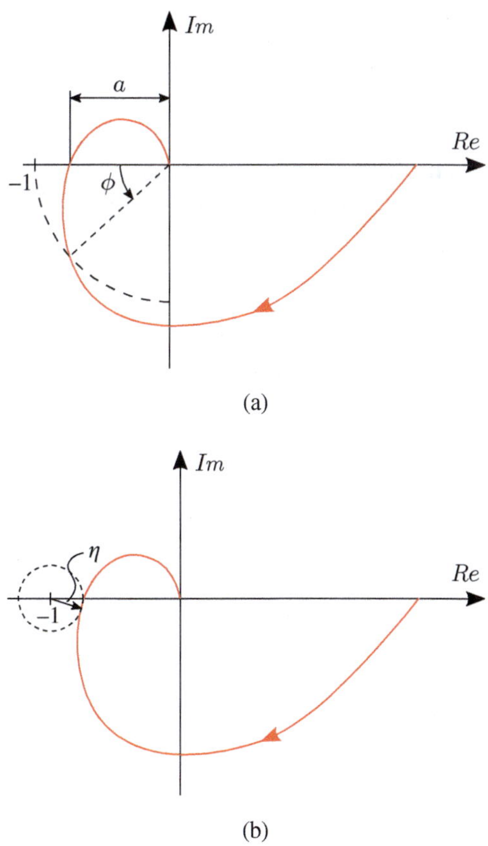

Figure 6.5 Stability margins and sensitivity peak. (a) Phase and gain margins and (b) sensitivity peak.

simple issue of closed-loop stability. In particular, it is usually desirable to obtain some quantitative measures of how far from instability the nominal loop is, i.e. to quantify *relative stability*. This can be readily achieved by introducing measures that describe the *distance* from the nominal open-loop frequency response to the critical stability point $(-1, 0)$.

Figure 6.5 shows two sets of relative stability indicators for the case when the open loop has no poles in the open RHP. Figure 6.5a describes the *gain margin*, M_g, and *the phase margin*, M_f. They are defined as follows:

$$M_g \triangleq -20\log_{10}(|a|), \tag{6.39}$$

$$M_f \triangleq \phi. \tag{6.40}$$

Thus, the gain margin indicates the additional gain that would take the closed-loop to the critical stability condition. The phase margin quantifies the pure phase lag that would achieve the same critical condition.

Figure 6.5b defines an alternative indicator for relative stability. Recall that a vector from $(-1, 0)$ to $G_o(j\omega_1)C(j\omega_1)$, for $\omega = \omega_1$, corresponds to $1 + G_o(j\omega_1)C(j\omega_1)$, i.e. to $|S_o(j\omega_1)|^{-1}$. Thus, the radius, η, of the circle tangent to the polar plot of $G_o(j\omega)C(j\omega)$ is the reciprocal of the *nominal sensitivity peak*. The larger the sensitivity peak is, the closer the loop will be to instability and the less tolerant it will be to modelling errors.

The sensitivity peak is an alternative indicator of relative stability to gain and phase margins.

6.9 From Polar Plots to Bode Diagrams

Stability margins can also be described and quantified by using Bode diagrams (see Chapter 4 for an introduction to Bode diagrams).

The gain margin can be determined by noting the gain when the phase is 180° lag. Similarly, the phase margin is the phase difference from 180° when the gain is 0 dB, see Figure 6.6.

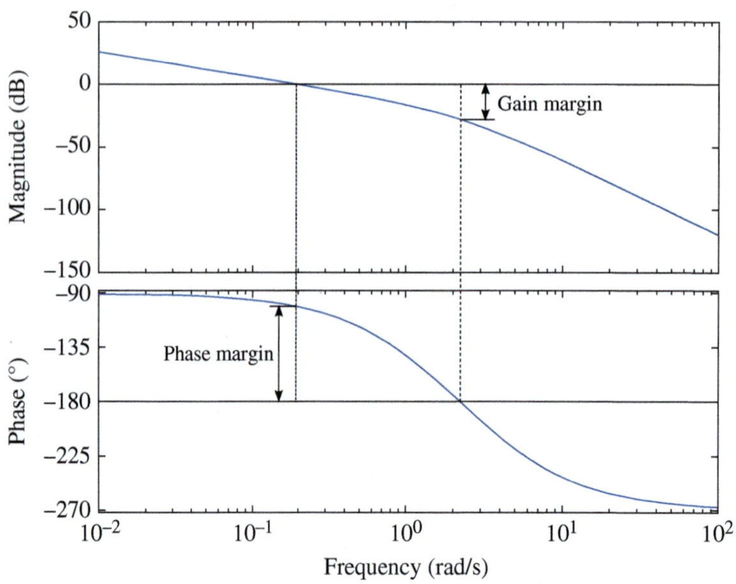

Figure 6.6 Stability margins in Bode diagrams.

6.10 Robustness

So far, only the effect that the controller has on the nominal closed-loop formed with the nominal model for the plant has been considered. In practice, one is invariably interested in, not only this nominal performance, but the true performance achieved when the controller is applied to the *true plant* which almost certainly differs from the nominal model. This is the so-called 'Robustness' issue. It will be shown below that the nominal sensitivities do indeed tell something about the true or achieved sensitivities.

6.10.1 Achieved Sensitivities

The achieved (or true) sensitivities when the controller $C(s)$ is applied to the 'true' model, $G(s)$ are given in Eqs. (6.42)–(6.45). Here, the transfer function of the true system is assumed to be given by:

$$G(s) = \frac{B(s)}{A(s)}, \tag{6.41}$$

which may differ from the 'nominal model', $G_o(s) = B_o(s)/A_o(s)$. The goal in the sequel will be to say something about $T(s)$ when some understanding of the possible differences that might exist between the nominal model, $G_o(s)$, and the 'true' system, $G(s)$, is known. That is, can something be said about achieved sensitivities, on the basis of knowledge of the nominal sensitivities, and some information about the likely model errors?

$$T(s) \triangleq \frac{G(s)C(s)}{1 + G(s)C(s)} = \frac{B(s)P(s)}{A(s)L(s) + B(s)P(s)}, \tag{6.42}$$

$$S(s) \triangleq \frac{1}{1 + G(s)C(s)} = \frac{A(s)L(s)}{A(s)L(s) + B(s)P(s)}, \tag{6.43}$$

$$S_i(s) \triangleq \frac{G(s)}{1 + G(s)C(s)} = \frac{B(s)L(s)}{A(s)L(s) + B(s)P(s)}, \tag{6.44}$$

$$S_u(s) \triangleq \frac{C(s)}{1 + G(s)C(s)} = \frac{A(s)P(s)}{A(s)L(s) + B(s)P(s)}. \tag{6.45}$$

Two descriptions for model errors are commonly used, namely, additive and relative model error. These errors are described by:

$$\text{Additive error} \triangleq G_o(s) + G_\epsilon(s), \tag{6.46}$$

$$\text{Relative Error} \triangleq G_\Delta(s) = \frac{G_\epsilon(s)}{G_o(s)}. \tag{6.47}$$

Note that

$$G(s) = G_o(s)[1 + G_\Delta(s)]. \tag{6.48}$$

6.10.2 Robust Stability

Insights into robustness issues can be obtained by relating the nominal and achieved sensitivity functions. In particular:

Theorem 6.4 ***(Result)*** Consider the nominal sensitivities $S_o(s)$ and $T_o(s)$ and a plant with Relative Model Error $G_\Delta(s)$. Then the achieved sensitivity functions S and T are given by

$$S(s) = S_o(s)S_\Delta(s), \tag{6.49}$$

$$T(s) = T_o(s)(1 + G_\Delta(s))S_\Delta(s), \tag{6.50}$$

$$S_i(s) = S_{io}(s)(1 + G_\Delta(s))S_\Delta(s), \tag{6.51}$$

$$S_u(s) = S_{uo}(s)S_\Delta(s), \tag{6.52}$$

$$S_\Delta(s) = \frac{1}{1 + T_o(s)G_\Delta(s)}. \tag{6.53}$$

$S_\Delta(s)$ is called the *error sensitivity*.

Proof: By direct substitution.

Next, a result is derived giving a sufficient condition for stability of the true closed-loop given that the nominal closed-loop is stable.

Sufficient conditions for a feedback loop to be robustly stable are stated in the following:

Theorem 6.5 ***(Robust stability)*** Consider a plant with nominal transfer function $G_o(s)$ and true transfer function given by $G(s)$. Assume that $C(s)$ is the transfer function of a controller that achieves nominal internal stability. Also assume that $G(s)C(s)$ and $G_o(s)C(s)$ have the same number of unstable poles. Then, a sufficient condition for stability of the true feedback loop obtained by applying the controller to the true plant is that

$$|T_o(j\omega)||G_\Delta(j\omega)| = \left|\frac{G_o(j\omega)C(j\omega)}{1 + G_o(j\omega)C(j\omega)}\right| |G_\Delta(j\omega)| < 1 \quad \forall\omega, \tag{6.54}$$

where $G_\Delta(j\omega)$ is the frequency response of the relative model error.

Proof: Recall that, by assumption, the nominal loop is stable and that $G(s)C(s)$ and $G_o(s)C(s)$ have the same number of unstable poles. This means that the real loop will be stable if and only if the Nyquist plot of $G(j\omega)C(j\omega)$ encircles the point $(-1,0)$ the same number of times (and in the same direction) that $G_o(j\omega)C(j\omega)$ does.

Also note that

$$G(s)C(s) = G_o(s)C(s) + G_\epsilon(s)C(s), \tag{6.55}$$

that is, the change in the open-loop transfer function is $G_\epsilon(s)C(s)$, where $G_\epsilon(j\omega)$ is the frequency response of the additive modelling error (AME).

Consider now Figure 6.7.

From that figure, it can be seen that a sufficient condition for the Nyquist plots to have the same number of encirclements, is

$$|G_\epsilon(j\omega)C(j\omega)| < |1 + G_o(j\omega)C(j\omega)| \quad \forall\omega. \tag{6.56}$$

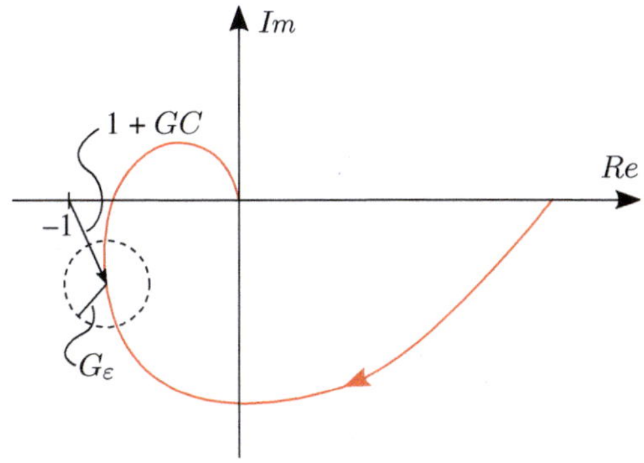

Figure 6.7 Nyquist plot for the nominal and the true loop.

Recalling that $G_\epsilon(s) = G_o(s)G_\Delta(s)$, then (6.56) is equivalent to

$$\frac{|G_\Delta(j\omega)G_o(j\omega)C(j\omega)|}{|1 + G_o(j\omega)C(j\omega)|} < 1. \tag{6.57}$$

This is equivalent to (6.54), upon using the definition of the nominal complementary sensitivity.

Remark 6.1 Equation (6.54) gives only a *sufficient* condition for robust stability, i.e. if the condition works then the true loop is guaranteed to be stable but if it fails, the true loop may be stable (although this is perhaps unlikely).

Remark 6.2 It is also possible to extend the result to the case when $G_\Delta(s)$ is unstable. All that is required is that one keep track of the appropriate number of encirclements to guarantee stability of the true system.

Remark 6.3 When applying the robust stability result, it is usual that $|G_\Delta(j\omega)|$ be replaced by some known upper bound, say $\epsilon(\omega)$. The sufficient condition can then be replaced by $|T_o(j\omega)\epsilon(\omega)| < 1, \forall\omega$.

Next the performance of the real loop is inferred from that of the nominal loop. This is often called the problem of *performance/robustness*. From Eqs. (6.49) to (6.53), it can be seen that the nominal performance will not be too different from the achieved performance if $S_\Delta(j\omega)$ is close to $1 + j0$ for all frequencies. It can be seen from Eq. (6.53) that this will hold if the frequency response $|T_o(j\omega)|$ 'rolls off' before the Relative Model Error $|G_\Delta(jw)|$ increases in magnitude to approach the value 1. This ensures that $|T_o(j\omega)G_\Delta(j\omega)| \ll 1$ at all frequencies.

6.11 Revision Questions

1. (a) Consider the system $T(s) = \frac{1}{s^2}$; Plot the locus of the closed-loop poles if a proportional controller of gain K is used.
   (b) Would this be a suitable control law? In not, why not? closed-loop
2. (a) Consider the system $T(s) = \frac{K}{(s+1)^3}$; Plot the root locus of the closed-loop poles as K goes from 0 to ∞.
   (b) Is this closed-loop system stable for all values of K?

3. Plot the Nyquist diagram for

$$T'(s) = \frac{1}{(s+1)^3}. \tag{6.58}$$

4. From your answer to Question 3, what is the phase when the Nyquist diagram crosses the negative real axis?
5. Where should $-\frac{1}{K}$ be on the Nyquist diagram for closed-loop stability?
6. Use your answers to Questions 3–5 to determine the maximum gain of a proportional controller to ensure closed-loop stability.
7. Consider a process having transfer function $G(s) = \frac{1}{(s-1)^2}$. A control law $C(s) = \frac{9(s-1)}{(s+5)}$ is proposed.
   (a) Evaluate the four nominal sensitivity functions: S_o, T_o, S_{io}, S_{uo}.
   (b) Do you think this is a satisfactory control law for this process? If not, why not?

Further Reading

This chapter is an abridged version of material covered in greater depth in the following books:

1 G.C. Goodwin, S.F. Graebe and M.E. Salgado, *Control system design*, Prentice Hall, 2000.
2 J.C. Doyle, B.A. Francis and A.R. Tannenbaum, *Feedback control theory*, Macmillan Publishing Company, 1992.
3 S. Skogestad and I. Postlethwaite, *Multivariable feedback control analysis and design*, John Wiley & Sons, Inc., New York, 1996.

7

Design of Control Laws for Single-Input Single-Output Linear Systems

7.1 Introduction

Chapter 6 introduced a number of tools that could be used to analyse the properties of single-input single-output closed-loop systems. Hence, a possible way that one could proceed to design a controller would be to simply hypothesise a possible control law and then analyse its stability and performance using the tools presented so far. However, this is clearly unsatisfactory since one could conceivably have to try an extremely large number of possible control laws before hitting on one that had the required properties. Hence, this chapter goes a step further and develops a number of direct design tools which closely link the desired closed-loop properties to the control law.

7.2 Closed-Loop Pole Assignment

It was shown in Chapter 6 that the nominal closed-loop characteristic polynomial is given by

$$A_{cl}(s) = A_o(s)L(s) + B_o(s)P(s), \tag{7.1}$$

where $A_o(s)$, $B_o(s)$, $L(s)$, and $P(s)$ are polynomials appearing in the denominator of the (nominal) plant, the numerator of the (nominal) plant, the denominator of the controller and the numerator of the controller, respectively.

Hence, a simple approach to control system design is, for given $A_o(s)$ and $B_o(s)$, to specify $A_{cl}(s)$ and then to solve Eq. (7.1) for the required polynomials $L(s)$ and $P(s)$. This leads to a set of simultaneous equations for the coefficients of the polynomials $L(s)$ and $P(s)$.

Additional features can be added. For example, if one wishes to enforce integral action, then it suffices to constrain $L(s)$ to have s as a factor, i.e. to write

$$L(s) = sL'(s), \tag{7.2}$$

and then to solve Eq. (7.1) for the coefficients in $L'(s)$ and $P(s)$.

A point to note is that, in order to have enough degrees of freedom to solve for $L(s)$ and $P(s)$, it is necessary to choose $A_{cl}(s)$ of sufficiently high degree.

For example, if a proper (but not strictly proper) control law is required, then the degree of $A_{cl}(s)$ should be chosen of degree $2n - 1$ where n is the degree of $A_o(s)$.

Practical Control System Design: Real World Designs implemented on Emulated Industrial Systems,
First Edition. Adrian Medioli and Graham Goodwin.

Companion website: www.wiley.com/go/medioli/practicalcontrolsystemdesign

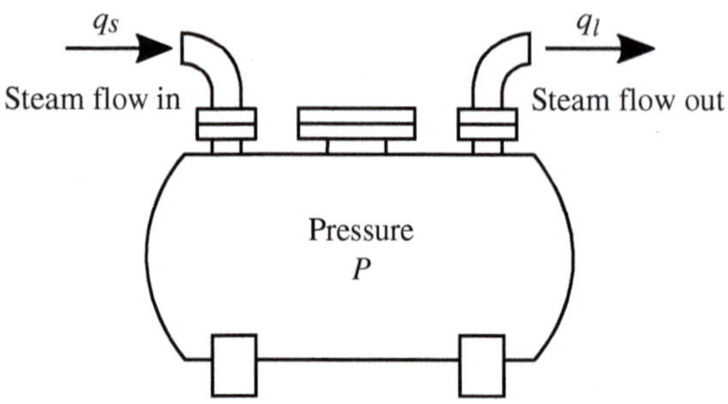

Figure 7.1 Steam receiver schematic.

7.2.1 Example: Steam Receiver

As the authors were writing in Section 7.2, they were simultaneously working on a Boiler Control problem in the Sugar Industry. This problem will be discussed in detail in Chapters 29 and 30. Part of this problem is related to the control of pressure in a Steam Receiver. A schematic diagram is shown in Figure 7.1.

A physical model was constructed from the full system and then various tests were conducted on that model. Details will be given in Chapter 29. A very simple linear model linking q_s to P turns out to be

$$\dot{P} = g(q_s - q_l), \tag{7.3}$$

where g is a function of the dimensions of the vessel. In other words, the rate of change of pressure is proportional to the difference between the inflow of steam, q_s, and the outflow of steam, q_e.

Hence, treating q_l as a disturbance and q_s as a controllable input, then the input–output transfer function becomes

$$\frac{P}{q_s} = \frac{g}{s} = G(s). \tag{7.4}$$

Part of the design was to implement a controller having the following transfer function:

$$C(s) = K_p + \frac{K_i}{s}, \tag{7.5}$$

(Actually, this is a simple PI control law.) The closed-loop characteristic polynomial is seen to be the numerator of

$$G(s)\,C(s) + 1 = M(s). \tag{7.6}$$

Hence, the closed-loop poles satisfy

$$\frac{g}{s}\left[K_p + \frac{K_i}{s}\right] + 1 = 0, \tag{7.7}$$

or

$$s^2 + gK_p s + gK_i = 0. \tag{7.8}$$

Say that it is desired that the closed-loop poles be two equal poles at $-a$. Then the desired closed-loop characteristic equation is

$$A_{cl}(s) = s^2 + 2as + a^2. \tag{7.9}$$

Equating coefficients in (7.8) and (7.9) shows that the appropriate choice of K_p (the proportional gain) and K_i (the integral gain) in the control law is

$$K_p = \frac{2a}{g},$$
$$K_i = \frac{a^2}{g}. \tag{7.10}$$

These values were tried and found to work as desired. This is a simple example of design by closed-loop pole assignment. Note that much deeper ideas related to this particular control problem will be studied in Chapter 30.

7.3 Using Root Locus

Another simple strategy, closely related to pole assignment, that can be used for control law design is to utilise root locus plots in innovative fashions to gain insight into the closed-loop performance.

7.3.1 Example: Double Integrator

As a simple illustration, the design given above will be repeated, but this time using root locus ideas.

It is to be noted that the plant has a single pole at the origin. Also, say that it is required that integral action be included. This means that an extra pole at the origin is needed in the control law. Thus, the design needs to account for two poles at the origin.

Say, that one wanted to use a simple integral control with gain K (i.e. as in Section 7.2.1 with $K = K_i$ and $K_p = 0$). Then the plant plus control law would have transfer function

$$L(s) = \frac{Kg}{s^2}. \tag{7.11}$$

The corresponding root laws for the closed-loop poles as K is varied from 0 to ∞ is shown in Figure 7.2.

Clearly the result shown in Figure 7.2 is unsatisfactory since the closed-loop system will have poles on the stability boundary (the imaginary axis) for all values of K.

The root locus rules imply that the zeros tend to attract the closed-loop poles. This suggests that one could deliberately include a zero in the control law leading to

$$C(s) = \frac{K(s+z)}{s}, \tag{7.12}$$

where $z > 0$. This changes the root locus diagram to the one shown in Figure 7.3.

This is much more satisfactory because the zero has clearly 'attracted' the closed-loop poles into the left-half plane.

All that remains is to adjust the values of z and K so that the location of the closed-loop poles accords with other design objectives.

Actually, one can use root locus diagrams in more sophisticated ways to gain design insights. This idea is illustrated in the next example which shows how one can better understand the impact of varying a parameter other than the overall gain.

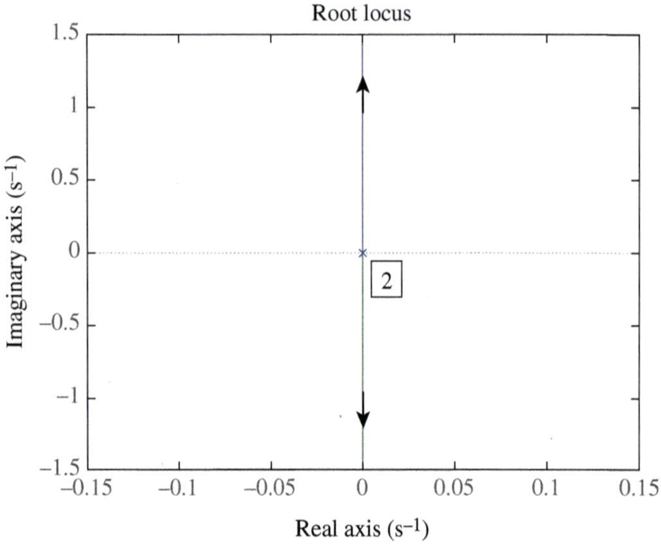

Figure 7.2 Steam receiver root locus of closed-loop poles where control law has one pole.

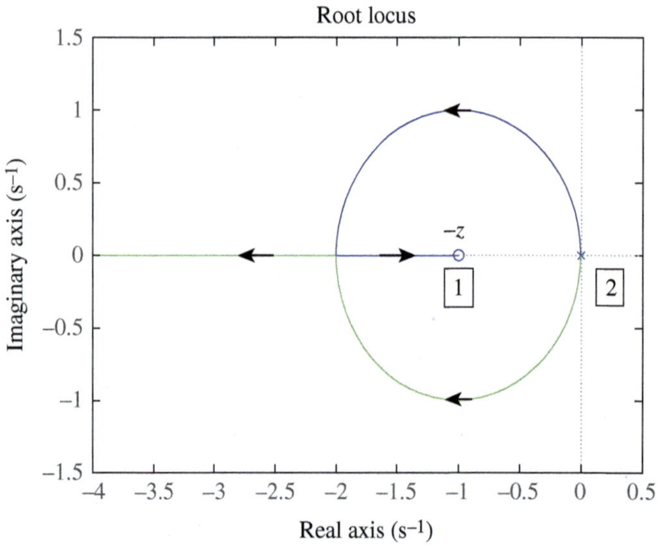

Figure 7.3 Steam receiver root locus of closed-loop where control law has one pole and one zero.

7.3.2 Example: Unstable Process

Consider a plant with transfer function $G_o(s)$ and a feedback controller with transfer function $C(s)$, where

$$G_o(s) = \frac{1}{(s-1)(s+2)} \quad \text{and} \quad C(s) = 4\frac{s+\alpha}{s}. \tag{7.13}$$

The idea is to examine the location of the closed-loop poles as α is varied from 0 to ∞. Notice that the closed-loop poles are the roots of

$$1 + 4\frac{s+\alpha}{s(s^2+s-2)} = \frac{s(s^2+s-2)+4s+4\alpha}{s(s^2+s-2)} = 0. \tag{7.14}$$

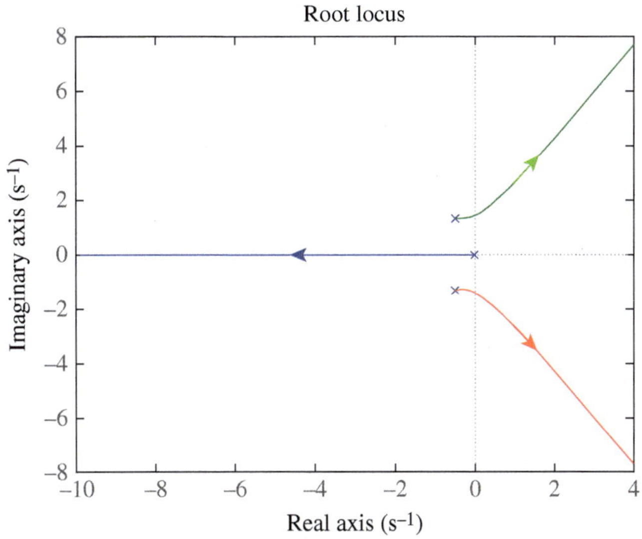

Figure 7.4 Locus for the closed-loop poles when the controller zero varies.

This equation can be rearranged so that it takes the form of (6.26) of Chapter 6. In particular, it can be rearranged so that the parameter α appears as a multiplier as follows:

$$s(s^2 + s + 2) + 4\alpha = 0, \tag{7.15}$$

Next, dividing by $s(s^2 + s + 2)$, then an equation of the form of (6.26) is achieved where

$$\lambda = 4\alpha \quad \text{and} \quad F(s) = \frac{1}{s(s^2 + s + 2)}. \tag{7.16}$$

The rules described in Chapter 6 (Section 6.6), lead to the following:

Rule 1: The root locus has three branches ($m = 0$ and $n = 3$).
Rule 4: The negative real axis belongs to the locus.
Rule 5: For α close to zero, that is, λ close to zero, the roots are at the poles of $F(s)$, that is, at $0, -0.5 \pm j0.5\sqrt{7}$.
Rule 7: When α tends to ∞, that is, when λ tends to ∞, the three roots tend to ∞, following asymptotes that intersect the real axis at $(\sigma, 0)$ with $\sigma = -\frac{1}{3}$. The angle of these asymptotes is π, $\pi/3$, and $2\pi/3$. This means that two branches go to the right-half plane.

The above rules allow one to sketch a root-locus diagram. This can also be obtained by using the MATLAB® command **rlocus**; the result is shown in Figure 7.4. A powerful MATLAB® environment, **rltool**, allows a versatile root-locus analysis, including shifting and adding of poles and zeros.

7.4 All Stabilising Control Laws

A powerful idea that can be used for control system design is the concept known as the Youla–Kucera parameterisation of all stabilising control laws. To introduce the idea, it is first assumed that the plant transfer function is $G_o(s)$ and that this transfer function is stable. It will be shown in the sequel that *every* possible linear stabilising control law can be implemented as in Figure 7.5.

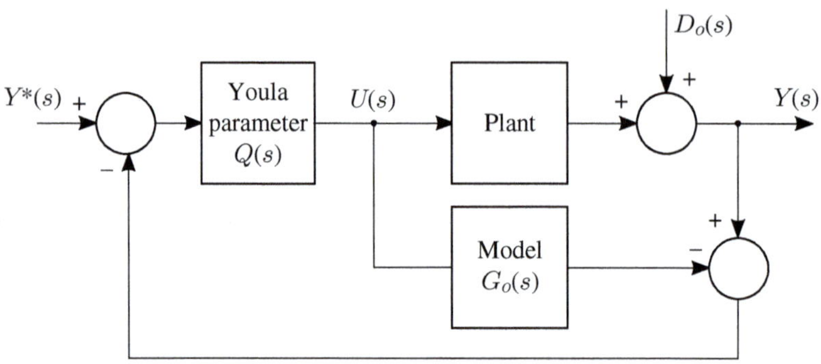

Figure 7.5 Youla parameterisation of all stabilising controllers.

From Figure 7.5, it follows that

$$U(s) = Q(s)\left[Y^*(s) - Y(s)\right] + Q(s)\,G_o(s)\,U(s), \tag{7.17}$$

or

$$U(s) = \frac{Q(s)}{1 - G_o(s)\,Q(s)}\left[Y^*(s) - Y(s)\right]. \tag{7.18}$$

Hence, Figure 7.5 is simply an alternative way of implementing the control law:

$$C(s) = \frac{Q(s)}{1 - G_o(s)\,Q(s)}. \tag{7.19}$$

There are two important properties that are associated with the parameterisation of the control law as in (7.19). These are stated and proved below.

Theorem 7.1 ***(Result)*** The parameterisation (7.19) has two key properties, namely, (i) all of the resultant closed-loop sensitivity functions are affine (i.e. linear plus a possible offset) in $Q(s)$ and (ii) a necessary and sufficient condition for closed-loop stability is that $Q(s)$ be a stable-transfer function.

Proof:

(i) The four closed-loop sensitivity functions are easily evaluated. The result is

$$T_o(s) = Q(s)\,G_o(s), \tag{7.20}$$

$$S_o(s) = 1 - Q(s)\,G_o(s), \tag{7.21}$$

$$S_{io}(s) = S_o(s)\,G_o(s), \tag{7.22}$$

$$S_{uo}(s) = Q(s). \tag{7.23}$$

These are affine in $Q(s)$.

(ii) It follows from Eq. (7.23) and the stability of $G_o(s)$ that a *necessary* condition for closed-loop stability is that $Q(s)$ be stable. Also, $Q(s)$ being stable is sufficient for all four transfer functions in (7.20), (7.21), (7.22) and (7.23) to be stable.

□□□

The above result is a powerful idea since it characterises *all* stabilising control laws. Of course one needs to be a little careful if one believes that the true plant is not precisely modelled by $G_o(s)$. Indeed, this will almost always be the case. Another caveat on the result is that the poles of $G_o(s)$

appear in the closed-loop system. They can be cancelled of course, but they will still impact the closed-loop performance.

To illustrate, say that the true plant transfer function is given by

$$G(s) = G_o(s)[1 + G_\Delta(s)], \tag{7.24}$$

where $G_\Delta(s)$ is the multiplicative model error.

It can then be shown that the true sensitivity function is then

$$S(s) = S_o(s) S_\Delta(s), \tag{7.25}$$

where

$$S_\Delta(s) = \frac{1}{1 + T_o(s) G_\Delta(s)}. \tag{7.26}$$

Hence, one needs to choose $Q(s)$ to ensure that $G_\Delta(s)$ does not destabilise the system. There is really nothing new since it all follows the same ideas as the robustness discussion in Section 6.10.2 of Chapter 6.

Another caveat on the representation in Figure 7.5 is that the open-loop poles (i.e. the poles of $G_o(s)$) appear in the closed-loop. Thus, it clearly must be necessary that the open-loop poles all be stable for the above to make sense. Actually, there may be other open-loop poles that one may wish to exclude from the closed-loop transfer function. For example, a pole having an excessively long transient time may be deemed 'undesirable'. Hence, the characterisation of all stabilising control laws will be revisited in Chapters 14 and 22 where a parameterisation that is applicable to cases where the open-loop plant contains 'undesirable' poles including those that are unstable will be studied.

7.5 Design Using the Youla–Kucera Parameterisation

Note that the Youla–Kucera parameterisation of all stabilising control laws leads to the results

$$T_o(s) = G_o(s) Q(s), \tag{7.27}$$

$$S_o(s) = 1 - G_o(s) Q(s). \tag{7.28}$$

An important point to note is that (7.27) and (7.28) are linear (actually affine) in $Q(s)$. This greatly simplifies the procedure of going from desired values for $T_o(s)$ and $S_o(s)$ to the required control law.

A common requirement is that one tracks the given reference signal and rejects disturbances. This is achieved by setting $T_o(s) \simeq 1$ and $S_o(s) \simeq 0$. This line of reasoning suggests that one should choose

$$G_o(s) Q(s) \simeq 1. \tag{7.29}$$

Of course, one needs to ensure that the transfer function $Q(s)$ is both proper and stable. Thus, a very simple design rule emerges, i.e. $Q(s)$ should be chosen as a stable proper approximation to the inverse of $G_o(s)$. This choice can be expressed by choosing

$$Q(s) = \hat{G}_o(s)^{-1} F(s), \tag{7.30}$$

where $\hat{G}_o(s)^{-1}$ is a stable approximation to the inverse of $G_o(s)$ and $F(s)$ is an extra filter used to modify the closed-loop behaviour. Note that (7.30) gives powerful support to the notion, introduced in Chapter 2 that control design is an inverse problem.

The above ideas are illustrated by several examples.

7.5.1 Example: Simple First-Order Model

Say that $G_o(s)$ is a first-order model of the form:

$$G_o(s) = \frac{k_0}{(s + \alpha_0)}, \alpha_0 > 0. \tag{7.31}$$

As in (7.30), the following choice is made

$$Q(s) = \hat{G}_o(s)^{-1} F(s), \tag{7.32}$$

where

$$\hat{G}_o(s)^{-1} = \left(\frac{s + \alpha_0}{k_0}\right), \tag{7.33}$$

$$F(s) = \frac{1}{(s\tau + 1)}. \tag{7.34}$$

The above design results in

$$T_o(s) = \frac{1}{(s\tau + 1)}. \tag{7.35}$$

Thus, the closed-loop nominal sensitivity function can be adjusted by choice of τ.

It is also interesting to find the equivalent control law. Using Eq. (7.19) the result is

$$\begin{aligned} C(s) &= \frac{Q(s)}{1 - G_o(s) Q(s)}, \\ &= \frac{\frac{s+\alpha_0}{k_0(s\tau+1)}}{1 - \frac{1}{(s\tau+1)}}, \\ &= \frac{1}{k_0}\left[\frac{s + \alpha_0}{s\tau}\right]. \end{aligned} \tag{7.36}$$

Hence, the above design actually leads to a PI controller of the form

$$C(s) = K_p + \frac{K_i}{s}, \tag{7.37}$$

where

$$K_p = \frac{1}{k_0 \tau}, K_i = \frac{\alpha_0}{k_0 \tau}. \tag{7.38}$$

7.6 Integral Action

It was shown in the example presented in Section 7.5.1 that integral action was imposed by simply choosing $Q(s)$ as a suitable approximation to the inverse of $G_o(s)$ with an exact inverse holding at $s = 0$.

Actually, this is a very general result. Irrespective of the true plant transfer function, it suffices that one chooses $Q(s)$ so that $G_o(s) Q(s)$ has unity gain at d.c. to achieve integral action, i.e. $Q(s)$ is an exact inverse to $G_o(s)$ at $s = 0$.

To show this, note that the equivalent control law is

$$C(s) = \frac{Q(s)}{1 - G_o(s)\, Q(s)}. \tag{7.39}$$

Hence, provided $Q(s)$ satisfies $G_o(0)\, Q(0) = 1$, then clearly, the denominator of $C(s)$ is zero when $s = 0$ (i.e. s is a factor of the denominator of the control law). The results will be given as an alternative interpretation in Section 22.12 of Chapter 22.

7.7 Anti-Windup

Actually, the Youla–Kucera parameterisation gives many other insights. To illustrate, note that a problem that occurs with integral action is that the integrator will continue to integrate if the tracking error cannot be brought to zero. The latter problem occurs very frequently in practical systems, e.g. when the input to a plant hits a saturation limit. Indeed, the second author of this book met this problem very early in his career. He had implemented a simple PI control law on a steam boiler but found that every time the controller was connected to the plant, it shut down (with a loud 'bang'). An older (and much wiser) engineer pointed out that the control law was running prior to being connected to the plant leading to integrator wind-up. Actually, the output of the integrator was sitting at the maximum number allowed by the computer when the controller was connected to the plant! A remedy to this problem is to stop the integrator when the input to the plant hits a saturation limit. This is called 'anti-windup' protection.

The Youla–Kucera parameterisation provides a simple way of implementing anti-windup. The key idea is to ensure that $G_o(s)$ [the model used in the controller] sees the same input as the real plant does.

This is readily achieved by embellishing the implementation in Figure 7.5 with the related implementation in Figure 7.6.

Since all of the transfer functions in Figure 7.5 are stable, then all signals remain bounded and wind-up does not occur. Another way of seeing this is that the integral in the equivalent control law is obtained via the feedback path through $G_o(s)$ and $Q(s)$. The inclusion of the saturation as shown in Figure 7.6 halts the integrator when saturation occurs since the loop through $G_o(s)$, $Q(s)$ is opened by action of the saturation.

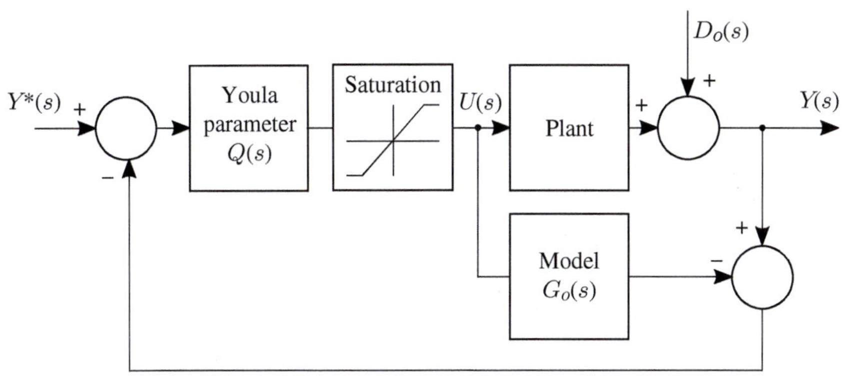

Figure 7.6 Youla parameterisation of all stabilising controllers with anti-wind-up.

7.8 PID Design

The example of Section 7.5 showed that a PI controller arises naturally for a first-order plant. Here, PID design is explored in slightly more depth. PID Control is simple and robust. Thus, it is used extensively for industrial control laws. Indeed, prior to the use of computers for control, almost all controllers were of PID type implemented in hydraulic or pneumatic form.

7.8.1 Structure

Consider the simple SISO control loop as shown in Figure 7.7.

The traditional expressions for PI and PID controllers can be described by their transfer functions, relating error $E(s) = Y^*(s) - Y(s)$ and controller output $U(s)$ as follows:

$$C_P(s) = K_p, \tag{7.40}$$

$$C_{PI}(s) = K_p\left(1 + \frac{1}{T_r s}\right), \tag{7.41}$$

$$C_{PD}(s) = K_p\left(1 + \frac{T_d s}{\gamma s + 1}\right), \tag{7.42}$$

$$C_{PID}(s) = K_p\left(1 + \frac{1}{T_r s} + \frac{T_d s}{\gamma s + 1}\right). \tag{7.43}$$

where T_r and T_d are known as the *reset time* and *derivative time*, respectively. Notice that a small filter has been included to render the derivative term a 'proper' controller.

As seen from (7.40) to (7.43), the members of this family include, in different combinations, three control modes or actions: *proportional (P), integral (I)*, and *derivative (D).*

Many tuning rules can be found in the literature dealing with PID design. However, caution must be exercised when applying these PID tuning rules, since there are a number of parameterisations used for PID structures. Equation (7.43) is known as the *standard form*. An alternative, *series form*, is as follows:

$$C_{series}(s) = K_s\left(1 + \frac{I_s}{s}\right)\left(1 + \frac{D_s s}{\gamma s + 1}\right). \tag{7.44}$$

Yet another form is the, so-called, parallel form which is as follows:

$$C_{parallel}(s) = K_p + \frac{I_p}{s} + \frac{D_p s}{\gamma s + 1}. \tag{7.45}$$

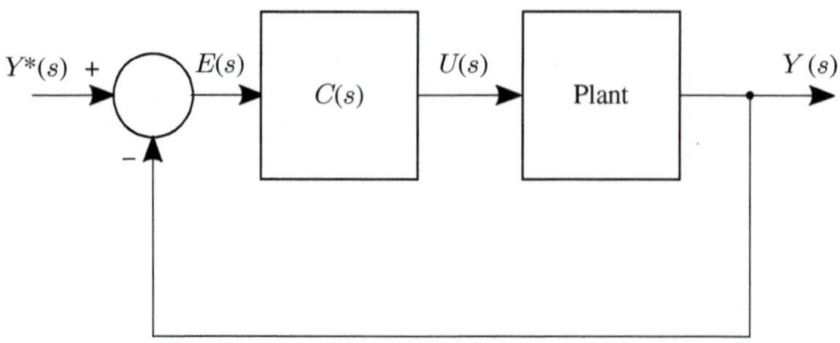

Figure 7.7 Basic feedback control loop.

Thus, terminology, such as *P-gain* is not uniquely defined and can refer to either K_s in (7.44), K_p in (7.45) or K_p in (7.43). It is therefore important to know which of these parameterisations is used. Also, if the PID is implemented in a different form, then one must transform the parameters appropriately.

PID tuning has historically been viewed in terms of the P, I, and D parameters. Although their impact on the closed-loop is far from independent of each other, their effect can be thought of heuristically as follows:

- **Proportional action:** This provides a contribution which depends on the instantaneous value of the control error. A proportional controller can control any stable plant provided the gain is chosen appropriately, but it provides limited performance and non-zero steady-state errors. This latter limitation is due to the fact that its frequency response is bounded for all frequencies.
  It has also been traditional to use the expression *proportional band* (PB) to describe the proportional action. The equivalence is

$$PB[\%] = \frac{100[\%]}{K_p}. \tag{7.46}$$

  The PB is defined as the error required (as a percentage of full scale) to yield a 100% change in the controller output.
- **Integral action:** This provides a controller output that is proportional to the accumulated error. This implies that it is a *slow* control mode. This characteristic is also evident in its low-pass frequency response. The integral mode plays a fundamental role in achieving perfect plant inversion at $\omega = 0$. This forces the steady-state error to zero in the presence of a step reference and/or constant output disturbances. The integral mode, viewed in isolation, has two major shortcomings: its pole at the origin is detrimental to loop stability. It also gives rise to the undesirable effect (in the presence of actuator saturation) known as *wind-up*. Methods for dealing with this problem were discussed in Section 7.7 (see Figure 7.6).
  The integral time, T_r, in Eqs. (7.41) and (7.43) can be thought of as the time taken for a unity error to cause the control signal to build up to the same value as given by the proportional part of the controller.
- **Derivative action:** This acts on the rate of change of the control error. Consequently, it is a *fast mode* which ultimately disappears in the presence of constant errors. It is sometimes referred to as a *predictive mode*, because of its dependence on the error trend. The main limitation of the derivative mode, viewed in isolation, is its tendency to yield large control signals in response to high-frequency control errors. Its implementation requires properness of the transfer functions, so a little filter is typically added to the derivative, as is evident in Eqs. (7.42) and (7.43). In the absence of other constraints, the additional time constant γ is normally chosen such that $0.1T_d \leq \gamma \leq 0.2T_d$. This constant is called the derivative time constant; the smaller it is, the larger the frequency range over which the filtered derivative approximates the exact derivative, with equality in the limit:

$$\lim_{\gamma \to 0} u_{PID}(t) = K_p e(t) + \frac{K_p}{T_r} \int_{t_0}^{t} e(\tau)\, d\tau + K_p T_d \frac{de(t)}{dt} \tag{7.47}$$

  The classical argument for choosing $\gamma \neq 0$ was to ensure that the controller be proper, to attenuate high-frequency noise. The latter point is illustrated in Figure 7.8, which shows that the filtered derivative approximates the exact derivative well at frequencies up to $\frac{1}{\gamma}$ (rad/s), but that it has finite gain at high frequencies, whereas an exact derivative has infinite gain.

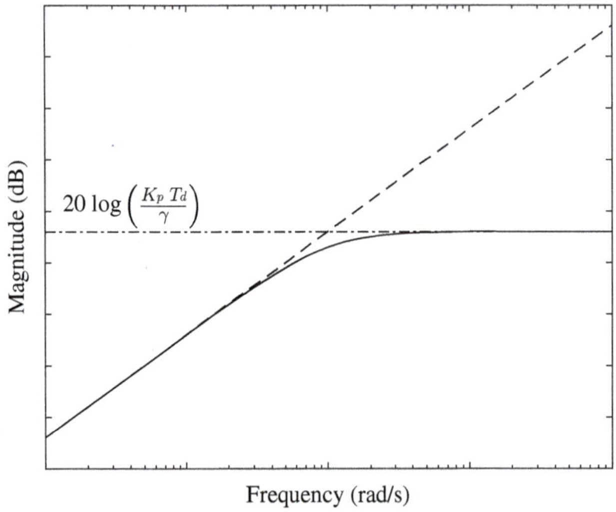

Figure 7.8 Bode magnitude plot of exact derivative (dashed) and filtered derivative (solid).

Because $\gamma \neq 0$ was largely seen as a necessary evil, i.e. as a necessary departure from a pure proportional, integral and derivative action, almost all industrial PID controllers set γ as a fixed fraction of T_d, rather than viewing it as an independent design parameter in its own right.

A heuristic explanation of derivative action is that $(1 + T_d s)$ in Eq. (7.42) can be thought of as cancelling a term of the form $\frac{1}{1+\tau s}$ in the plant transfer function. Hence, under ideal conditions, one might begin by setting T_d equal to nominal time constant of the plant.

7.8.2 Using the Youla–Kucera Parameterisation for PID Design

Since the Youla–Kucera parameterisation gives all stabilising control laws, it should not surprise the reader that a PID controller will result from assuming a particular model structures. Indeed, it has already been shown in the example of Section 7.5 that a PI controller results from assuming a first-order plant transfer function. Similarly a PID law results from assuming a second-order plant transfer function. Indeed, this, in part, explains the wide spread use of PID controllers since many real-world systems can be described (at least approximately) by simple first-order or second-order models.

7.9 Empirical Tuning

There exists a vast literature on empirical tuning rules. These are typically for PID control laws based on measurements made on the real plant. The classical techniques are often preferred by practitioners, so the core ideas are reviewed below for one of the most commonly deployed classical tuning methods.

7.10 Ziegler–Nichols (Z–N) Oscillation Method

This procedure is valid only for open-loop stable plants. It is carried out by means of the following steps.

Table 7.1 Ziegler–Nichols tuning, using the oscillation method.

	K_p	T_r	T_d
P	$0.50K_c$		
PI	$0.45K_c$	$\frac{P_c}{1.2}$	
PID	$0.60K_c$	$0.5P_c$	$\frac{P_c}{8}$

- Set the true plant under proportional control, with a very small gain.
- Increase the gain until the loop starts oscillating.
- Record the controller critical gain $K_p = K_c$ and the oscillation period of the controller output, P_c.
- Adjust the controller parameters according to Table 7.1. This table refers to the parameterisation of (7.43).

Observe that the underlying model being obtained in the experiment is only *one point on the frequency response*, namely, the one corresponding to a phase equal to $-\pi$ (rad) and a magnitude equal to K_c^{-1}, because the Nyquist plot for $K_pG(j\omega)$ crosses the point (−1,0) when $K_p = K_c$.

The settings in Table 7.1 were suggested by Ziegler and Nichols, who aimed to achieve an underdamped response to a step for those plants having an approximate model has the form

$$G_o(s) = \frac{K_o e^{-s\tau_o}}{\nu_o s + 1}, \tag{7.48}$$

where $\nu_o > 0$.

7.10.1 Example: Third-Order Plant

Consider a plant with a model given by

$$G_o(s) = \frac{1}{(s+1)^3}. \tag{7.49}$$

Goal is to find the parameters of a PID controller by using the Z–N oscillation method.

The critical gain K_c and critical frequency ω_c are first computed. These values must satisfy

$$K_c G_o(j\omega_c) = -1 \iff K_c = -(j\omega_c + 1)^3. \tag{7.50}$$

From this equation, the values $K_c = 8$ and $\omega_c = \sqrt{3}$ are obtained. Hence, the critical period is $P_c \approx 3.63$. Now using the settings in Table 7.1, leads to

$$K_p = 0.6K_c = 4.8; \; T_r = 0.5P_c \approx 1.81; \; T_d = 0.125P_c \approx 0.45. \tag{7.51}$$

The derivative mode will be attenuated with a fast pole with time constant $\gamma = 0.1T_d = 0.045$. Thus, the final loop-transfer function becomes

$$G_o(s)C(s) = K_p \frac{(T_d + \gamma)s^2 + (1 + \frac{\gamma}{T_r})s + \frac{1}{T_r}}{s(\gamma s + 1)(s + 1)^3} = \frac{52.8s^2 + 109.32s + 58.93}{s(s + 22.2)(s + 1)^3}. \tag{7.52}$$

The corresponding closed-loop response actually exhibits significant overshoot that might be unacceptable in some applications. Nonetheless Z–N tuning provides a starting point for finer tuning. The reader is invited to check that $T_d = 1$ provides better performance.

Remark 7.1 A shortcoming of this method is that it requires that the plant be forced to oscillate; this can be dangerous and expensive.

A modern way of obtaining the critical gain and frequency is to use relay autotuning. This method also essentially provides an estimate of K_c and ω_c.

7.11 Two Degrees of Freedom Design

So far in this chapter, only the design of a single element has been considered, namely, the controller transfer function $C(s)$ (or equivalently $Q(s)$ in the Youla–Kucera parameterisation). These are termed one-degree-of-freedom architectures.

Frequently, however, extra measurements are available. For example, one may have a direct measurement of a disturbance. This extra measurement can be used in a *two-degree-of-freedom* (two-d.o.f.) architecture, as shown in Figure 7.9.

7.12 Disturbance Feedforward

The extra measurement $\eta(t)$ can be used to proactively compensate for the effect of a disturbance before its impact is registered on the output response.

To show how this works, consider the system shown in Figure 7.9.

The standard feedback solution to disturbance compensation leads to

$$y(t) = \frac{\overline{D}(s)}{1 + G(s)C(s)}\eta(t) + \frac{G(s)C(s)}{1 + G(s)C(s)}y^*(t). \tag{7.53}$$

In the presence of a changing disturbance, the response can never be made perfect due to the difficulties of designing a suitable value for $S_o(s) = \frac{1}{1+G(s)C(s)}$. It will be shown in Chapters 12 and 13 that delays in $G(s)$ are particularly problematic and work against achieving 'tight' control.

On the other hand, say that $\overline{D}(s)$ and, $G(s)$ are accurately modelled and that $\eta(t)$ is measured. Then it makes sense to change the control architecture to that shown in Figure 7.9 with $H(s) \neq 0$.

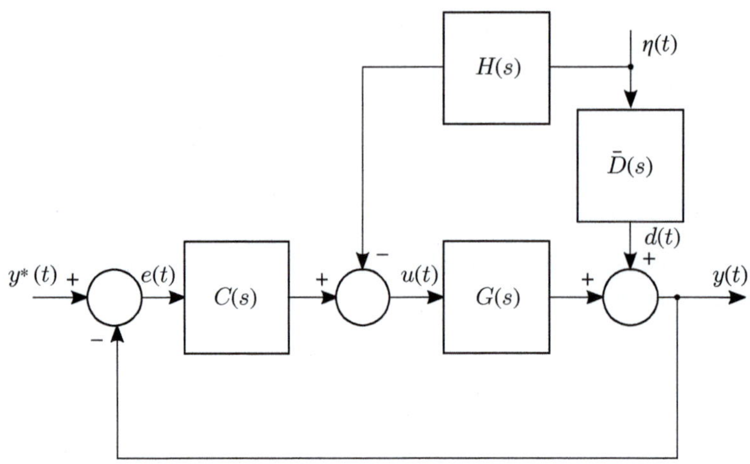

Figure 7.9 Disturbance feedforward structure.

From Figure 7.9, it can be seen that

$$y(t) = \frac{\overline{D}(s) - H(s)G(s)}{1 + G(s)C(s)}\eta(t) + \frac{G(s)C(s)}{1 + G(s)C(s)}y^*(t). \tag{7.54}$$

It then follows that the choice

$$H(s) \simeq G(s)^{-1}\overline{D}(s), \tag{7.55}$$

will lead to near-perfect nulling of the effect of the disturbance on the output.

The above idea is particularly advantageous in some cases, e.g. when the plant, $G(s)$ has a large delay and $\overline{D}(s)$ has a similar delay. It will be shown in Chapter 12, that the presence of a delay makes the design of 'tight' feedback control particularly problematic. On the other hand, if $G(s)$ and $\overline{D}(s)$ have a similar delay, then $H(s)$ as in (7.55) will be delayed free leading to almost perfect disturbance rejection.

7.13 Revision Questions

1. Consider the system $G(s) = \frac{1}{s(s+2)}$.
   (a) What is the closed-loop characteristic polynomial if a proportional control law is used with gain K?
   (b) What proportional controller gain would give closed-loop poles at -1?
2. Consider the following RLC circuit

Figure 7.10 Question 3 RLC circuit.

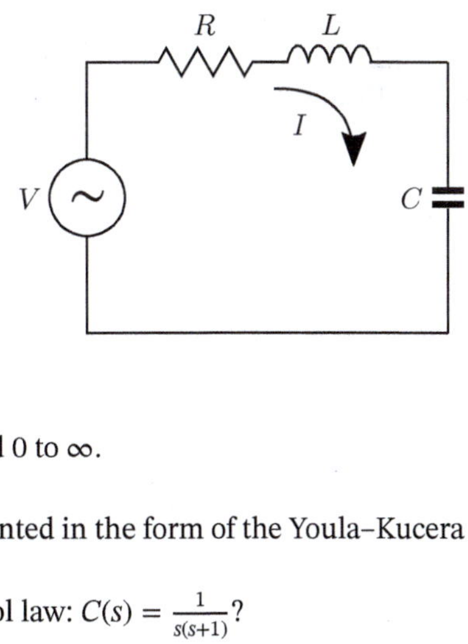

   (a) What is the transfer function for V to I?
   (b) Plot the locus of the poles of the system as R is varied 0 to ∞.
   (c) Is the system always stable?
3. Consider a plant $G(s) = \frac{3}{(s+1)}$ with a control law implemented in the form of the Youla–Kucera parameterisation.
   (a) What is the value of $Q(s)$ corresponding to the control law: $C(s) = \frac{1}{s(s+1)}$?
   (b) Is the system closed-loop stable?
4. Consider the same plant as in Question 3. What transfer function, $Q(s)$, is needed to implement a PI control law $C(s) = \left[1 + \frac{0.1}{s}\right]$?
5. For the system of Question 3, what is the equivalent control law if $Q(s)$ is chosen as

$$Q(s) = \frac{1}{s+3}. \tag{7.56}$$

6. Consider the system having transfer function

$$G(s) = \frac{1}{(s+1)^3}. \tag{7.57}$$

   (a) Choose $Q(s) = \frac{(s+1)^3}{(0.1s+1)^3}$. What are the closed-loop sensitivity and complementary sensitivity functions?
   (b) Plot the closed-loop step response for a unit step in the reference.
7. Consider the same plant as in Question 6. Use the PID settings given in Table 7.1. Plot the closed-loop step response for a unit step change in the reference signal.
8. Consider a plant having transfer function

$$G(s) = \frac{e^{-s}}{s+1}. \tag{7.58}$$

   Say one uses the Youla–Kucera parameterisation with $Q(s) = \left[\frac{s+1}{0.1s+1}\right]$. What is the closed-loop transfer function?
9. (a) For the system in Question 8, what is the transfer function from an input disturbance to the output?
   (b) Say that the input disturbance is a sine wave of frequency π rad/s. What is the approximate magnitude of the output response?
10. Say the input disturbance could be measured. Why would it help to use a two d.o.f. control law?

Further Reading

The material presented in this chapter is described in greater detail in the following book:

1 G.C. Goodwin, S.F. Graebe and M.E. Salgado, *Control system design*, Prentice Hall, 2000.

8

Laboratory 2: Position Control of Electromechanical Servomechanism

8.1 Introduction

An electromechanical servomechanism used in a door-opening system was previously studied in Chapter 5. In that chapter, the parameters in a simple model were estimated. For convenience, the laboratory schematic is repeated as Figure 8.1.

In the current chapter, the effect of various feedback control structures will be examined on the same servomechanism. In particular, the effect of position and position plus velocity feedback will be explored.

8.2 Proportional Feedback

Here, the class of closed-loop transfer functions that can be achieved by position feedback alone are considered.

The transfer function from y^* to y for the closed-loop in Figure 8.2 can be calculated as follows:

$$e = y^* - y,$$
$$y = GV_{in},$$
$$V_{in} = Ce,$$

where G and C denote plant and controller transfer functions, respectively. Hence, as usual,

$$\frac{y}{y^*} = \frac{GC}{1+GC} = T. \tag{8.1}$$

Here, the input–output transfer function takes the following form (see Chapter 5).

$$G = \frac{K_m}{s(\tau s + 1)}. \tag{8.2}$$

In this section, it has been decided that proportional control will be used. Hence, $C = K_P$.

Substituting the known form of the nominal plant transfer function, G, into the closed-loop expression yields

$$T = \frac{K_m K_P}{s(\tau s + 1) + K_m K_P}. \tag{8.3}$$

Practical Control System Design: Real World Designs implemented on Emulated Industrial Systems, First Edition. Adrian Medioli and Graham Goodwin.

Companion website: www.wiley.com/go/medioli/practicalcontrolsystemdesign

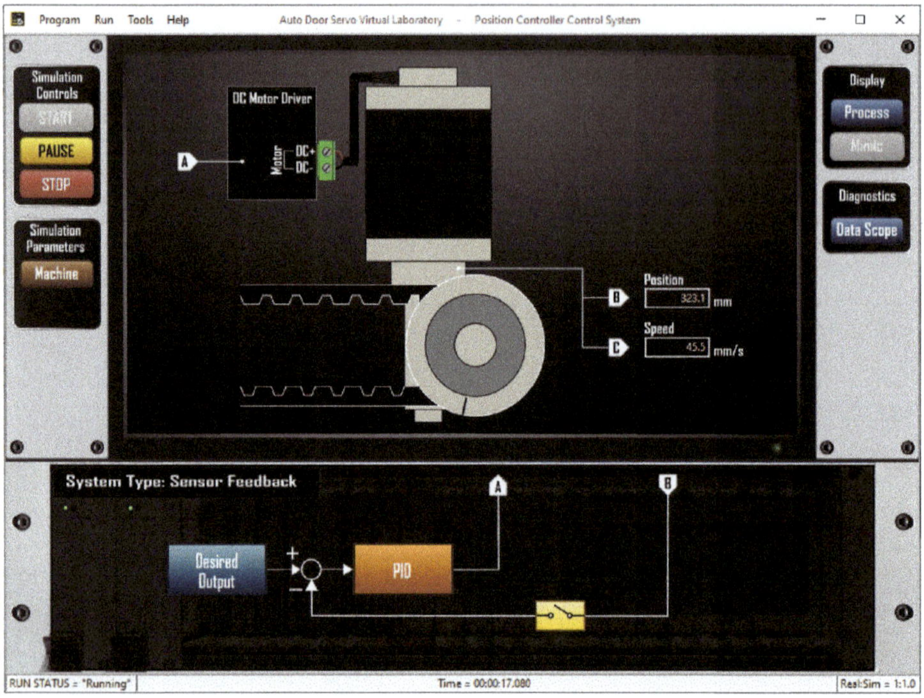

Figure 8.1 The servomechanism.

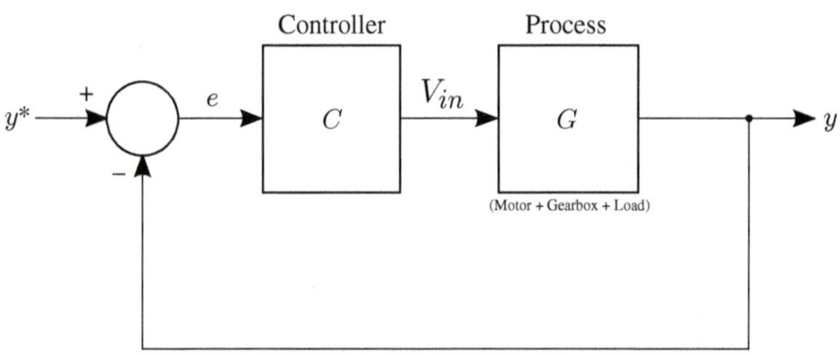

Figure 8.2 Block diagram.

Thus, the closed-loop transfer function between y^* and y takes the following generic form:

$$\frac{y(s)}{y^*(s)} = \frac{\omega_n^2}{s^2 + 2\zeta\omega_n s + \omega_n^2}. \tag{8.4}$$

Using (8.3), the variables ω_n (natural frequency) and ζ (damping ratio) can be expressed in terms of the plant parameters, K_m and τ and the controller gain, K_P, as follows.

$$\omega_n = \sqrt{\frac{K_m K_P}{\tau}}, \tag{8.5}$$

$$\zeta = \frac{1}{2}\sqrt{\frac{1}{K_m K_P \tau}}. \tag{8.6}$$

It can thus be seen that increasing the controller gain, K_P, increases the natural frequency (ω_n) (i.e. speeds the system up) but, at the same time, decreases the damping ratio (ζ) (i.e. makes the system more oscillatory).

Hence, there is an inherent trade-off associated with this particular design. This is a result of the use of the very restrictive control architecture, namely, a simple proportional gain. Alternative controllers will be explored in the sequel.

The virtual laboratory will first be used to study the closed-loop performance when position-only feedback is utilised.

8.2.1 Experiment: Testing a Proportion only Control Law

The reader should use the measured values of K_m and τ (found in Chapter 5) to compute values of controller gain (K_P) required to give damping ratios (ζ) for the closed-loop system of 0.4 and 0.7, respectively.

It is suggested that the reader test the closed loop for a step reference change for these two values of K_P and compare with the anticipated response from your thumbnail calculation.

The reader should find that the speed of response is severely compromised by the need to keep the overshoot small. The reader is asked to revisit Section 5.3.1 of the book.

8.3 Using Proportional Plus Derivative Feedback

It was found in Section 8.2 that using proportional feedback alone was restrictive in the sense that any attempt to increase the closed-loop natural frequency had the effect of simultaneously decreasing the damping ratio. This is a consequence of the restrictive control law. This problem can be mitigated by adding extra flexibility to the control law, e.g. by introducing velocity feedback. This is achieved by changing the controller to a PD controller form.

Thus, consider

$$C(s) = K_P + K_D s. \tag{8.7}$$

The following special choice is made:

$$K_D = K_P \tau. \tag{8.8}$$

The loop-transfer function then becomes

$$L(s) = G(s)C(s) = \frac{K_m K_P(\tau s + 1)}{s(\tau s + 1)} = \frac{K_m K_P}{s}. \tag{8.9}$$

Hence, the closed-loop-transfer function reduces to

$$T(s) = \frac{K_m K_P}{\frac{K_m K_P}{s} + 1} = \frac{K_m K_P}{s + K_m K_P}. \tag{8.10}$$

Hence, by adjusting K_P, then, in principle, the response between y^* and y can be made as fast as desired since the closed-loop response is (theoretically) a first-order system having time constraint:

$$\tau_{cl} = \frac{1}{K_m K_P}. \tag{8.11}$$

8.3.1 Experiment: Testing a PD Control Law

By adjusting K_P and K_D (i.e. not restricting to the choice (8.8)), it can be easily shown that essentially any (second order) closed-loop response can be obtained. This is shown below. For general K_P and K_D, it follows that:

$$\begin{aligned} T(s) &= \frac{\frac{K_m(K_P + K_D s)}{s(\tau s + 1)}}{1 + \frac{K_m(K_P+K_D s)}{s(\tau s+1)}}, \\ &= \frac{K_m(K_P + K_D s)}{\tau s^2 + (1 + K_m K_D)s + K_m K_P}, \\ &= \frac{\frac{K_m(K_P + K_D s)}{\tau}}{s^2 + (1 + K_m K_D)(\frac{1}{\tau})s + K_m K_P}. \end{aligned} \tag{8.12}$$

Hence any closed-loop-transfer function can, in principle, be achieved by choice of K_P and K_D. To test this idea, the controller can be implemented using the PID block. To avoid having an improper controller, a small filter should be added to the derivative term to yield

$$C = K_P + \frac{K_D s}{(\gamma s + 1)}. \tag{8.13}$$

It is suggested that one choose $\gamma = 0.01 < 0.1$.

The reader should experiment with different values of K_P and K_D and record the resultant closed-loop step responses.

8.4 Tachometer Feedback

As a side comment, it is observed that the PD controller implements "derivative action" by differentiating the position signal. Some servo systems are actually equipped with an extra measurement of output velocity by using a Tachometer. If this is available, then this can be used instead of the Derivate term in the PD controller.

8.5 PID Design

The inclusion of an integral term is next explored via a PID controller. The reader may wonder why there is any need for the integral term. After all, there is already an integrator inherent in the servo itself. This is a very interesting question and depends on whether or not one has output or input disturbances and what form the disturbances take. The two cases are explained below.

8.5.1 Output Disturbances

Recall from Section 6.3 of Chapter 6 that the closed-loop-output response for an output disturbance, d_o, is

$$y = S(s)d_o, \tag{8.14}$$

where $S(s)$ is the sensitivity function given by

$$S(s) = \frac{1}{1 + G(s)C(s)}. \tag{8.15}$$

In the case of the problem under study, it follows that:

$$G(s) = \frac{K_m}{s(\tau s + 1)}. \tag{8.16}$$

Hence,

$$S(s) = \frac{s(\tau s + 1)}{s(\tau s + 1) + K_m C(s)}, \tag{8.17}$$

where $C(s)$ is the controller transfer function.

Clearly, $S(0) = 0$ even when $C(s)$ does not have an extra integrator. Hence, zero steady-state error is obtained in closed-loop for a *constant* output disturbance. However, the situation with input disturbances is different as explained in Section 8.5.2.

8.5.2 Input Disturbance

The output response to an input disturbance, d_i is different from that of an output disturbance. In particular, from (6.15) of Chapter 6

$$y = S_i(s)d_i, \tag{8.18}$$

where

$$S_i(s) = \frac{G(s)}{1 + G(s)C(s)}. \tag{8.19}$$

Again using (8.16)

$$S_i(s) = \frac{K_m}{s(\tau s + 1) + K_m C(s)}. \tag{8.20}$$

It can then be seen that the s in the numerator of $S(s)$ arising from the plant is not present in $S_i(s)$. Indeed, the transfer function $S_i(s)$ has finite DC gain given by

$$S_i(0) = \frac{1}{C(0)}, \tag{8.21}$$

where $C(0)$ is the DC gain of the controller.

Alternatively, if a PD controller is used, then

$$S_i(0) = \frac{1}{K_P}. \tag{8.22}$$

Hence, a constant input disturbance will lead to an off-set in the output response.

It follows from (8.22) that the impact of input disturbances for the servo systems decreases as the controller gain K_P increases but the error response is never zero.

This suggests that for input disturbances, the design needs to go further. Thus, an integral term is added leading to a PID controller of the form:

$$C(s) = K_P + \frac{K_I}{s} + K_D s; \; K_I \neq 0. \tag{8.23}$$

In this case,

$$S_i(0) = 0, \tag{8.24}$$

as required.

A similar situation arises when an output disturbance is not a constant. For example, it may be that the disturbance is best modelled as a ramp at the output. The Laplace transform of a ramp is of the form $\frac{\alpha}{s^2}$. Hence, even if $S(s)$ has a single s in the numerator, this is insufficient to yield zero steady-state error. Thus, again one would need to include an extra integrator in the control law, even when the plant has an integrator in its transfer function.

8.5.3 A Simple Design Procedure

As shown below, it is actually possible to adjust the PID gains to achieve essentially any desired closed-loop response. It is suggested that the reader re-examine Section 7.3 of Chapter 7. To illustrate the idea, say one chooses K_P, K_I, K_D to cancel the open-loop pole and place the two other closed-loop poles at $-\alpha^*$. To achieve this, it suffices to choose $C(s)$ to be of the form:

$$C(s) = \frac{K_P(\tau s+1)\left(s+\frac{\alpha^*}{2}\right)}{s}. \tag{8.25}$$

Then, after cancelling the term $(\tau s + 1)$, the loop-transfer function becomes

$$G(s)C(s) = \frac{K_m K_P\left(s+\frac{\alpha^*}{2}\right)}{s^2}. \tag{8.26}$$

The resulting closed-loop characteristic polynomial is readily seen to be

$$\begin{aligned} A_{cl}(s) &= s^2 + K_m K_P\left(s+\frac{\alpha^*}{2}\right), \\ &= s^2 + K_m K_P s + \frac{K_m K_P \alpha^*}{2}. \end{aligned} \tag{8.27}$$

Hence to achieve two closed-loop poles at α^*, it suffices to choose $K_m K_P = 2\alpha^*$ or $K_P = \dfrac{2\alpha^*}{K_m}$. This yields a closed-loop characteristic polynomial:

$$\lambda(s) = s^2 + 2\alpha^* s + (\alpha^*)^2 = (s+\alpha^*)^2, \tag{8.28}$$

as required.

8.5.4 Experiment: Testing a PID Control Law

It is suggested that the reader test the design found in Section 8.5.3. In particular, a PID controller can be implemented having transfer function

$$C(s) = \frac{K_P\left[\tau s^2 + \left(\frac{\tau\alpha^*+1}{2}\right)s + \frac{\alpha^*}{2}\right]}{s}. \tag{8.29}$$

The controller $C(s)$ can be manipulated into the form it appears in the laboratory beginning with (8.29):

$$C(s) = \frac{K_P\left[\tau s^2 + \left(\frac{\tau\alpha^*}{2}+1\right)s + \frac{\alpha^*}{2}\right]}{s}, \tag{8.30}$$

$$= K_P\left(\frac{\tau\alpha^*}{2}+1\right) + \frac{K_P\alpha^*}{2s} + K_P\tau s. \tag{8.31}$$

Hence, the PID gains in the laboratory should be chosen as $K_P\left(\frac{\tau\alpha^*}{2}+1\right)$, $\frac{K_P\alpha^*}{2}$ and $K_P\tau$, respectively, where $K_P = \frac{2\alpha^*}{K_m}$.

It is suggested that the reader test the PID on the servo system for different values of α^*.

Also, the reader could try the previous PD design with a constant input disturbance and compare it with the results obtained with the alternative PID design.

8.6 Revision Questions

1. Consider the same system as studied in this chapter and ignore the limits on travel.
   (a) If a proportional control law with gain K_p is used what is the steady-state error to a constant output disturbance of magnitude d?
   (b) Repeat (a) for a ramp-type output disturbance, $d(t)$, where

$$d(t) = ct. \tag{8.32}$$

2. Repeat Question 1 for a PI controller of the form $C(s) = K_P + \frac{K_I}{s}$.
3. Can you design a control law for this plant that has zero steady-state error to a constant input disturbance whilst ensuring that the closed-loop response does not overshoot to a step change in the reference signal? [Note: This was a design task set to a young engineer working in the Chemical Industry.]
4. In this chapter, control laws have been studied for the door-opening system that provides zero tracking errors in the presence of step or ramp-type output disturbances. What changes in the control law are necessary to achieve zero steady-state error for a ramp change in desired position?
5. Continuing the tracking theme, what changes would be needed to have zero steady-state error if there is a sinusoidal disturbance of frequency ω_0? (The reader may feel this would not occur for a door opener. Whilst this is true, the idea of rejecting a sinusoidal disturbance occurs in many other areas especially power electronics.)
6. Say that the desired closed-loop poles are all at $-\alpha^*$. Set up a suitable proper (but not strictly proper) control law incorporation $\left(s^2+\omega_0^2\right)$ in the denominator and write down the pole assignment identity required to solve for the control law coefficients.
7. How would you change your answer to Question 6 if a strictly proper control law was needed?
8. Write the control law of Question 7 in state-space form. (This is the form in which it would be implemented.)
9. Repeat Question 8 for the control law of Question 6.

Further Reading

1 G.C. Goodwin, S.F. Graebe and M.E. Salgado, *Control system design*, Prentice Hall, 2000.

9

Laboratory 3: Continuous Casting Machine: Linear Considerations

9.1 Introduction

This chapter explores the application of simple linear control system design to a machine that continuously casts molten steel into a strip. This study was conducted by the authors and their colleagues in industry over a two-year period. The original motivation for the study was that the real-world system exhibited sustained oscillations which led to poor product quality. Many hypotheses were advanced to explain these oscillations. Indeed, the company spent large sums of money studying various hypothesised causes of this problem. Interestingly, it was known that similar machines elsewhere in the world exhibited similar oscillations.

The laboratory will be visited twice. On this first occasion, the use of simple versions of P, I and PI control will be examined. A later laboratory (Chapter 26) will return to the question of the oscillations.

9.2 The Physical Equipment

A continuous caster is a process in which molten liquid steel is solidified continuously into a rectangular bar. The molten steel is poured into a mould which is a rectangular-shaped container, open at each end (see Figure 9.1). A cast begins when the mould, which has a rectangular cross-section, is closed off at the bottom, so as to build up the molten steel, and is accompanied by intense cooling using water jets. When the steel in the mould starts to solidify, it is drawn slowly out from the bottom of the mould, as a solid piece. As the steel is drawn out from the bottom of the mould it leaves a space at the top. This space is continually filled by molten steel poured into the mould from a container called a tundish. The process results in a long bar of steel, with a rectangular cross-section, being drawn out of the bottom of the mould while a stream of molten steel is poured into the top. In traditional steel mills the bar would be drawn out for a number of metres by which time it would be completely solidified. This solid bar would then be cut into manageable pieces called blooms which can then be further processed, e.g. by rolling or forging.

In a continuous casting machine, the cast material can be directly fed to a hot rolling mill, i.e. the intermediate step of forming blooms is avoided.

As can be imagined, continuous casting requires a number of processes to run in unison. The failure to balance any one process could lead to an extremely dangerous and expensive accident. An example would be if solidifying steel was drawn out too quickly from the mould. This would expose the molten steel in the mould which would then flow out in an uncontrolled fashion. The lives of operators could be threatened by such an event.

Practical Control System Design: Real World Designs implemented on Emulated Industrial Systems, First Edition. Adrian Medioli and Graham Goodwin.

Companion website: www.wiley.com/go/medioli/practicalcontrolsystemdesign

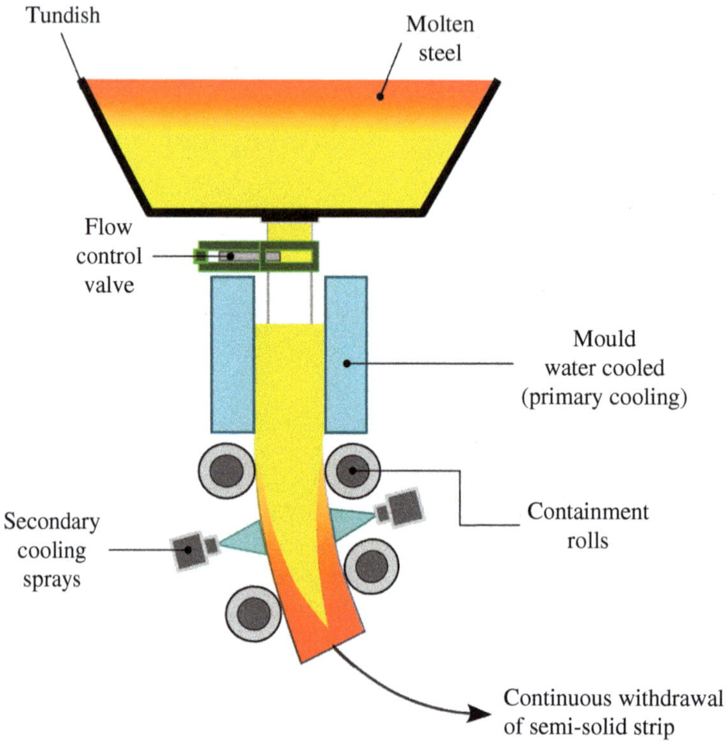

Figure 9.1 Continuous caster schematic.

It is assumed here that the cooling system and withdrawal mechanisms have been properly tuned and work safely and hence it suffices to study the details of the system controlling the pouring of the molten steel into the mould.

As used in the BHP Steel mill in Newcastle, Australia in the 1990s (see Figure 9.3), the continuous caster comprises a reservoir of steel (the Tundish) directly above the mould (see Figure 9.1). The flow of the molten steel is controlled by a slide gate valve (SGV). An important issue with the delivery of the molten steel is that the height of steel in the mould should be kept nearly constant (or, at least, regulated to a specific time-varying set-point). If the height of steel is allowed to fluctuate too widely then contamination of the steel can occur from flux material, e.g. slag, being embedded into the surface of the steel. The removal of these impurities is an expensive extra process. The impurities are also capable of causing structural weakness in the products made from the system.

The liquid steel enters the mould via a submerged entry nozzle. The nozzle has an outer refractory lining to deal with the extreme conditions. However, the flux material on the surface of the molten steel eventually destroys this lining. In order to maximise the nozzle life, the mould height set-point is periodically changed. Thus, the control system needs to be able to facilitate these changes.

A screenshot of the virtual system emulating the continuous caster is shown in Figure 9.2. Note that the correct operation of the real system is extremely critical. A 'break-out' of the steel from the caster can cause millions of dollars of damage and threaten human life. Thus, to experiment on a real machine of this type, industry requires management approval at the highest level and would usually be impossible.

The virtual laboratory accurately reflects the design of a real-life control system carried out on the continuous caster at BHP's Newcastle Australia steel works. This control application was

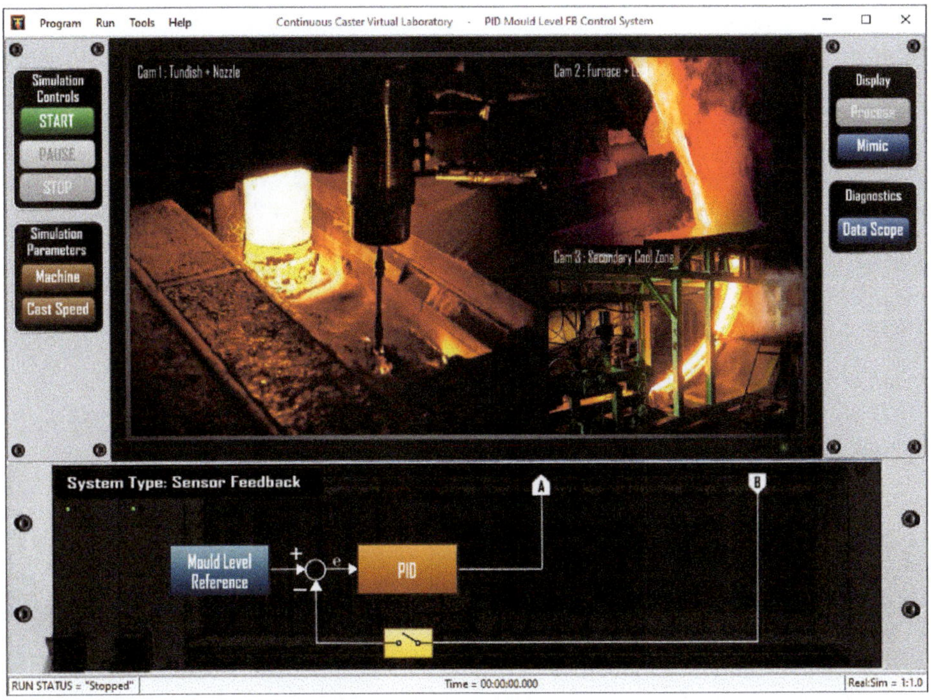

Figure 9.2 Screenshot of continuous caster laboratory.

motivated by a desire to improve the quality attributes of the cast material. Post-implementation audits showed a significant improvement in quality with an estimated saving to the company of US$ 1.5 million annually. Thus, the payback period for the control system design and implementation was of the order of one to two months. It is suggested that the reader delay reading references [1] and [2] until the experiment described in Chapter 26 has been completed. Photographs of the real system are shown in Figures 9.3–9.5.

9.3 Modelling of Continuous Casting Machine

The key components of the associated model of the process will next be described.

The model linking the level of molten steel in the mould, h, the mould cross-sectional area, A_m, the flow of steel into the mould f_{in} and the flow of steel out of the mould f_{out} is first derived.

A mass balance of the mould gives

$$A_m \frac{dh(t)}{dt} = f_{in} - f_{out}$$

or alternatively

$$h(t) = \frac{1}{A_m} \int (f_{in} - f_{out})\, dt.$$

Assuming that the mould has uniform cross-sectional area, A_m (mm^2), then

$$f_{out} = A_m d(t),$$

where $d(t)$ is the casting speed (mm/s).

Figure 9.3 Overall view of the continuous caster at BHP, Newcastle, Australia.

Figure 9.4 The steel is pouring through the slide gate valve.

Figure 9.5 The continuously cast steel passing through the secondary cooling chamber.

The flow into the mould is proportional to the area of the SGV opening, A_v (mm^2), and the square root of the steel level in the Tundish. Assuming that the latter is held roughly constant by other means, then

$$f_{in} = K_1 A_v.$$

It is assumed, for the moment, that the valve area, A_v (mm^2), is directly related to the valve position $u(t)$. Then,

$$f_{in} = K_1 K_2 u(t).$$

It turns out that this simple relationship will be perfectly adequate for initial control system design (i.e. for the nominal model). In Chapter 26, a more realistic model relating $u(t)$ to the SGV area will be explored.

For the purpose of control system design, a 'nominal' model will be used essentially as described above. Note that this models the mould as a simple integrator. Letting $K_v = K_1 K_2 / A_m$ results in the block diagram shown in Figure 9.6 for the open-loop plant.

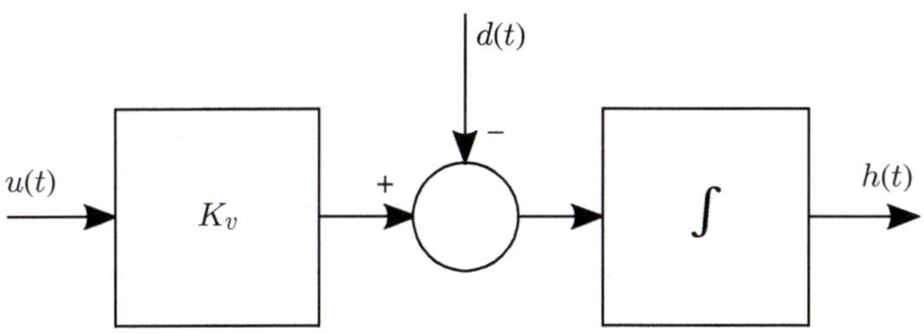

Figure 9.6 Simplified open-loop model.

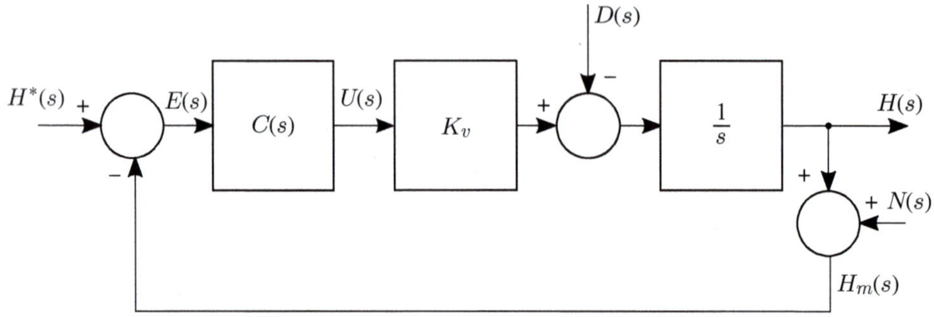

Figure 9.7 Closed loop.

9.4 Proportional Control

Next control system design is explored for the continuous caster. First, a simple proportional controller will be used, as shown schematically in Figure 9.7. In this case, the controller is $C(s) = K_p$ (i.e. proportional only). The level sensor introduces noise, $N(s)$, to the measured level signal $H_m(s)$.

An expression for the closed-loop relationship between $H(s)$ and $H^*(s)$, $D(s)$ and $N(s)$ is next developed. The answer will be expressed in the usual format:

$$H(s) = T(s)H^*(s) + S_i(s)D(s) - T_s(s)N(s),$$

where $T(s)$ and $S_i(s)$ are the complementary and input disturbance sensitivity functions, respectively.

Note that

$$H(s) = \frac{1}{s}(K_p K_v E(s) - D(s)).$$

$$E(s) = H^*(s) - H(s) - N(s).$$

Hence,

$$H(s) = \frac{1}{s}K_p K_v (H^*(s) - H(s) - N(s) - D(s)).$$

Rearranging gives,

$$H(s)(s + K_p K_v) = K_p K_v H^*(s) - K_p K_v N(s) - D(s)$$

or

$$H(s) = \frac{K_p K_v}{s + K_p K_v} H^*(s) - \frac{K_p K_v}{s + K_p K_v} N(s) - \frac{1}{s + K_p K_v} D(s).$$

Thus,

$$H(s) = \frac{1}{\frac{1}{K_p K_v}s + 1} H^*(s) - \frac{1}{\frac{1}{K_p K_v} + 1} N(s) - \frac{\frac{1}{K_p K_v}}{\frac{1}{K_p K_v}s + 1} D(s). \tag{9.1}$$

Finally it can be seen that

$$T(s) = \frac{1}{\frac{1}{K_pK_v}s + 1}.$$

$$S_i(s) = -\frac{\frac{1}{K_pK_v}}{\frac{1}{K_pK_v}s + 1}.$$

Note that the measured mould level is given by

$$H_m(s) = H(s) + N(s), \tag{9.2}$$

where $N(s)$ denotes measurement noise.

Also note that the disturbance is the output flow from the mould which is a function of the casting speed. This results in a ramp-type disturbance at the output due to the integrator in the plant.

9.5 Response to Set-Point Changes

The analysis is commenced by assuming that the casting speed is controlled and hence that the outflow disturbance, modelled by $D(s)$, is of the form α/s, i.e. a constant outflow. It is also assumed, for the moment, that the sensor is noise free, i.e. $N(s) = 0$. The effect that the controller gain, K_p, has on the response to set-point changes is then explored.

By inspection of Eq. (9.1), the closed-loop-time constant is given by

$$\tau = \frac{1}{K_pK_v}. \tag{9.3}$$

9.6 Experiments

9.6.1 Experiment: Model Parameter Estimation

The first task will be to determine the closed-loop-time constant and use it to calculate the plant gain K_v.

1. Start the virtual laboratory program.
2. Note that it would be extremely dangerous to operate this system in open loop. Hence, this experiment will be performed 'in closed-loop', i.e. with feedback between the system outputs and inputs. To ensure this is the case, confirm that the switch, in the mould level feedback path, is in the closed state.
3. Next, configure the controller by clicking on the PID block and setting the proportional term K_p to a gain of 2.0 whilst ensuring that K_i and K_d are zero.
4. Use the desired output (i.e. the mould level) block to apply a square wave signal with Offset = 80, Peak amplitude = 5.0, Frequency = 0.05 Hz and Noise SD = 0. Note that you will need to change the mould level reference as explained in Section 9.2.
5. Set the casting speed to be constant by clicking on the 'Casting Speed' button and selecting the constant tab and setting the value to 12 mm/s.
6. Using the Data Logger, observe the mould level reference y^* (tag 'y*'), the measured mould level (tag 'y_m') and the casting speed (tag 's_m').
7. Examine the measured mould level response and calculate the time constant of the plant and K_v.

9.6.2 Low Gain Feedback

Next the impact of changing the controller gain will be explored.

1. Start the virtual laboratory program.
2. This experiment will be performed 'in closed-loop', i.e. with feedback between the system outputs and inputs. To ensure this is the case, confirm that the switch, in the mould level feedback path, is in the closed state.
3. Next configure the controller by clicking on the PID block and setting the proportional term K_p to a gain of 2.0 whilst ensuring that K_i and K_d are zero.
4. Use the desired output (i.e. the mould level) block to apply a square wave signal with Offset = 80, Peak amplitude = 5.0, Frequency = 0.05 Hz and Noise SD = 0. Note that the need to change the mould level reference was explained in Section 9.2.
5. Set the casting speed to be constant by clicking on the 'Casting Speed' button and selecting the constant tab and setting the value to 12 mm/s.
6. Using the Data Logger, observe the mould level reference y^* (tag 'y*'), the measured mould level (tag 'y_m') and the casting speed (tag 's_m').
7. Examine the responses for different values of K_p say 2.0, 5.0 and 10.0 and comment on its effect of the responses.

It may surprise the reader that the output has an offset from the desired output. The reader is asked to pause and question how this can be since the plant contains an integrator. More will be said about this issue in Section 9.9.

9.6.3 High Gain Feedback

1. Examine the responses for higher values of K_p say 100.0 and 200.0 and note the effect on the transient responses.

The reader should observe that the transient response time stops decreasing. The reader is asked to question why this can happen.

9.7 Effect of Measurement Noise

Measuring the mould level in practice is a difficult process. Clearly, one cannot use contact measuring devices – these would melt. Hence some form of proximity measurement is used – these tend to be noisy. To model this effect, measurement noise will be added to the system. The following relationships then hold

$$H(s) = T(s)\,(H^*(s) - N(s)) - S_i(s)D(s), \tag{9.4}$$

$$H_m(s) = T(s)H^*(s) + S(s)N(s) - S_i(s)D(s), \tag{9.5}$$

where $S(s)$ is the output disturbance sensitivity function. It can then be seen that as one increases $T(s)$ towards 1, the effect of disturbances decreases but the effect of noise increases until the output has the full noise impressed upon it.

9.7.1 Experiment: Measuring the Impact of Measurement Noise

The virtual laboratory will next be used to determine the effect of measurement noise.

1. Start the virtual laboratory program.
2. The experiment will be performed 'in closed-loop', i.e. with feedback between the system outputs and inputs. To ensure this is the case, confirm that the switch, in the mould level feedback path, is in the closed state.
3. Next, configure the controller by clicking on the PID block and setting the proportional term K_p to a gain of 2.0 whilst ensuring that K_i and K_d are zero.
4. Use the desired output (i.e. the mould level) block to apply a square wave signal with Offset = 70, Peak amplitude = 5.0, Frequency = 0.05 Hz and Noise $SD = 0$. Note that you will need to change the mould level reference as explained in Section 9.2.
5. Set the casting speed to be constant by clicking on the 'Casting Speed' button and selecting the constant tab and setting the value to 12 mm/s.
6. Add noise to the mould level measurement by clicking on the 'Machine' button on the 'Simulation Parameters' section of the lab. Once open click on the 'Use mould level noise' checkbox.
7. Using the Data Logger, observe the mould level reference y^* (tag 'y*'), the measured mould level (tag 'y_m') and the casting speed (tag 's_m').
8. Examine the responses for different values of K_p say 2.0, 4.0 and 10.0.

Note that varying K_p has little effect on the noise that appears on the measured output. The reader is asked to think about this observation.

9.8 Pure Integral Control

Although the plant has an integrator, it was seen in Section 9.6.2 that a constant outflow input disturbances leads to as non-zero tracking error. Hence, an alternative controller, namely, pure integral control, as shown schematically in Figure 9.7, will be used where $C(s) = \frac{K_i}{s}$.

The closed-loop-transfer function from $H^*(s)$ to $H(s)$ will be derived when pure integral control is used. Note that

$$E(s) = H^*(s) - H(s).$$

$$H(s) = \frac{K_i K_v}{s^2} E(s).$$

$$s^2 H(s) = K_i K_v E(s).$$

$$s^2 H(s) = K_i K_v (H^*(s) - H(s)).$$

$$(s^2 + K_i K_v) H(s) = K_i K_v H^*(s).$$

$$H(s) = \frac{K_i K_v}{(s^2 + K_i K_v)} H^*(s).$$

The reader is asked to ponder why this result is unsatisfactory.

9.8.1 Experiment: Testing Pure Integral Control

The above conclusions are examined using the virtual laboratory. (Note that these experiments could never be performed on the real plant because the resulting performance would endanger the lives of operators!)

1. Start the virtual laboratory program.
2. This experiment will be performed 'in closed-loop', i.e. with feedback between the system outputs and inputs. To ensure this is the case, confirm that the switch, in the mould level feedback path, is in the closed state.
3. Next configure the controller by clicking on the PID block and setting the integral term K_i to a gain of 1.0 whilst ensuring that K_p and K_d are both zero.
4. Use the desired output (i.e. the mould level) block to apply a square wave signal with Offset = 80, Peak amplitude = 5.0, Frequency = 0.05 Hz and Noise SD = 0.
5. Set the casting speed to be constant by clicking on the 'Casting Speed' button and selecting the constant tab and setting the value to 12 mm/s.
6. Remove the mould level measurement noise if it was applied previously by clicking on the 'Machine' button on the 'Simulation Parameters' section of the lab. Once open click on the 'Use mould level noise' checkbox to deselect it.
7. Using the Data Logger, observe the mould level reference y^* (tag 'y*'), the measured mould level (tag 'y_m') and the casting speed (tag 's_m').
8. Examine the responses for a gain K_i of say 1.0 and comment on its effect of the responses.

9.9 PI Control

Based on the results in Sections 9.4 and 9.8, it seems that proportional control gives excellent transient performance but leads to significant steady-state errors in the presence of input disturbances. There are two equivalent explanations for this phenomenon. On the one hand, if the disturbance is modelled at the input, then the input disturbance sensitivity function has non-zero DC gain. On the other hand, if the disturbance is transferred to the output, then it appears as a ramp. Then although the normal sensitivity function has an s in the numerator, the Laplace transform of the ramp disturbance is of the form α/s^2. This suggests that integral control is desirable. However, as observed by the results of Section 9.8, pure integral control is unsatisfactory when applied to this application since it leads to oscillatory closed-loop behaviour. This motivates the next step where the combination of proportional and integral control is shown schematically in Figure 9.7 with $C(s) = K_p + \frac{K_i}{s}$.

For this controller, the transfer function relationships between $H(s)$ and $H^*(s)$ and $D(s)$ (for simplicity we take $N(s) = 0$), is derived as follows:

Note that

$$U(s) = \left(K_p + \frac{K_i}{s}\right) E(s).$$

$$E(s) = H^*(s) - H(s).$$

$$H(s) = \frac{1}{s}\left(K_v U(s) - D(s)\right).$$

$$H(s) = \frac{1}{s}\left[K_v \left(K_p + \frac{K_i}{s}\right) E(s) - D(s)\right].$$

$$H(s)s^2 = K_v(K_p s + K_i)(H^*(s) - H(s)) - D(s)s.$$

$$H(s)(s^2 + K_v K_p s + K_v K_i) = (K_v K_p s + K_v K_i)H^*(s) - D(s)s.$$

$$H(s) = \frac{K_v K_p s + K_v K_i}{s^2 + K_v K_p s + K_v K_i} H^*(s) - \frac{s}{s^2 + K_v K_p s + K_v K_i} D(s). \tag{9.6}$$

Using the above results, the effect of different values of K_i and K_p will be explored on

1. Steady-state Errors
2. The natural frequency and damping of the closed-loop system

Recall that,

$$H(s) = \frac{K_v K_p s + K_v K_i}{s^2 + K_v K_p s + K_v K_i} H^*(s) - \frac{s}{s^2 + K_v K_p s + K_v K_i} D(s).$$

Hence, even with a constant input disturbance, $H(s) = H^*(s)$ in steady state i.e. the steady-state error is zero, provided $K_i > 0$ and $K_p > 0$.

Comparing terms in the denominators with a standard second order system, leads to

$$s^2 + K_v K_p s + K_v K_i = s^2 + 2\zeta\omega_n s + \omega_n^2.$$

Thus, the natural frequency is given by

$$\omega_n = \sqrt{K_v K_i}, \tag{9.7}$$

and damping by

$$\zeta = \sqrt{\frac{K_v}{4K_i}} K_p. \tag{9.8}$$

Thus, one can arbitrarily assign ω_n and ζ by choosing K_p and K_i appropriately.

9.9.1 Experiment: Testing PI Control

Using the understanding gained from the above analysis design a PI controller. The responses should not exhibit ringing as this would mix more flux onto the surface of the caste steel. Thus, take $\zeta = 1$. The PI controller should not drive the valve onto it's limits. N.B. You may assume that $K_v = 0.28338$.

Hence set $\zeta = 1$ and choose various values of ω_n. The reader is asked to design K_p and K_i for $\omega_n = 1.0$ 0.7, 0.5 and 0.1 rad/s, respectively.

1. Start the virtual laboratory program.
2. This experiment will be performed 'in closed-loop', i.e. with feedback between the system outputs and inputs. To ensure this is the case, confirm that the switch, in the mould level feedback path, is in the closed state.
3. Use the desired output (i.e. the mould level) block to apply a square wave signal with Offset = 80, Peak amplitude = 5.0, Frequency = 0.05 Hz and Noise SD = 0.
4. Set the casting speed to be constant by clicking on the 'Casting Speed' button and selecting the constant tab and setting the value to 12 mm/s.
5. Remove the mould level measurement noise if it was applied previously by clicking on the 'Machine' button on the 'Simulation Parameters' section of the lab. Once open click on the 'Use mould level noise' checkbox to deselect it.

6. Using the Data Logger, observe the mould level reference y^* (tag 'y*'), the measured mould level (tag 'y_m') and the casting speed (tag 's_m').
7. Examine the responses for the different values of K_p and K_i that you designed above and select the set that best meets the design criteria. Note that you may experience 'Breakout' of the Mould. This would be catastrophic in practice. This underlines the importance of testing designs in simulation before applying them to a real plant.

9.9.2 Experiment: Testing the Response to Varying Casting Speed

Using the best K_p and K_i values, the design will next be tested when the casting speed is varied by performing the following steps.

1. Use the desired output (i.e. the mould level) block to apply a constant reference of Value = 70 and Noise SD = 0.
2. Set the casting speed to be a square wave by clicking on the 'Casting Speed' button, selecting the square wave tab and setting the Offset = 12, Peak amplitude = 2.0, Frequency = 0.02 Hz and Noise SD = 0.
3. Examine the responses and comment on its effect of the responses.

9.10 Feedforward Control

It has been observed in Section 9.9, that PI control results in excellent steady-state and transient response due to a change in the reference mould level. However, it still suffers a transient error when a change in the disturbance (i.e. the casting speed) occurs. Next, it is shown how feedforward control can be used to reduce these transient effects.

At this point, it is suggested that the reader revisits Section 7.12.

It is known that a change in casting speed will require a change in the valve opening to increase the inflow of steel and thus maintain the mould level to its reference. When a change in casting speed occurs (i.e. a disturbance), an error is introduced in the mould level, to which the feedback controller reacts by adjusting the valve. If the disturbance can be directly measured then this information can be used to move the valve in a pre-emptive fashion and thus avoid the transient error from occurring in the mould level. This is disturbance feedforward and is illustrated in Figure 9.8.

For Figure 9.8, with a PI controller, the transfer function relationships between $H(s)$ and $H^*(s)$ and $D(s)$ can be derived. For simplicity, take $N(s) = 0$.

Then from Figure 9.8, it follows that:

$$U(s) = \left(K_p + \frac{K_i}{s}\right) E(s) + K_{ff}D(s).$$

Recall that $D(s)$ has been modelled as an input disturbance. Thus, a casting speed of 12 mm/s is represented as a disturbance having Laplace transform $12/s$. Also, note that

$$E(s) = H^*(s) - H(s).$$

$$H(s) = \frac{1}{s}\left[K_v U(s) - D(s)\right].$$

$$H(s) = \frac{1}{s}\left[K_v\left(K_p + \frac{K_i}{s}\right) E(s) + K_v K_{ff}D(s) - D(s)\right].$$

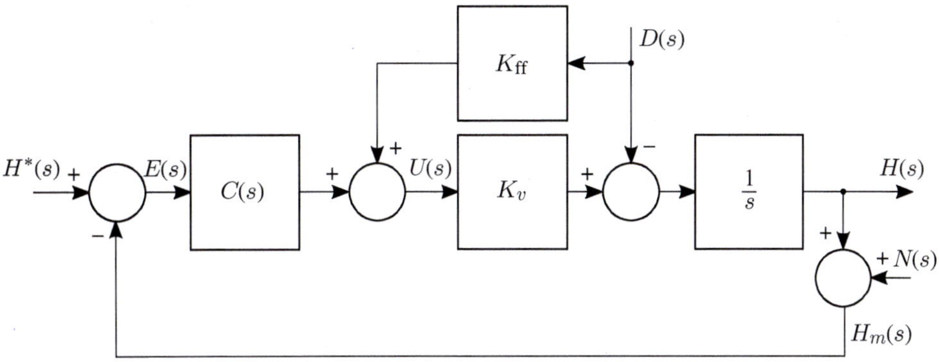

Figure 9.8 Control loop with disturbance feedforward.

$$H(s)s^2 = K_v\left(K_p s + K_i\right)(H^*(s) - H(s)) - (1 - K_v K_{ff})sD(s).$$

$$H(s)(s^2 + K_v K_p s + K_v K_i) = (K_v K_p s + K_v K_i)H^*(s) - (1 - K_v K_{ff})sD(s).$$

$$H(s) = \frac{K_v K_p s + K_v K_i}{s^2 + K_v K_p s + K_v K_i} H^*(s) - \frac{s(1 - K_v K_{ff})}{s^2 + K_v K_p s + K_v K_i} D(s). \tag{9.9}$$

The reader is asked to think about what value of K_{ff} would eliminate any transient due to a change in the casting speed.

Next say that a 10% error is made in measuring the casting speed. The effect that this has on the transient response due to a casting speed change will be studied when feedforward is applied.

Let $K_{ff} = \alpha \frac{1}{K_v}$ where α is a scaling factor, substituting gives

$$H(s) = \frac{K_v K_p s + K_v K_i}{s^2 + K_v K_p s + K_v K_i} H^*(s) - \frac{s(1 - \alpha)}{s^2 + K_v K_p s + K_v K_i} D(s).$$

For 10% overestimation, $\alpha = 1.1$ and

$$H(s) = \frac{K_v K_p s + K_v K_i}{s^2 + K_v K_p s + K_v K_i} H^*(s) + \frac{0.1s}{s^2 + K_v K_p s + K_v K_i} D(s).$$

For 10% underestimation, $\alpha = 0.9$

$$H(s) = \frac{K_v K_p s + K_v K_i}{s^2 + K_v K_p s + K_v K_i} H^*(s) - \frac{0.1s}{s^2 + K_v K_p s + K_v K_i} D(s).$$

Thus, if α is not 1, the feedforward does not perfectly remove the effects of the disturbance, though it does still attenuate its effect. The transient response is still the same, but the amplitude is reduced. Note also the sign change with over and under estimation of the casting speed.

9.10.1 Experiment: Testing Feedforward Control

The above ideas will be explored using the virtual laboratory.

Feedforward control will be studied with step changes in casting speed to experimentally verify the theory.

Designing for an ideal value, the feedforward gain, K^*_{ff}, can be determined. This value can be implemented in the feedforward gain block (Note: $K_v = 0.28338$). Next, robustness is explored by incorrectly setting the value of K_{ff} (say due to imprecise knowledge of K_v).

1. Start the virtual laboratory program.
2. This experiment will be performed 'in closed-loop', i.e. with feedback between the system outputs and inputs. To ensure this is the case, confirm that the switch, in the mould level feedback path, is in the closed state.
3. Select a new controller architecture which includes feedforward by selecting the Program > Load Controller > PID Mould Level FB + FF menu item. Set the 'Dither' to zero.
4. Use the desired output (i.e. the mould level) block to apply a constant reference of Value = 70 and Noise SD = 0.
5. Set the casting speed to be a square wave by clicking on the 'Casting Speed' button, selecting the square wave tab and setting the Offset = 12, Peak amplitude = 2.0, Frequency = 0.02 Hz and Noise SD = 0.
6. Remove the mould level measurement noise if it was applied previously by clicking on the 'Machine' button on the 'Simulation Parameters' section of the lab. Once open click on the 'Use mould level noise' checkbox to deselect it.
7. Using the Data Logger, observe the mould level reference y^* (tag 'y*'), the measured mould level (tag 'y_m') and the casting speed (tag 's_m').

9.10.2 Experiment: Testing Sensitivity to the Feedforward Gain

1. Examine the responses for the values of $K_{\text{ff}} = 0.9K^*_{\text{ff}}$ and $K_{\text{ff}} = 1.1K^*_{\text{ff}}$ and record your observations.

9.11 Revision Questions

1. Assume that the mould associated with the laboratory in this chapter has dimensions 400 × 600 mm. If the casting speed is 12 mm/s, how many metric tonnes of steel are cast per hour?
2. In Section 9.6.3, it was found that the response time did not keep reducing when the controller gain was increased. Why is this the case?
3. In Section 9.7, it was found that increasing the control law gain led to greater sensitivity to measurement noise. Explain this in terms of the fundamental limits connecting S and T.
4. In Section 9.8, it was suggested that a pure integral controller is unsatisfactory. Why?
5. Consider the closed-loop response given in Eq. (9.6) in the text. Show that K_p and K_i can be chosen to yield closed-loop poles at any desired location.
6. Given your answer to Question 5, why would it not make sense to design K_p and K_i to put the closed-loop poles in the far left-half complex plane.
7. Section 9.10.2 described an experiment to test the sensitivity of the feedforward loop to errors in the gain.
   (a) Why is feedforward control sensitive to gain errors?
   (b) Would you use feedforward alone, i.e. without feedback?

Further Reading

The design procedure described in this virtual laboratory essentially mirrors the initial steps used to solve the real problem in industry. It appears that good performance is readily achievable using the control architectures studied in this chapter. Note, however, that in practice, an oscillation (of about ±10 mm) was found to spontaneously develop and required operator intervention to stop. The investigation into the cause of this problem and a remedy are described in Chapter 26.

Further details of this practical control system design problem can be found in the following references. However, the reader may wish to delay consulting these references until the laboratory in Chapter 26 has been performed since the references give answers to some of the questions posed in Chapter 26.

1 S.F. Graebe, G.C. Goodwin and G. Elsley, "Control design and implementation in continuous steel casting", *IEEE Control Systems Magazine*, 15(4):64–71, 1995.
2 M.S. Jenkins, B.G. Thomas, W.C. Chen and R.B. Mahapatra, "Investigation of Strand Surface Defects Using Mold Instrumentation and Modelling", *77th Steelmaking Conference*, Chicago, Illinois, USA, Iron and Steel Society, Warrendale, Pennsylvania, USA, pp. 337–345, 1994.

10

Laboratory 4: Modelling and Control of Fluid Level in Tanks

10.1 Introduction

Tanks of various types are frequently used in industry to provide a place where different reactions can occur. Alternatively, tanks can be used as a buffer between different unit operations so that a variable 'incoming flow' can be averaged out so as to present a more uniform 'outgoing flow' to the next unit operation. The residence time in the tank can be exploited to give 'time' for other operations, e.g. pH neutralisation. Thus, the control of tanks is a very common problem met in industry.

In this virtual laboratory, a system having two tanks will be studied. The first tank can be controlled as a single unit. Alternatively, the two tanks can be coupled together by an interconnecting pipe. The latter configuration makes the dynamics more complex than would be seen on a single tank.

The virtual laboratory emulation is shown in Figure 10.1 and a schematic of the system is shown in Figure 10.2.

The block diagram, shown in Figure 10.3, describes a two-tank system in terms of its input and output.

For the purpose of the laboratory, the gain associated with the pump (which includes the DC motor, pump and associated drive circuit) will be considered a single component of the coupled tank apparatus.

10.2 The Controllers

In this laboratory, you will design closed-loop control systems of the form shown in Figure 10.4.

In this first laboratory, a single tank system will be studied. Later the two tank system, will be studied.

10.3 Physical Modelling

A physical model of the system will first be developed using mass balance. Consider tank T1, with valve V2 closed and V1 open, then

$$\frac{dV_1}{dt} = Q_{1i} - Q_{1o},$$

Practical Control System Design: Real World Designs implemented on Emulated Industrial Systems, First Edition. Adrian Medioli and Graham Goodwin.

Companion website: www.wiley.com/go/medioli/practicalcontrolsystemdesign

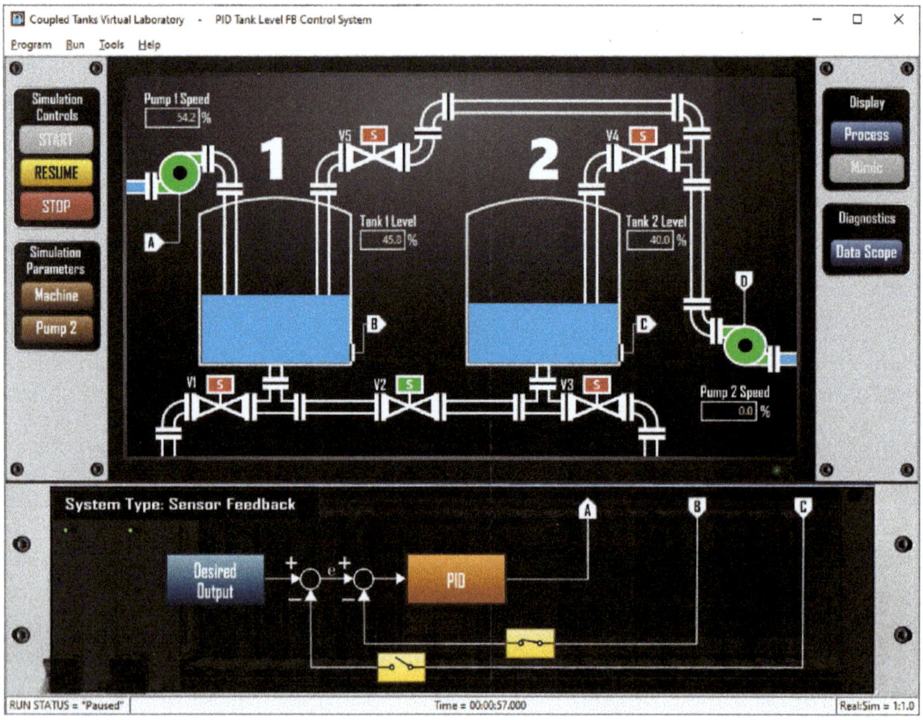

Figure 10.1 Coupled tanks virtual lab interface.

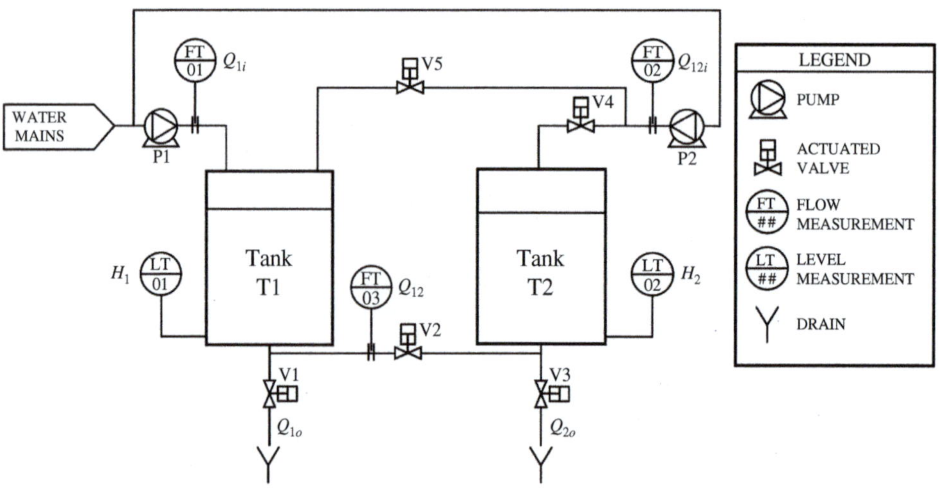

Figure 10.2 Coupled tanks schematic.

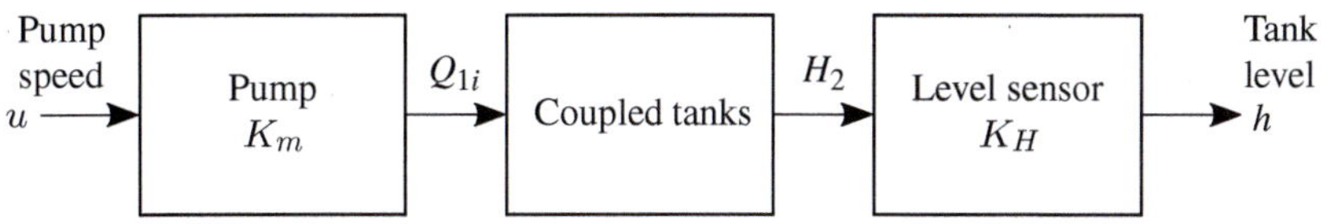

Figure 10.3 Coupled tanks block diagram.

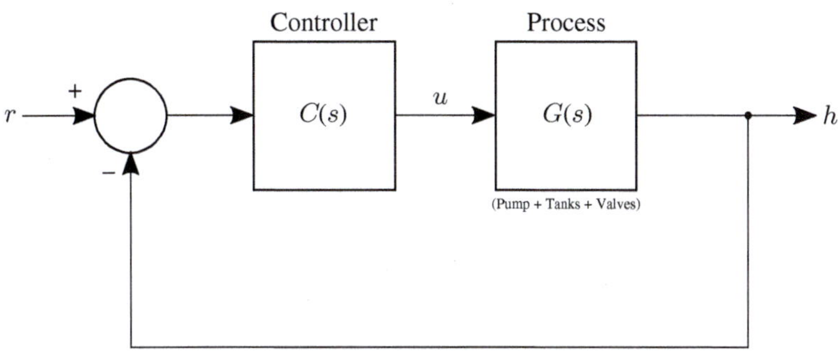

Figure 10.4 Coupled tanks control block diagram.

which can be rewritten as

$$A_1 \frac{dH_1}{dt} = Q_{1i} - Q_{1o}, \tag{10.1}$$

where

V_1 = volume of liquid in tank T1,
H_1 = level of liquid in tank T1,
A_1 = cross-sectional area of tank T1,
Q_{1i} = flow into tank T1,
Q_{1o} = flow out of tank T1.

The pump flow is assumed to be linearly related to the input voltage[1] u_1,

$$Q_{1i} = K_m u_1 \text{ for } u_1 > 0.15. \tag{10.2}$$

where K_m is the pump's flow constant. The level sensors return a percentage reading indicating the current height of fluid in the tank. Thus,

$$h = K_H H, \tag{10.3}$$

where H is the level in metres and K_H is the level sensor gain.

The flow out of the tanks may be modelled using the following equations

$$Q_{1o} = c_1 \sqrt{H_1 + \varepsilon_1}, \tag{10.4}$$

where the constants, c_1 and ε_1, are related to the discharge coefficients of the piping and valves.

The reader may wonder where the square root and ε_1 in (10.4) come from. The square root comes from assuming that the output is well-modelled as a turbulent flow. Physical models for this suggest a square root relationship.

$$Q_o = C\sqrt{H}. \tag{10.5}$$

1 Below approximately 0.15 V, the 'cut in' voltage, there is no flow.

The reader is also referred to Section 3.2.4 where a simple model for flow out of a tank was developed.

The additional term ϵ_1 comes from the geometry of the tank outlets. The physical process consists of the outlet of each tank followed by a length of pipe that discharges into the top of a reservoir. The actual discharge point is some distance below the floor of the tank. Applying Bernoulli's Equation to the system, it can be shown that the extra level (pressure head), which exists below the sensor's zero level, gets 'bundled' into the flow equations. Had the outlets been simply holes on the side wall, then Eq. (10.5) would more accurately describe the flow. This effect is fairly pronounced in this apparatus due to the small ratio of the outlet level to the tank level.

Equation (10.4) is valid for tank levels above 5%. The relationships were determined based on commonly used models for the type of physical apparatus which this laboratory replicates.

The goal of this virtual laboratory is to achieve 'tight' closed-loop control of the level of the fluid in the tanks. Note, however, that tanks are often used as buffers between unit operations. In this case, tight control may not be desirable since tight level control means that the output flow would follow the input flow variations. When a tank is needed as a buffer, then one either uses relatively 'loose' control or one deliberately tries to match the output flow to the average input flow over some period of time.

Using the above equations, it can be seen that the relationship between pump P1 input voltage, u_1, and level sensor output voltage, h_1, is given by

$$\frac{A_1}{K_H}\frac{dh_1}{dt} + c_1\sqrt{\frac{h_1}{K_H} + \varepsilon_1} = K_m u_1.$$

This is a nonlinear dynamical system. Using the techniques described in Chapter 3, an approximation linear model valid at an operating point can be obtained.

Using deviation variables $\tilde{u}_1 = u_1 - \overline{u}_1$ and $\tilde{h}_1 = h_1 - \overline{h}_1$, leads to the following linear model around the steady state operating point $(\overline{u}_1, \overline{h}_1)$.

$$\frac{\tilde{h}_1(s)}{\tilde{u}_1(s)} = \frac{K_m K_H r_1}{A_1 r_1 s + 1}, \tag{10.6}$$

where

$$r_1 = \frac{2\sqrt{\frac{\overline{h}_1}{K_H} + \varepsilon_1}}{c_1}. \tag{10.7}$$

The above transfer function describes the dynamic response of h_1 to small changes in u_1 around the steady-state operating point $(\overline{u}_1, \overline{h}_1)$.

Given the linearised model described above the next step is to determine its parameters. One approach would be to obtain the values of the parameters from component data sheets, fluid dynamic equations and specialised experiments. Another approach is to treat the system as a simple first-order system (or lag) of the form:

$$\frac{\tilde{h}_1(s)}{\tilde{u}_1(s)} = \frac{K}{\tau s + 1},$$

and determine K and τ from step tests performed experimentally. The latter approach will be adopted.

10.3.1 Experiment: Estimating Plant Gain and Time Constant

1. Perform an 'open-loop' test, by ensuring that the level feedback signals from the tanks, as shown on the system diagram, are both open.
2. Since a Signal Generator 1 will be used to drive the pump P1 directly, the PID controller gains are set to $K_p = 1, K_i = 0$, and $K_d = 0$.
3. Open valve V1 and close valves V2–V5. (Click on schematic and make sure Valve 1 is green and the remainder of the valves are red.)
4. Ensure the Data Scope is running. For the following experiments, you will need a large timescale (s/div)[2] to capture the full-step response.
5. Apply a constant input voltage u_1, to pump P1 to obtain a desired operating level. Aim for a level near the top of the tank, say 90. Note, with the speed up factor of 10 : 1 used in the simulator, it should take about one minute for the level to settle to the steady-state value.
6. Apply a small step to the speed of the pump. Again the system will take some time for the level to settle to the new steady-state value. It is suggested you might try starting with an input flow of 8. Then step it to 6. The gain is $\Delta h_1/\Delta u_1$ and the time constant may then be determined.
7. Do two more step tests and determine the plant dynamics at 60 and 40 steady-state height.
8. Compare the experimentally determined parameters for the different operating points. Does the trend make sense based on the physical modelling described earlier.

- The reader should see that both the gain and time constant increase with the tank level, in accordance with Eqs. (10.6) and (10.7).

10.4 Closed-Loop Level Control for a Single Tank

Using the parameters identified in Section 10.3, different closed-loop control strategies will be explored to regulate the level of fluid in a single tank. The control system aims to make h_1 follow a reference signal which is here denoted r, supplied from Signal Generator 1. The control system should maintain h_1 approximately equal to r despite 'disturbance' flows from pump P2 entering tank T1.

10.4.1 Proportional Only Control

The properties of proportional-only control will first be explored.

10.4.2 Experiment: Testing Proportional Control

The aim will be to control the level at around the 60 full range, i.e. $h_1 = 6.0$ V.

- Set valves V1, V5 open and V2–V4 closed. Also, set the sensor noise to zero.
- B1 feedback switch should be on, i.e. H1.
- The PID block should have $K_p = 1, K_i = 0$, and $K_d = 0$ all other options unchecked.
- Connect the Data Scope to observe r, h_1, the error and the controller output.
- Set r equal to 6.0 V.
- Let the tank settle and observe h_1, steady-state error and the controller output. Repeat for K_p equal 10 and 40.

---

2 Suggest 20 s/div.

- With $K_p = 1$ adjust r so that h_1 settles near 6.0 V. Then step r by 2 V and estimate the closed-loop time constant τ_{cl}. Repeat for $K_p = 10$.
- With $K_p = 1$ adjust r so that h_1 settles near 6.0 V. Recalling the cut in voltage of no flow, step pump P2 input voltage from 0.15 to 2 V. Estimate the disturbance response time constant and gain. Repeat for $K_p = 10$.

10.4.3 Integral Only Control

Again, the linearised transfer function for the single tank, integral only control, i.e. $C(s) = \frac{K_i}{s}$ will be explored:

The transfer function from r to h is

$$\frac{GC}{1+GC},$$

Substituting G and C gives

$$\frac{\frac{K}{\tau}K_i}{s^2 + \frac{1}{\tau}s + \frac{K}{\tau}K_i}, \tag{10.8}$$

The goal is to achieve a given closed-loop transfer function of the form $\frac{\omega_n^2}{s^2 + 2\zeta\omega_n s + \omega_n^2}$. Hence, equating coefficients gives

$$\omega_n = \sqrt{\frac{K}{\tau}K_i}.$$

$$\zeta = \frac{1}{2\sqrt{K\tau K_i}}.$$

$$K_i = \frac{1}{4K\tau\zeta^2}.$$

For under damping $\zeta < 1$, critically damping $\zeta = 1$ and over damping $\zeta > 1$:

$$K_i < \frac{1}{4K\tau} \quad \text{gives over damped,}$$

$$K_i = \frac{1}{4K\tau} \quad \text{gives critically damped,}$$

$$K_i > \frac{1}{4K\tau} \quad \text{gives under damped.}$$

It can be seen that, as the integral gain increases, the damping decreases and the response becomes more oscillatory.

10.4.4 Experiment: Testing Integral Control

Apply the previously derived expressions to determine a value of K_i to achieve a critically damped system. Use the model determined from the step test around 60 full.

Set r to 6 V. Set the K_p value in the PID block to zero and set the K_i term to the calculated value. Ensure the Bumpless Parameter Changes box is set in the PID block.[3]

3 Bumpless update means the state of the integral term in the controller is adjusted so the controller output initially remains the same with the new controller parameters applied. Without this facility the integral state is set to zero and the output will jump.

Wait until the tank settles to steady-state and then apply a 1 V step on r and observe the response. What is the steady-state error? Increase the integral gain K_i by a factor of 10, how does the response change?

It concluded from the above test that having integral control only leads to a poor compromise between speed of response and damping. In Section 10.4.5, an alternative study of PI control will be studied which gives a better trade-off.

10.4.5 Proportional Plus Integral Control

In this design, the controller parameter will be chosen to cancel the open-loop pole.

For a first-order plant, the design results in the following values for control law gains:

$$K_p = \frac{\tau}{\beta K} \qquad \text{proportional constant,}$$

$$K_i = \frac{1}{\beta K} \qquad \text{integral constant,}$$

where K is the plant gain, τ is the time constant in seconds and β is the desired closed-loop time constant in seconds. The parameter τ is the single tuning parameter.

In closed-loop the rise time, i.e. the time taken for the output to change from 10 to 90 of the steady-state value, is $T_r \approx 2.2\beta$.

10.4.6 Experiment: Testing PI Control

Design all the following PI controllers based on your model at the 60 full operating point. The controllers are generated based on a desired closed-loop time constant, β. Design controllers with β equal to τ, 0.1τ and 0.01τ. Observe and discuss the effects of decreasing the desired closed-loop time constant on

- reference tracking (try some large steps, e.g. 2 V)
- input disturbance rejection
- controller output (note that the pump input is limited to 0–10 V).

10.4.7 Experiment: Alternative PI Controller

Next a design is studied in which the open-loop plant pole is not cancelled. In the following design equations do not cancel the slow plant pole (if $\beta < \tau$). Use these design equations to generate a PI controller for $\beta = 0.1\tau$ leads to

$$K_p = \frac{2\tau - \beta}{K\beta} \qquad K_i = \frac{\tau}{K\beta^2}.$$

Examine

- the reference tracking (try some good-sized steps, e.g. 2 V)
- input disturbance rejection
- the controller output (note that the pump input is limited to 0–10 V)

10.5 Closed-Loop Level Control of Interconnected Tanks

The control of the two-tank system is next explored by using only pump P1, and setting V1 closed, V3 open and opening the interconnecting valve V2. Note that the flow out of tank 1 is now governed by the height difference between tanks 1 and 2. Thus, the linear model takes the form:

$$A_1 \frac{dh_1}{dt} = -k(h_1 - h_2) + K_m u_1, \tag{10.9}$$

where u_1 is the pump and $K_m u_1$ is the flow into tank 1.

Similarly, from tank 2.

$$A_2 \frac{dh_2}{dt} = k(h_1 - h_2) - Bh_2, \tag{10.10}$$

where Bh_2 describes the flow out of tank 2.

Note that, in steady-state, it is necessary that there be a difference in height between the two tanks.

In the experiment, the two levels h_1 and h_2 will be measured, and attention will be focused on controlling the level h_2 using the pump into tank 1. The transfer function model for h_1 is thus derived.

Using Laplace transforms of (10.9) and (10.10) and assuming the tanks have equal area $A = A_1 = A_2$ yields

$$(s + \alpha)H_1(s) = \alpha H_2(s) + \beta U_1(s), \tag{10.11}$$

$$(s + \alpha + \gamma)H_2(s) = \alpha H_1(s), \tag{10.12}$$

where $\alpha = \frac{k}{A}, \beta = \frac{K_m}{A}, \gamma = \frac{B}{A}$.

Rearranging (10.12) in terms of $H_1(s)$ and substituting into (10.11) yields

$$(s + \alpha)(s + \alpha + \gamma)H_2(s) = \alpha(\alpha H_2(s) + \beta U_1(s)) \tag{10.13}$$

or

$$\left(s^2 + (2\alpha + \gamma)s + \alpha\gamma\right) H_2(s) = \alpha\beta U_1(s). \tag{10.14}$$

Thus, the total system behaves as an (approximate) second-order linear system.

It is suggested that the reader carry out experiments to find the various parameters in the above model (This experiment is similar to the experiments performed earlier in Section 10.3.). Next, a control law will be designed for this second-order system.

This part of the laboratory will be left for our readers to explore on their own.

10.6 Revision Questions

1. In Section 10.3, it was suggested that the outflow of the tank is roughly the square root of the level of liquid in the tank. Do you think this nonlinearity will have a large impact on the performance of a closed-loop control system?
2. In practice, tanks are often used as a buffer between stages of a process. Why does this mean that it is not desirable to use a high-gain feedback controller?
3. Can you suggest a better solution?
4. Why would a tank be a useful device if one wishes to neutralise an incoming flow with an elevated pH?
5. Build a simple nonlinear model for pH neutralisation in a tank. Assume that the tank holds V (m^3) of liquid and that the inflow is f (m^3/s). Assume that the outflow is also f (m^3/s) rate. (Hint: Use ion concentration as the state variable.)
6. If there is no control law, what is the steady-state ion concentration on the outflow?

7. Consider the following conditions:
   - The inflow rate jumps from f_a to f_b at time 0.
   - The incoming ion concentration increases, at the same time, from C_i^a to C_i^b.
   - There is a tight (high-gain) level control system.

   What is the time constant of the exit ion concentration?
8. Use your answer to Question 7 to explain why a large tank is desirable. (Hint: assume that the incoming ion concentration is a sine wave of frequency ω.)
9. What does pH actually measure?
10. Propose a controller to add a strong base to neutralise the solution in the tank if the solution pH is measured.

Further Reading

1 P.K. Kushwaha and V.K. Giri, *Control strategies for water level control of two tank system*, IJBSTR Publication, 2013.

2 T.V. Phung and V. Tzouanas, *Design and control of a twin tank water process*, American Society for Engineering Education, 2012.

3 C. Urrea and F. Páez, "Design and comparison of strategies for level control in a nonlinear tank", *Process*, 9:735, 2021.

11

Laboratory 5: Wind Power (Mechanical Components)

11.1 Introduction

Significant international pressure is building to move from traditional energy sources, e.g. coal-and oil-fired power stations, to renewal energy sources, e.g. solar, hydro and wind power. Pressure to move in this direction comes from many sources. Amongst other things, it is driven by the desire/necessity to reduce green house emissions and to deliver energy to an ever-increasing world population.

This laboratory is based on one source of renewable energy, namely, wind-power. Extracting energy from the wind leads to many interesting control challenges. This laboratory studies the mechanical aspects of wind turbine control, including yaw control, angular velocity control and pitch control. Chapter 20 will study the electrical aspects of this problem.

The laboratory[1] is based, in part, on the experience gained by the authors with the design and control of several small wind generators. A commercial-scale wind generator is shown in Figure 11.1.

The laboratory uses the Wind Power Virtual Laboratory program. The wind power generator in the laboratory is based on a Vestas V39 500 kw generator. Figure 11.2 shows an anemometer together with the plot of simulated wind velocities. Figure 11.3 shows the main screen of the program. Other screens show the controllers for yaw control, pitch control, generator control and AC/DC/AC converter.

Some physical parameters of the wind power generator are as follows:

- Blade radius 20 m
- Gear Box ratio 50 : 1
- Inertia referred to the generator 6.3×10^6 kg m^2
- Friction coefficient: 1.2×10^3 kg m^2/s
- Maximum rated wind speed: 12 m/s.

11.2 Yaw Control

A relatively simple feedback control loop is typically used to point the turbine into the wind direction. This is called the yaw controller. When the turbine faces the wind then the rotation plane of the blades is perpendicular to the wind, (as in Figure 11.4), and the yaw angle error $e_y = 0$. The yaw controller aims to minimise e_y, since the power extracted from the wind is proportional to $|\cos e_y|$.

1 Many of the ideas were supplied by Professor Galina Mirzaeva from the University of Newcastle.

Practical Control System Design: Real World Designs implemented on Emulated Industrial Systems,
First Edition. Adrian Medioli and Graham Goodwin.

Companion website: www.wiley.com/go/medioli/practicalcontrolsystemdesign

Figure 11.1 Photo of wind turbines.

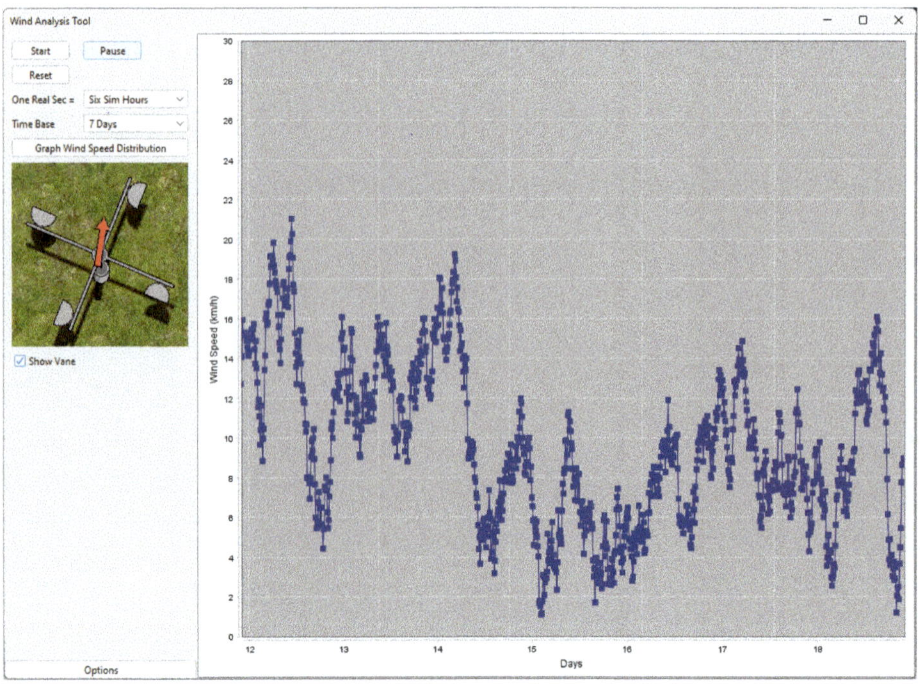

Figure 11.2 Screenshot of anemometer and wind speed.

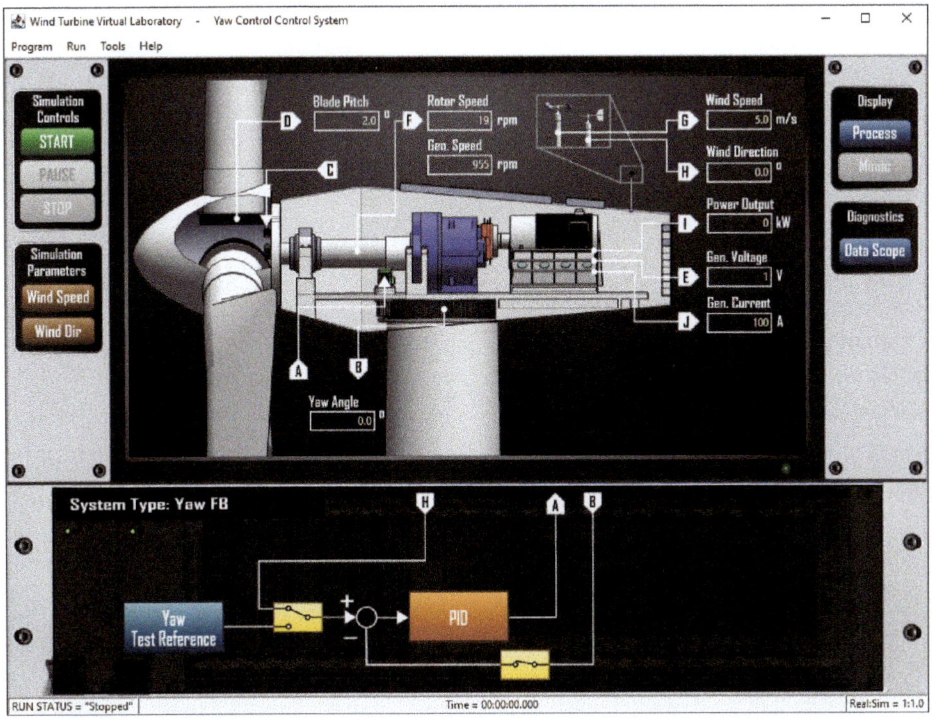

Figure 11.3 Main screen.

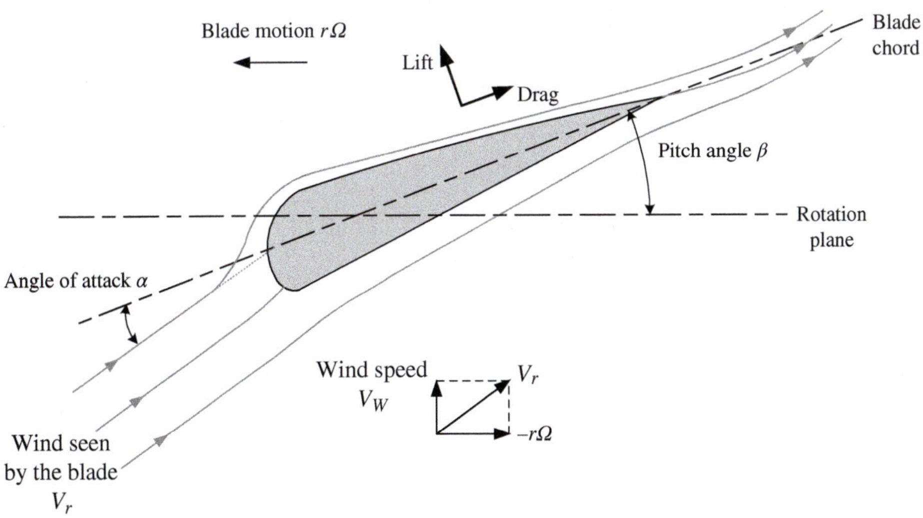

Figure 11.4 Blade cross-sectional view.

The yaw control loop is usually formulated as a simple servo control problem. This is essentially the same problem as was studied earlier in Chapters 5 and 8. Assuming that the actuator for the yaw control is a DC servo motor, the following are the equations describing its behaviour.

$$V_a = R_a i_a + L_a \frac{di_a}{dt} + K_f \omega_y, \tag{11.1}$$

$$T_y = K_f i_a, \tag{11.2}$$

where V_a and i_a are the motor armature voltage and current, respectively.

R_a and L_a are the armature resistance and inductance, respectively.

K_f is the motor field constant.

ω_y is yaw angular velocity.

T_y is yaw torque (produced by the motor).

The mechanical model is described by

$$I_y \frac{d\omega_y}{dt} + D_y \omega_y = T_y, \tag{11.3}$$

$$\frac{d\theta_y}{dt} = \omega_y, \tag{11.4}$$

where T_y is yaw torque (supplied by the motor).

I_y is the moment of inertia about a vertical axis.

D_y is a damping term.

θ_y is the orientation angle.

The block diagram of the yaw control is shown in Figure 11.5, where C_y represents the yaw controller.

The time constant of the yaw motor electrical circuit is much smaller than the mechanical time constant of the yaw rotational dynamics of the turbine.

This leads to the following simplified model:

$$\begin{aligned} i_a &= \tfrac{V_a - E_b}{R_a} && \text{(Motor Current)},\\ E_b &= K_f \omega_y && \text{(Back emf)},\\ T_y &= K_f i_a && \text{(Torque)},\\ I_y \dot{\omega}_y + D_y \omega_y &= T_y && \text{(Yaw Dynamics)}. \end{aligned}$$

The above equations are next used to obtain a simple model linking ω_y and V_a. Combining the above equations leads to

$$I_y \dot{\omega}_y + D_y \omega_y = T_y = K_f i_a = K_f \left[\frac{V_a - E_b}{R_a} \right], \tag{11.5}$$

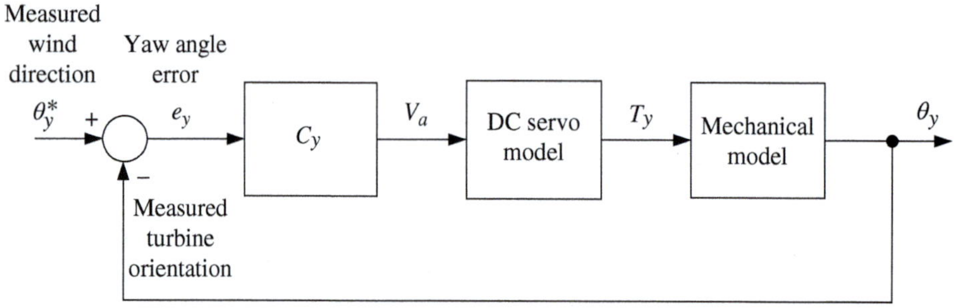

Figure 11.5 Block diagram of yaw control.

$$= K_f \left[\frac{V_a - K_f \omega_y}{R_a} \right].$$

Rearranging (11.5) leads to

$$I_y \dot{\omega}_y + \left[D_y + \frac{K_f^2}{R_a} \right] \omega_y = \frac{K_f}{R_a} V_a. \tag{11.6}$$

Manufacturing data can be used to determine I_y and D_y. For the purpose of the laboratory, the following values are approximately true for the yaw dynamics:

$$I_y = 3000 \text{ kg m}^2 \quad \text{and} \quad D_y = 1000 \text{ kg m}^2/\text{s}$$

By neglecting the term $\frac{K_f^2}{R_a}$, then an estimate of the time constant of the yaw dynamics can be obtained. The approximate result is

$$\tau_y \approx \frac{I_y}{D_y}, \tag{11.7}$$

$$= 3 \text{ seconds}.$$

11.2.1 Experiment: Estimating the Yaw Time Constant

Start the Wind Power Generator laboratory. Go to the section marked 'Program' and select 'Yaw Control'.

Use the Yaw Test Reference to put a step change in voltage into the yaw motor and plot the speed response. You will use this to estimate the time constant.

Note that the time taken to reach 63% of the final velocity can be used to estimate the time constant. Also, estimate the gain of the system. You should find that the yaw velocity time constant is approximately 3 seconds and the gain is approximately $0.2\,{}^\circ\text{s}^{-1}\text{V}^{-1}$.

11.2.2 Design of Yaw Controller

The yaw controller is usually designed to have a relatively low bandwidth so as not to mechanically stress the structure.

A PD controller will be designed to give a first-order closed-loop response for the Yaw angle having a time constant of T seconds (where T is a user choice).

The model transfer function is

$$\frac{\theta_y(s)}{V_a(s)} = \frac{g}{\left(\tau_y s + 1\right) s} = G(s), \tag{11.8}$$

where from Section 11.2 the following approximate values hold:

$$g \simeq 0.2\,{}^\circ\text{s}^{-1}\text{V}^{-1}, \tag{11.9}$$

$$\tau_y \simeq 3 \text{ seconds}. \tag{11.10}$$

(The reader should use the values obtained using the experiment suggested in Section 11.2.1. The results given above are purely indicative.)

A PD controller takes the form:

$$V_a(s) = \left(K_p + K_d s\right) \left[\theta_y^*(s) - \theta_y(s)\right]. \tag{11.11}$$

The following choice is made

$$\left(K_p + K_d s\right) = K \left[\tau_y s + 1\right] = C_y(s). \tag{11.12}$$

Note that the zero in the control law approximately cancels the pole of the plant.

Then the closed-loop transfer function is

$$\frac{\theta_y(s)}{\theta_y^*(s)} = \frac{C_y(s)\,G(s)}{1+C_y(s)\,G(s)}, \tag{11.13}$$
$$= \frac{\frac{Kg}{s}}{1+\frac{Kg}{s}},$$
$$= \frac{Kg}{s+Kg}.$$

Hence to achieve a closed-loop time constant of T seconds, then it suffices to choose K so that

$$Kg = \frac{1}{T}. \tag{11.14}$$

Hint: Use the 'Yaw Step Reference' to add a square wave reference of ± 10. Also, make sure the Yaw angle is 180° before you start.

11.2.3 Experiment: Testing the Yaw Controller

Implement the controller given in (11.11) and investigate the effect of changing K on the closed-loop time constant.

The response of the yaw controller to step changes in the wind angle should be as in Eq. (11.14), i.e. the response should be linear with a time constant, T. However, the actual response will differ from the ideal (see for example Figure 11.6). You may also find that the actual response deviates further from the ideal as K increases. Note that the voltage into the motor is heavily constrained. Begin with $K_p = 0.5, K_d = 1.5$. This should yield about a 10-second closed-loop time constant.

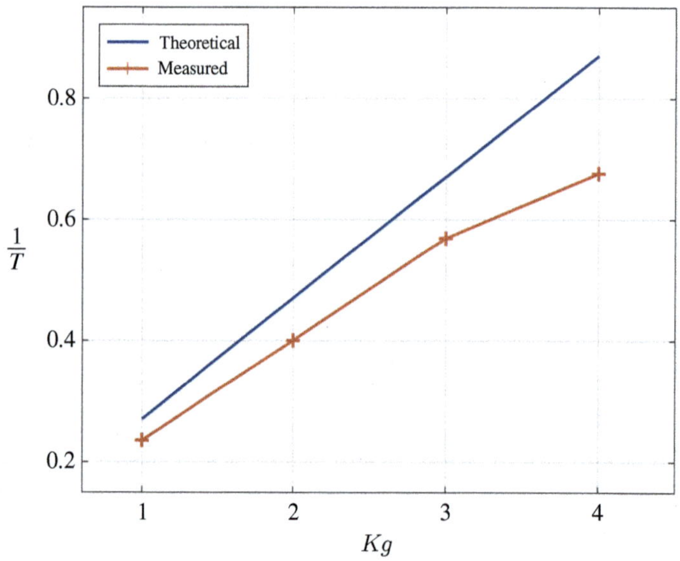

Figure 11.6 One over time constant versus gain.

11.3 Rotational Velocity Control

The blades in a wind turbine act like a wing of an aircraft. Figure 11.4 shows a horizontal cross-sectional sketch of a turbine blade when its tip points into the sky.

Heuristically, it can be seen that the wind that passes over the top of the aerofoil has further to travel and this produces a pressure drop and, in turn, this leads to 'lift' and 'drag' forces. It is desirable to have a high ratio. The shape of the aerofoil is crucial. It is also possible to adjust the blade pitch angle β between the plane of rotation of the blade chord. The impact of doing this is to change the angle of attack α, as shown in Figure 11.4. This would slow down, and possibly even stall, the blade rotation. This can be used to protect the turbine when the wind speed is too high.

The maximum power that can be extracted from the wind is described by Betz Law. This law was formulated in 1919 and is presented in Theorem 11.1 below. The law is based on conservation of mass and momentum.[2]

Theorem 11.1 ***(Betz Law)*** Consider the schematic representation of air flow for a wind turbine as shown in Figure 11.7.

Then the following properties hold:

(i) $v = \frac{v_1+v_2}{2}$

(ii) Power $= \frac{1}{4}\rho A v_1^3 \left(1 + r_v - r_v^2 - r_v^3\right)$, where $r_v = \frac{v_2}{v_1}$.

(iii) Maximum power occurs when $r_v = \frac{1}{3}$

(iv) Maximum power $= \left[\frac{16}{27}\right] \frac{1}{2}\rho A v_1^3$.

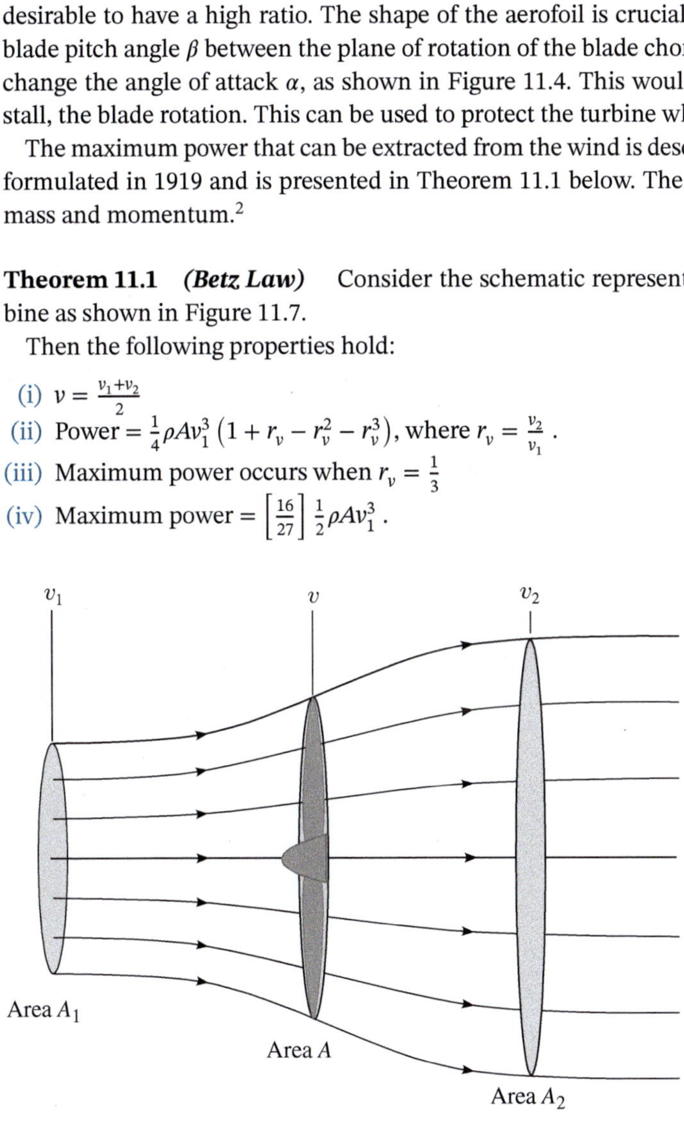

Figure 11.7 Air flow for air turbine ($v_1 > v > v_2$).

2 This result is given for completeness. However, readers who are primarily interested in control can move on to the rest of the laboratory.

Proof:

(i) Assume that the air density, ρ is constant. Then consider a period of time T. The mass of air entering the turbine in time T is $\rho A v T$. This air undergoes deceleration from velocity v_1 to v_2. Hence, the acceleration is $\frac{(v_1 - v_2)}{T}$. Then,

$$\begin{aligned} \text{Force} &= \text{Mass} \times \text{Acceleration} \\ &= \rho A v T \left(\frac{v_1 - v_2}{T} \right) \\ &= \rho A v \left(v_1 - v_2 \right) \\ &= F. \end{aligned} \tag{11.15}$$

Power is rate of doing work. Hence, power is force times velocity, i.e. using (11.15)

$$\begin{aligned} \text{Power} &= Fv \\ &= \rho A v^2 \left(v_1 - v_2 \right). \end{aligned} \tag{11.16}$$

However, this power must be equal to the decrease in kinetic energy.
The kinetic energy, KE_1, of the air stream at Area A_1 delivered over the time period T is given by

$$\begin{aligned} KE_1 &= \frac{1}{2} m v_1^2 \\ &= \frac{1}{2} \rho A v T v_1^2. \end{aligned}$$

The kinetic energy, KE_2, of the air stream at Area A_2 delivered over the time period T is

$$\begin{aligned} KE_2 &= \frac{1}{2} m v_2^2 \\ &= \frac{1}{2} \rho A v T v_2^2. \end{aligned} \tag{11.17}$$

Hence, the rate of change of kinetic energy is

$$\begin{aligned} \frac{d(KE)}{dt} &= \frac{KE_1 - KE_2}{T} \\ &= \frac{1}{2} \rho A v T \frac{\left(v_1^2 - v_2^2 \right)}{T} \\ &= \frac{1}{2} \rho A v \left(v_1^2 - v_2^2 \right). \end{aligned} \tag{11.18}$$

Equating (11.17) and (11.18) gives

$$v^2 \left(v_1 - v_2 \right) = \frac{1}{2} v \left(v_1^2 - v_2^2 \right). \tag{11.19}$$

Solving (11.19) for v yields

$$v = \frac{1}{2} \left(v_1 + v_2 \right). \tag{11.20}$$

This completes the proof of part (i).

(ii) Substituting (11.20) into (11.16) yields

$$\begin{aligned}\text{Power} &= \frac{1}{4}\rho A(v_1+v_2)^2\,(v_1-v_2)\\ &= \frac{1}{4}\rho A\,(v_1^3+2v_1^2v_2+v_1v_2^2-v_1^2v_2-2v_1v_2^2-v_2^3)\\ &= \frac{1}{4}\rho A v_1^3\,(1+r_v-r_v^2-r_v^3),\end{aligned} \tag{11.21}$$

where $r_v = \frac{v_2}{v_1}$.

This completes the proof of part (ii).

(iii) Differentiating (11.21) with respect to r_v yields the following condition for a maximum

$$1-2r_v-3r_v^2=0. \tag{11.22}$$

Hence, maximum power is achieved when r_v takes the following value:

$$r_v=\frac{1}{3}. \tag{11.23}$$

This completes the proof of part (iii)

(iv) Finally, substituting (11.23) into (11.21) yields

$$\begin{aligned}\text{Max Power} &= \frac{1}{4}\rho A v_1^3\left[\frac{32}{27}\right]\\ &= \left[\frac{16}{27}\right]\frac{1}{2}\rho A v_1^3.\end{aligned} \tag{11.24}$$

□□□

The above law states that no turbine can extract more than $16/27 = 0.59$ of the kinetic energy in the wind. Practical systems achieve between about 0.45 and 0.5 of the kinetic energy. In practice, the mechanical power satisfies a form of (11.24), namely,

$$P=\frac{1}{2}C_p\rho A v_1^3, \tag{11.25}$$

where ρ is the air density, A the swept area (πr^2) and v_1 is the upstream wind velocity. The coefficient C_p in (11.25) is a function of both (λ) The tip-speed ratio (TSR) and the pitch angle of the blades (β). Hence, it can be expressed as $C_p(\lambda, \beta)$. A key factor is the TSR, λ, where

$$\lambda=\frac{\text{Blade tip speed}}{\text{Wind speed}}=\frac{(\text{angular velocity})\times 2\pi r}{v_1}, \tag{11.26}$$

where r is the blade radius.

A rule of thumb is that maximum power is extracted when the TSR is approximately 7 for a 3-blade rotor. (The laboratory uses a TSR of 8.) At every wind speed, this corresponds to a different optimal rotational velocity for the turbine, as shown in Figure 11.8. Adjustment of the turbine rotational velocity to keep TSR constant and hence maximise the extracted power is called maximum power point tracking (MPPT).

One option for MPPT is to use the electrical torque available from the generator attached to the shaft to roughly maintain the TSR at 7. (See Figure 11.9; (Note that a later laboratory (Chapter 20) will study how this torque is generated.)

Referring to Figure 11.9, the following notation is used:

- ω^* is the desired turbine rotational velocity for MPPT
- C_ω represents the turbine speed controller
- T^* is a reference torque supplied to the electrical system
- T_e is electromagnetic torque produced by the electrical system
- T_m is mechanical torque due to friction.

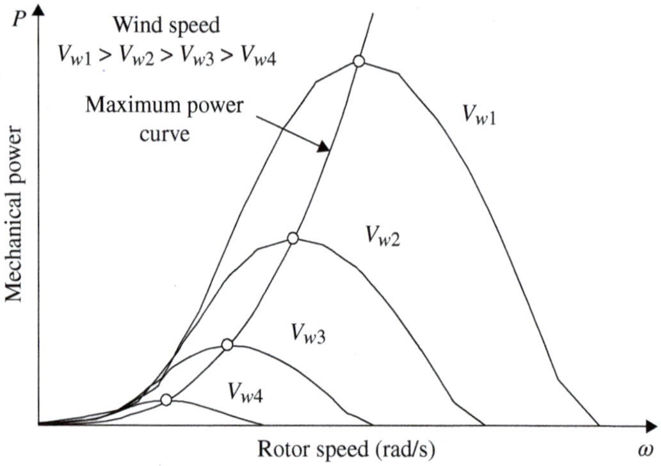

Figure 11.8 Tracking of maximum power depending on the wind speed.

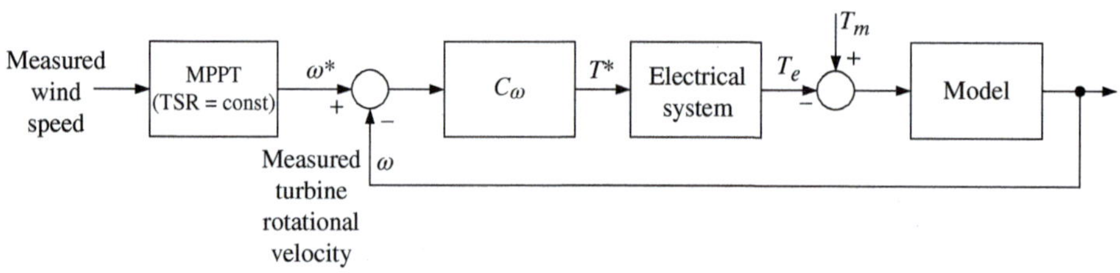

Figure 11.9 Block diagram of rotational velocity control.

The mechanical model can then be approximately described as:

$$I\frac{d\omega}{dt} + B\omega = \left(T_m - T_e\right), \tag{11.27}$$

where

I is the inertia of the turbine/gearbox combination. (11.28)

B is the turbine/gearbox friction coefficient. (11.29)

Mechanical torque in (11.27), for a given TSR and pitch angle, is proportional to ω^2, i.e.

$$T_m = K(\lambda, \beta)\omega^2, \tag{11.30}$$

where $K(\lambda, \beta)$ is a proportionality coefficient depending on the blade geometry, TSR (λ) and pitch angle (β). (See Revision Question 3 given at the end of the chapter.)

Because of the relatively high inertia, it is typical to design the bandwidth of the system to be relatively low, say, a few rad/s. On the other hand, the bandwidth of the electrical system is very high. Therefore, when modelling the rotational velocity control of Figure 11.9, it can be assumed that, the achieved electrical torque, T_e, is equal to the requested torque, T^*. (Chapter 20 will study the electrical system that leads to T_e.)

Note that, when the dynamics of the electrical system are neglected, then the transfer function between, T^*, and the measured angular velocity in rad/s is simply the torque,

$$G(s) = \frac{1}{(Is + B)}. \tag{11.31}$$

The reader may notice that the size of B relative to I is very small. Hence, the 'plant' is almost a pure integrator. However, it is still necessary to include an integrator in the controller to achieve zero steady-state errors due to the presence of input disturbances. (See Revision Question 2 in the exercises at the end of the chapter.)

11.3.1 Experiment: Testing the Rotational Velocity Control Law

Load the rotational velocity controller.

Begin with a wind speed of 8 m/s.

The reader should convert this into a desired rotational shaft speed for the generator in rad/s. To test the closed loop try a step in wind velocity from 8.0 to 7.5.

Both a P and PI controller will be designed and tested. The reader may care to start with the following values.

- P design try: $K_p = 249{,}000$.
- What do you estimate the closed-loop constant to be?
- PI design try: $K_p = 240{,}000$, $K_i = 2400$.

These values may seem large but this is because physical units have been used when describing the variables. Also, note that the traces show rotational speed in rad/s, whereas the mimic shows rotational speed in rpm. (Note rpm $\simeq$ 10× rad/s.)

An important point is that one must never exceed the rated speed of the turbine ω^{max}. Hence, the set-point for the speed is based on TSR until the maximum speed is reached, at which point the demand for increased speed saturates as shown in Figure 11.10.

11.4 Pitch Control

The effect of blade pitch control is next studied. It has already been seen that pitch angle effects the power coefficient. Therefore, pitch angle can be used to fine-tune the power coefficient $C_p(\lambda, \beta)$ to extract maximum power from the wind. Pitch control usually becomes active when the speed of the turbine rises above the rated wind speed to limit the generated power to its rated value. This idea is illustrated in Figure 11.11.

There is also a certain 'cut in' speed below which the turbine does not produce enough power to justify its running costs. In the wind speed following region from ($v_{cut\ in}$ to v_{rated}) the turbine produces power according to equation (11.29).

At rated wind speed, the turbine generator produces the power for which its electrical components are rated. Above this power, the turbine should not be operated. By controlling the pitch

Figure 11.10 Saturation of rotational speed set point.

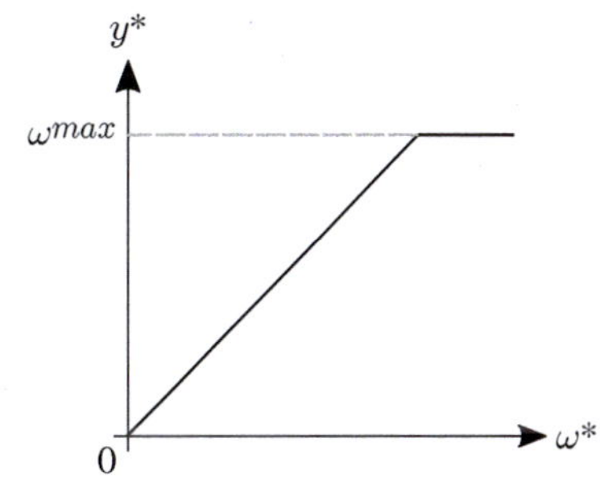

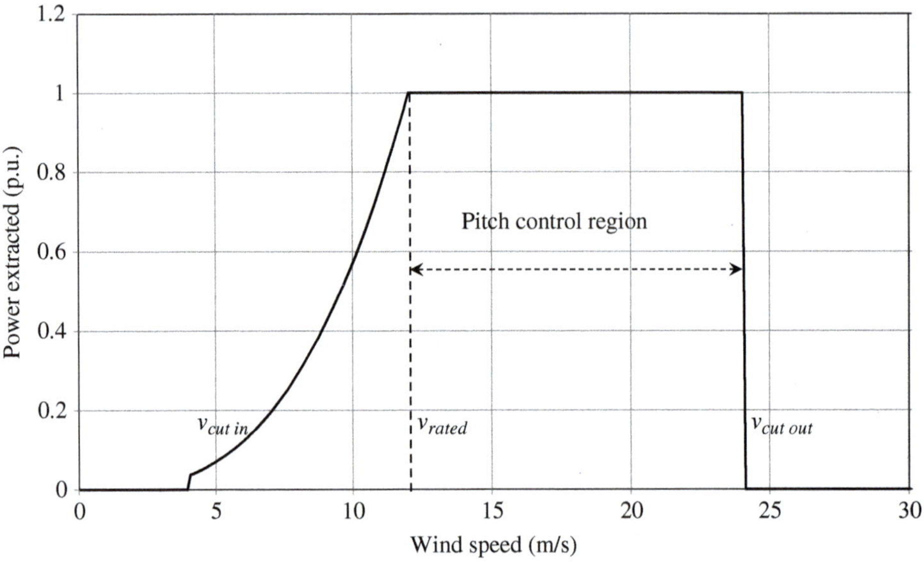

Figure 11.11 Power control regions.

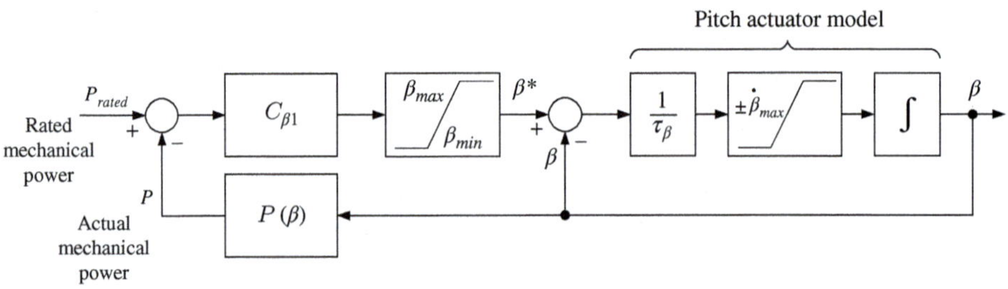

Figure 11.12 Block diagram of pitch control.

angle, the turbine efficiency can be reduced and the power kept constant from v_{rated} to $v_{cut\ out}$. At 'cut out' wind speed, the turbine blades are turned to a stall position and the turbine is shut down. Operation above that speed becomes dangerous for the mechanical structure.

Pitch angle control is usually based on the rotational velocity or generated power rather than the wind speed due to better accuracy (see diagram in Figure 11.12). When the turbine rotational velocity is below the rated value the reference pitch angle (β^*) is set to the nominal position β_{min}. When the turbine rotational velocity is above the rated value, then the pitch controller $C_{\beta 1}$ generates an increased reference pitch angle, which is limited by its maximum (stall) value (β_{max}).

The pitch servo model is similar to the 'DC servo model' for the yaw control in Figure 11.5. The mechanical model for the pitch actuator is satisfactorily represented by a rate-of-change limiter for the pitch angle β ($\pm\dot{\beta}_{max}$), time constant τ_β and an integrator, as in Figure 11.12.

11.5 Experiment: Testing the Pitch Controller

Set the rotational velocity controller to $K_p = 4$, $K_i = 2$. Then carry out the following steps:

(1) Increase the wind speed to rated speed (12 m/s). Measure generated power and turbine speed.

(2) Increase the wind speed to 15 m/s. Measure generated power and turbine speed without the pitch controller.
(3) Turn on the Pitch Controller by choosing C_β as a Proportional controller with $K_p = 1$. Measure the generated power and turbine speed.

11.6 Revision Questions

1. Consider a system modelled as in Eq. (11.31). Let $C(s)$ denotes the control law transfer function
   (a) Obtain an expression for the output speed, y, as a function of the desired speed, y^*, an input disturbance d_i and an output disturbance d_0.
   (b) Say that a proportional control law with gain K_p is used. Determine the steady-state value of y as a function of y^*, d_i, d_0, K, I, B and K_p when y^*, d_i and d_0 are constant.
   (c) Evaluate the answer found in part (b) for B small.
2. Use your results in Question 1 to argue why a PI controller is desirable for this system.
3. Equation (11.25) claims that the power extracted from the wind is proportional to the wind velocity cubed. whereas Eq. (11.30) states that the Mechanical Torque is proportional to the rotational velocity squared. How are these results related?
4. What is the theoretical relationship between wind speed and extracted power in the region 5–12 m/s. Verify that this is consistent with Figure 11.10.

Further Reading

1 A. Grauers, "Efficiency of three wind energy generator systems", *IEEE Transactions on Energy Conversion*, 11(3):650–657, 1996.

2 M.M. Hand, K.E. Johnson, L.J. Fingersh, and A. D. Wright, "Advanced Control Design and Field Testing for Wind Turbines at the National Renewable Energy Laboratory", *World Renewable Energy Congress VIII*, Denver, Colorado, USA, 2004, available www.nrel.gov/docs/fy04osti/36118 .pdf.

3 L.H. Hansen, P.H. Madsen, F. Blaabjerg, H.C. Christensen, U. Lindhard and K. Eskildsen, "Generators and Power Electronics Technology for Wind Turbines", *Annual Conference of the IEEE Industrial Electronics Society*, Denver, Colorado, USA, 2001.

4 S.S. Murthy, B. Singh, P.K. Goel and S.K. Tiwari, "A Comparative Study of Fixed Speed and Variable Speed Wind Energy Conversion Systems Feeding the Grid", *International Conference on Power Electronics and Drive Systems*, Bangkok, Thailand, 2007.

5 L.Y. Pao and K.E. Johnson, "A Tutorial on the Dynamics and Control of Wind Turbines and Wind Farms", *Proceedings of the American Control Conference*, St. Louis, Missouri, USA, 2009, pp. 2076–2089.

6 H. Stiesdal, Bonus Info Autumn 1999, "The Wind Turbine Components and Operation", Bonus Energy, 1999, available http://www.windmission.dk/workshop/BonusTurbine.pdf.

7 E. Valsera-Naranjo, A. Sumper, O. Gomis-Bellmunt, A. Junyent-Ferre and M. Martinez-Rojas, "Design of a Pitch Control of a Wind Turbine to Improve System Frequency Response", *2009 13th European Conference on Power Electronics and Applications*, Barcelona, Spain, 2009.

8 J. Zhang, M. Cheng, Z. Chen and X. Fu, "Pitch Angle Control for Variable Speed Wind Turbines", *International Conference on Electric Utility Deregulation and Restructuring and Power Technologies*, Nanjing, China, 2008.

Part III

More Complex Linear Single-Input Single-Output Systems

This part continues the theme of Part II but addresses slightly more complex issues. Chapter 12 describes theory necessary to understand systems having a time delay. Chapter 13 applies the ideas described in Chapter 12 to the problem of Centre Line Gauge Control in a rolling mill. Chapter 14 presents specific results for systems that are open-loop unstable. Chapter 15 studies the control of a rocket in two-dimensional flight. Chapter 16 summarises important theory that describes the sensitivity trade-offs that occur in the frequency domain.

Practical Control System Design: Real World Designs implemented on Emulated Industrial Systems, First Edition. Adrian Medioli and Graham Goodwin.

Companion website: www.wiley.com/go/medioli/practicalcontrolsystemdesign

12

Time Delay Systems

12.1 Introduction

Pure time delays are a common occurrence in practical control systems. (Delays are frequently caused by the transport of material from one point to another or result from delays in measurements.) Because of their importance, this chapter is devoted to developing special methods aimed at the control of systems having a pure delay.

A pure time delay introduces phase shift into signals. Moreover, this phase shift increases linearly with frequency. Unfortunately, phase shift is a major hindrance to high performance control system design. For example, a simple examination of the Nyquist stability result (Chapter 6) shows that phase shift rotates the Nyquist diagram in a clockwise direction. This reduces the phase margin and will ultimately lead to instability.

This chapter will give some initial ideas on the impact of delays on control system design. These ideas will be further examined in Chapter 13 where a real-world example of a control problem with significant delay will be presented.

12.2 Transfer Function Analysis

The transfer function of a pure delay was studied in Sections 4.6.6 and 4.6.7. The transfer function corresponding to a delay of T_d seconds is

$$G_d(s) = e^{-sT_d}. \tag{12.1}$$

To obtain the corresponding frequency response, one simply replaces s by $j\omega$ yielding

$$G_d(j\omega) = e^{-j\omega T_d}. \tag{12.2}$$

It can be seen that this transfer function has gain 1 and phase given by

$$\angle G_d(j\omega) = -j\omega T_d. \tag{12.3}$$

It can thus be concluded that phase increases linearly with frequency. Moreover, the phase shift is made more pronounced for larger delays.

Practical Control System Design: Real World Designs implemented on Emulated Industrial Systems,
First Edition. Adrian Medioli and Graham Goodwin.

Companion website: www.wiley.com/go/medioli/practicalcontrolsystemdesign

12.3 Classical PID Design Revisited

The ideas described in Chapters 7 and 9 can, in principle, be immediately applied to systems containing a delay. However, it is unfortunately a fact of life that the quality of the performance will be deteriorated by the impact of the phase lag produced by the delay.

12.4 Padé Approximation

One simple idea that can be used to think about control system design for plants having a pure delay is to approximate the delay by a rational transfer function. One such approximation that is commonly used is a first-order Padé approximation as follows:

$$e^{-sT_d} \approx \frac{1 - \frac{T_d}{2}s}{1 + \frac{T_d}{2}s}. \tag{12.4}$$

Note that this transfer function has gain 1 at all frequencies and a phase lag that increases with frequency as in the case for a delay. (See Question 1 given in the Revision Questions at the end of the chapter.)

The approximation (12.4) can often give insight into the control problem for a time-delayed system. However, there are more direct methods for dealing with delayed systems. One such technique is to use the Youla–Kucera parameterisation of all stabilising control laws. This idea will be explored below.

12.5 Using the Youla–Kucera Parameterisation

It is suggested that the reader reviews Section 7.5 before proceeding.

A pure delay is a linear element. It will thus not surprise readers to know that a useful trick when dealing with a pure delay is to ensure that the model used in the Youla–Kucera parameterisation also contains a model for the delay. Hence, the appropriate Youla–Kucera parameterisation form of the control law is as shown in Figure 12.1, where T_d denotes the delay and $G_o^*(s)$ is the

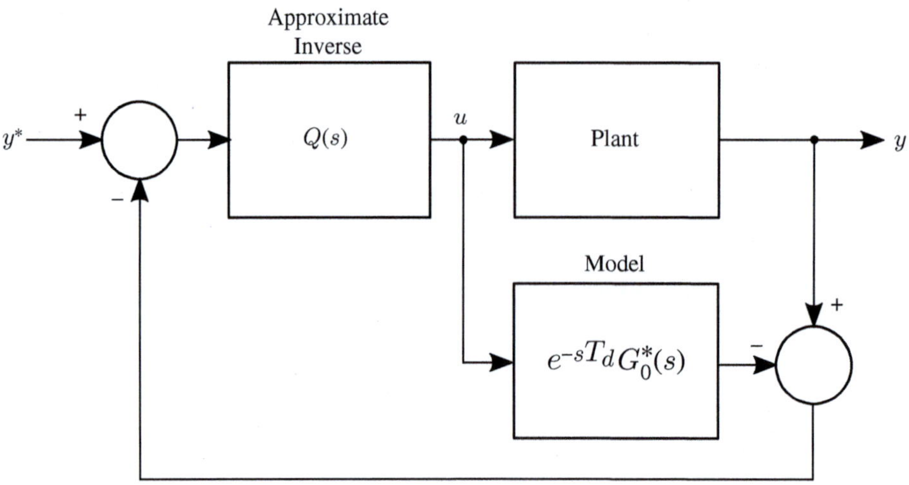

Figure 12.1 Youla–Kucera delay block diagram.

plant model with the delay removed. A source of difficulty when controlling systems with pure delay is that one cannot invert the delay. Hence, $Q(s)$ will typically be an inverse for the plant *without delay*.

12.6 Smith Predictor

Actually the implementation shown in Figure 12.1 has a historical interpretation. Specifically, the ideas go back to an original idea by Otto Smith which now goes under the name 'Smith Predictor'. The only slight caveat is that Smith used a particular form to implement the transfer function $Q(s)$. The idea will be described below. It is to be noted that the delay cannot be inverted. Hence, the best one could hope for would be to choose $Q(s)$ as an approximation to the inverse of $G_o^*(s)$. Actually, Smith suggested that the inverse in $Q(s)$ be constructed by using a high-gain controller applied to $G_o^*(s)$ as shown in Figure 12.2. This leads to an equivalent $Q(s)$ given by

$$Q(s) = \frac{C(s)}{1 + G_o^*(s)\, C(s)}, \tag{12.5}$$

where $G_o^*(s)$ is a nominal model of the plant *with the delay removed*. Hence, this is equivalent to implementing the proper stable inverse, as used in the transfer function $Q(s)$, by placing a 'high gain' feedback controller around a model of the plant with delay removed. Of course, this still has the flavour of inversion. Indeed, it can be seen from Eq. (12.5) that if $C(s)$ is 'large', then $Q(s) \approx G_o^*(s)^{-1}$.

The Smith predictor idea leads to the final control implementation as shown in Figure 12.2.

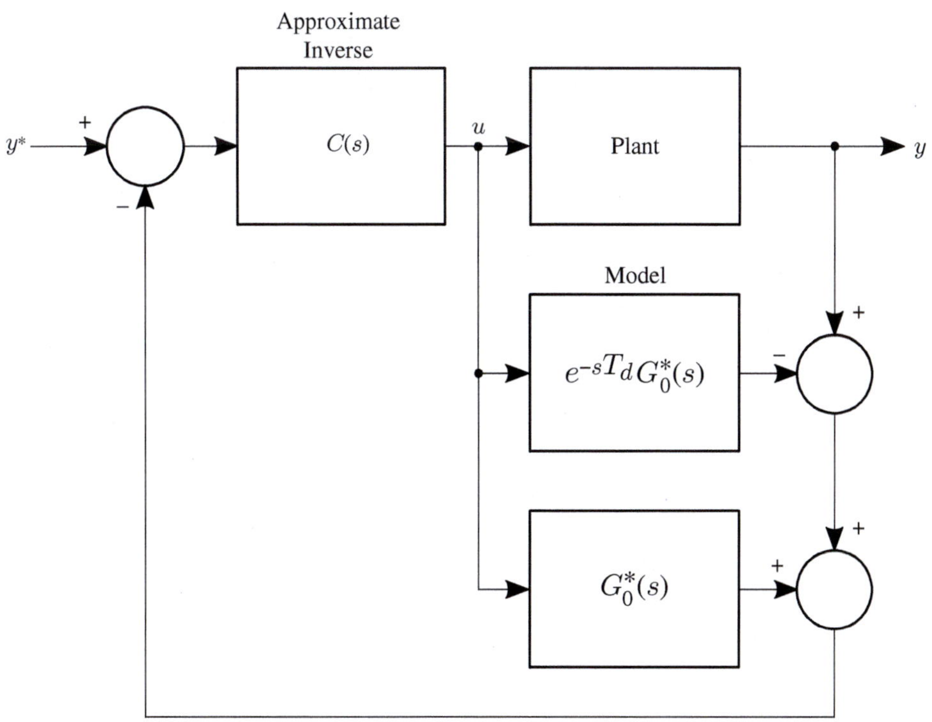

Figure 12.2 Smith predictor block diagram.

12.7 Modern Interpretation of Smith Predictor

A modern interpretation of the Smith predictor is found in a design technique known as model predictive control (MPC). The core ingredients in MPC are (i) the use of a Model and (ii) a *Prediction* (as the name suggests). Further ideas regarding MPC will be presented in Section 22.13.

Note that these ideas are also inherent in the Smith predictor. There is no doubt that MPC is a powerful tool for complex control problems. However, it is recommended to readers that they always begin with simpler architectures such as those described above and in the sequel. One can move onto MPC when the deficiencies, if any, of simpler ideas have been fully explored.

12.8 Sensitivity Trade-Offs

Assuming that the plant model is perfect, it follows that the Smith Predictor Architecture leads to:

$$\begin{aligned}\frac{y(s)}{y^*(s)} &= \frac{C(s)}{1+C(s)\,G_o^*(s)}G_o^*(s)\,e^{-sT_d},\\ &= T_o(s)\,e^{-sT_d},\end{aligned}\tag{12.6}$$

where $T_o(s)$ is the nominal complementary sensitivity for the loop including $G_o^*(s)$ and $G_o^*(s)$ is the nominal plant model with delay removed.

Hence, if the nominal (undelayed) complimentary sensitivity function is, $T_o(s)$, then the achieved complementary sensitivity function, $\overline{T}_o(s)$, is given by

$$\overline{T}_o(s) = T_o(s)\,e^{-sT_d}.\tag{12.7}$$

This all looks very simple. However there are two important caveats to the above design idea.

(1) The design can be sensitive to model errors since it depends on the cancellation of the two upper paths in Figure 12.2. Hence, a good model is needed. Analysis of the impact of model error is given in Section 12.9.
(2) One needs to be aware that whilst designing $T_o(s)$ may seem simple, the disturbance sensitivity function may contain surprises.

    In particular,

$$\begin{aligned}\overline{S}_o(s) &= 1-\overline{T}_o(s),\\ &= 1-T_o(s)\,e^{-sT_d}.\end{aligned}\tag{12.8}$$

    The phase shift inherent in the e^{-sT_d} term in the above expression can cause $|\overline{S}_o(s)|$ to be large even though $|\overline{T}_o(s)|$ is near 1. The reason is that $\overline{T}_o(s)$ and $\overline{S}_o(s)$ are actually complex numbers when considered as a function of frequency.

    The problem is illustrated in Figure 12.3 where $|\overline{T}_o| = 1$ but $|\overline{S}_o| > 1$.

Actually, the authors of this book came across a practical case of the above difficulty when designing a closed-loop control law for insulin injections aimed at blood glucose regulation for Type 1 diabetes. The authors found that they could easily design an acceptable complementary sensitivity function, i.e. $\overline{T}_o(s)$, for this system but the sensitivity function always seemed poor. The cause of the difficulty turned out to be due to a significant pure time delay in the transfer function linking injected insulin to the blood glucose response.

Further insights into the impact of pure time delays on sensitivity trade-offs will be presented in Section 16.5.2 when Bode's sensitivity integrals will be studied.

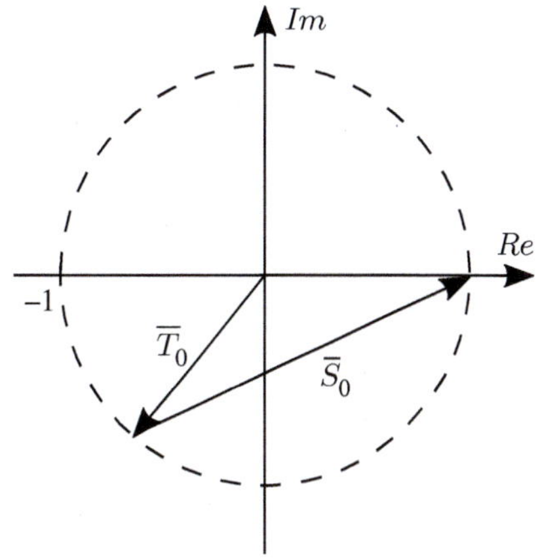

Figure 12.3 Impact of delay.

12.9 Theoretical Analysis of Effect of Delay Errors on Smith Predictor

For the purpose of this section it is assumed that the true plant is

$$G(s) = \frac{Ke^{-sT_d}}{\tau s + 1}. \tag{12.9}$$

The model takes the form

$$\hat{G}(s) = \frac{\hat{K}e^{-s\hat{T}_d}}{\hat{\tau} s + 1}, \tag{12.10}$$

where $\hat{K}$, $\hat{T}_d$, and $\hat{\tau}$ are approximate values for K, T_d, and τ, respectively.

The open-loop pole at $-\frac{1}{\hat{\tau}}$ will be cancelled in the design and integral control will be used leading to nominal closed-loop complementary sensitivity of

$$\overline{T}_o(s) = \left(\frac{1}{\beta s + 1}\right) e^{-sT_d}. \tag{12.11}$$

It turns out that Smith predictors are most sensitive to errors in the delay. Hence, the effects of errors in modelling the delay will be explored. It is assumed that the values of $\hat{K}$ and $\hat{\tau}$ are correct. (Of course, these parameters are also estimates and hence in practice may also differ from the 'true' values.) It is assumed that the nominal model has a time delay $\hat{T}_d$ whilst the actual time delay is T_d.

The sufficient condition for robust stability described in Section 6.10 is that the true closed-loop system will be stable if $|T_o(j\omega)||G_\Delta(j\omega)| < 1 \ \forall \omega$ where $T_o(j\omega)$ is the nominal complementary sensitivity based on the nominal model and where $G_\Delta(j\omega)$ is the multiplicative modelling error defined as

$$G_\Delta(s) = \frac{G(s) - \hat{G}(s)}{\hat{G}(s)}.$$

In this definition, $G(s)$ is the true plant and $\hat{G}(s)$ is the nominal plant model.

It can be shown (see Question 2 in Revision Questions) that the magnitude of $T_o(j\omega)G_\Delta(j\omega)$ satisfies:

$$|T_o(j\omega)||G_\Delta(j\omega)| = \frac{2\left|\sin\left(\frac{\omega(\hat{T}_d - T_d)}{2}\right)\right|}{\sqrt{(\omega\beta)^2 + 1}}. \tag{12.12}$$

12.10 Revision Questions

1. (a) Compare the Bode diagram of $e^{-s\tau}$ and $\frac{1-\frac{\tau}{2}s}{1+\frac{\tau}{2}s}$.
   (b) For what range of frequencies is the second transfer function a good approximation to the first?
2. (a) For the system in this chapter, use the robust stability result discussed above to confirm Eq. (12.12). Hence obtain an expression for the error in $\hat{T}_d - T_d$ which is sufficient to ensure closed-loop stability.
3. Use the following bounds to obtain a simple relationship between β and $|\hat{T}_d - T_d|$ to guarantee stability.

$$(|a| + 1)^2 > a^2, \tag{12.13}$$

$$|\sin\theta| < \theta. \tag{12.14}$$

4. A system is known to have a long-time delay. The closed-loop response appears to have an oscillation imposed on it. You suspect the system might be unstable. What would you look for on the controller output to confirm your suspicion?
5. The authors of this book were once presented with a control system having the gain and phase characteristics are shown in Figure 12.4. (Actually, this was for an autonomous agriculture vehicle.)

   What do you believe the model to be?

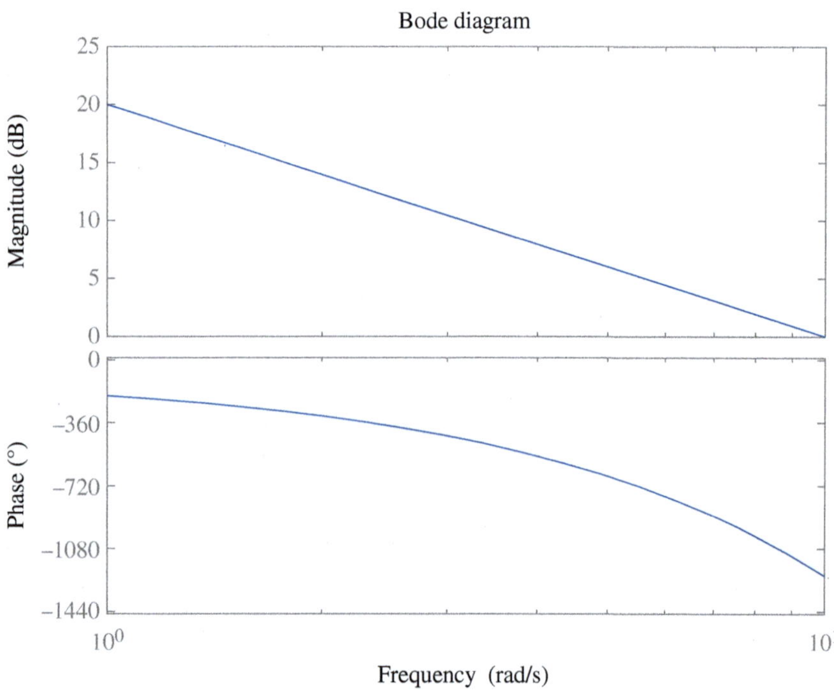

Figure 12.4 Bode plot of real-world system.

Further Reading

1 L. Dugard and E.I. Verriest, *Stability and control of time-delay systems*, Springer-Verlag, 1998.
2 M.-H. Khooban and T. Dragicevic, *Control strategies for time-delay systems*, Elsevier, 2021.
3 S. Yi, P.W. Nelson and A.G. Ulsoy, *Time-delay systems: analysis and control using the Lambert W function*, World Scientific, New Jersey, 2010.

13

Laboratory 6: Rolling Mill (Transport Delay)

13.1 Introduction

This chapter is the first in a series of chapters dealing with the control of centre line gauge control in a Rolling Mill. A photo of a rolling mill is given in Figure 13.1.

In principle, this is an easy problem, namely, if one wants the exit thickness to become smaller, then one simply needs to push the work rolls of the rolling mill closer together. However, there is a fundamental issue with this solution because the exit thickness can only be measured a few metres from the roll gap. This introduces an inherent measurement delay. A typical mill strip speed is 30 km/h, i.e. 8.3 m/s. Hence, if the distance between the roll gap and the measuring device is 2 m, then this introduces a measurement delay of 0.25 seconds. This is very large in comparison with the speed at which the roll gap can be adjusted. Based on the observations made in Chapter 12, this is bound to lead to control difficulties. This laboratory will study the impact of the delay on closed-loop performance for this real-world system.

13.2 The Physical System

Rolling Mills are of major technological significance in modern society. Perhaps surprisingly, they are extremely high technology pieces of equipment with strip speeds that can exceed 100 km/h and tolerances of 0.1%. Essentially, the aim of rolling is to reduce the thickness and change the metallurgical properties of metal strip as it passes through a pair of driven parallel cylindrical rolls. The high external roll forces cause a permanent plastic deformation of the strip, decreasing its thickness and producing a corresponding increase in length. The strip may be either *hot* or *cold* rolled.[1]

Rolled sheet metal is an important material in modern manufacturing. Rolled products are found everywhere in our daily lives. For example, rolled aluminium products can be found in many applications as follows:

- Foil less than 0.2 mm thick is used in packaging such as foil containers and wrapping, in electrical applications, building insulation and in printing industries.
- Steel 0.2–6 mm thick has a wide variety of uses in construction industries such as sidings and roofing. Sheets are also used in transport, for automotive body panels, airframes and boat hulls.
- Plate over 6 mm thick is used in airframes, military vehicles and construction components for buildings and bridges.

1 For example, steel plate (typically >1.5 mm) is hot rolled and steel sheet (typically <0.1 mm) or aluminium foil is cold rolled.

Practical Control System Design: Real World Designs implemented on Emulated Industrial Systems,
First Edition. Adrian Medioli and Graham Goodwin.

Companion website: www.wiley.com/go/medioli/practicalcontrolsystemdesign

Figure 13.1 Single stand 4 high cold rolling mill.

For example, in the year 2000, 3.44 million tonnes of aluminium-rolled products were consumed in western Europe.[2]

Rolling mills have been in use since the fifteenth century, however, increasing performance demands have led to greater sophistication in rolling mill technology and associated control. Forces can be of the order of 2000 tonnes, strip speeds up to 120 km/h and (in the aluminium industry), thickness as low as 0.05 mm and tolerances of 5 μm. (This represents a tolerance of 0.1% which is 'high tech' by any standard.) The mechanical systems, materials, sensors, actuators and control systems required to achieve this have undergone substantial improvement over time.

Of particular relevance to this virtual laboratory is the problem of centreline thickness control in rolling mills. This problem has been the subject of much investigation. Through the development of new sensors and actuators combined with the appropriate application of new control architectures, the fundamental performance limits of earlier architectures have been overcome and tolerances reduced by 3 orders of magnitude (from 10% to 0.1%)!

2 European Aluminium Association's web site www.eaa.net.

13.3 Modelling

13.3.1 Description of the Process

While numerous rolling mill configurations exist, they are all comprised of one or more stands.[3]

A stand consists of a large cast housing within which a pair of horizontal rolls is mounted. These work rolls 'squeeze' the metal so as to achieve a reduction in thickness. Using smaller work rolls reduces the size of the contact region with strip and thus allows smaller roll forces and torques to be used for a given thickness reduction. It also facilitates better cooling. Larger diameter idling rolls (backup rolls) are usually mounted above and below the work rolls to restrict the amount of roll bending which occurs as the metal passes through the rolls. Such a stand is referred to as 4 high. While this configuration is typical, some mills have 6-high roll stacks or even 20 rolls in triangular cluster formation.

Single-stand mills usually operate by reversing, i.e. multiple passes are usually needed to attain the desired final thickness. Each pass involves rolling a coil from the uncoiling mandrel to the coiling mandrel as fast as possible subject to achieving the desired thickness reduction.

For simplicity, this virtual laboratory will focus on a single stand with the mill running at a fixed nominal operating speed.

13.3.2 Sensors and Actuators

Control is typically affected by changing the relative gap between the work rolls with a hydraulic cylinder[4] acting on the backup roll bearing. Usually, the bearing position is measured with respect to the support frame (the so-called unloaded 'roll-gap' position). The separation of the work rolls is not equal to this measurement because of the significant elastic deformation of the mill stand components. A load cell measures the roll force applied to the strip. Typically, X-ray thickness gauges provide continuous measurements of the strip's input and output thickness at locations some distance from the stand. The input and output strip velocities may also be measured.

Figure 13.2 shows the typical sensor and actuator signals present in a single stand.

13.3.3 Disturbances

The most severe disturbances in terms of off-specification products, are those originating in the incoming strip.[5] Input strip thickness and hardness variations can have particularly severe waveforms that can be approximated by step functions.

13.3.4 Aims of the Control System

The schematic of the rolling mill laboratory is shown in Figure 13.3

The primary aim is to maintain the mill exit thickness within specification without jeopardising the satisfactory operation of the mill. In particular, looping and tearing must be avoided. Considering the extreme forces present and the strip's momentum and speed, one can readily imagine the consequences of the strip being torn or looping due to poor control. Even with fault detection

---

3 Multi-stand mills are also known as tandem mills.

4 Or motor-driven screw in older systems.

5 Set up errors, hydrodynamic backup roll bearing speed effects and input strip surface roughness are also significant but are beyond the scope of this laboratory.

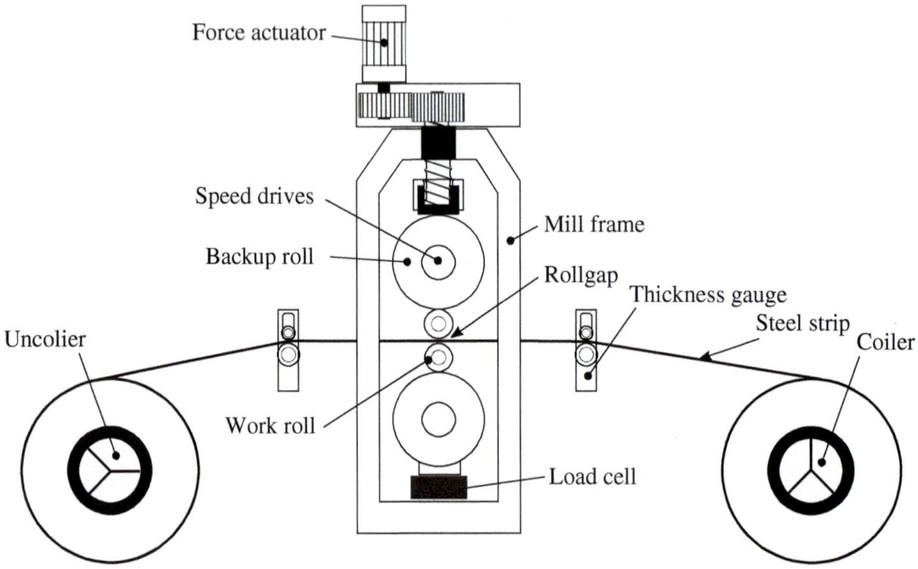

Figure 13.2 Typical measurements and control signals in a single stand.

systems taking immediate corrective action a break in the strip or looping can, and does, result in spectacular, expensive and widespread damage.

13.4 Building a Model

13.4.1 The Mill Frame

The mill frame and the rolls themselves are not perfectly rigid. So if the gap between the rolls (relative to the frame) is set to σ, called the passive (or unloaded) roll gap, then entry of the strip causes the mill to stretch and the gap to increase to σ_a, the active roll gap. Since the elastic recovery of the strip is small enough to be ignored, the exit thickness of the strip, h, is approximately equal to σ_a.

Approximating the stand as a spring,[6] the force of the strip on the rolls is given by

$$F_1 = M(h - \sigma). \tag{13.1}$$

The constant M denotes the 'spring stiffness'. In rolling mill technology, this constant is usually called the 'Mill Modulus'.

13.4.2 Strip Deformation

The force required to deform the strip is known as the roll force and is a function of input and output strip thickness. The following simplified relationship is used to describe the deformation force:

$$F_2 = W(H - h), \tag{13.2}$$

where W is a constant related to the material properties of the strip, H is the input strip thickness and h is the exit thickness.

6 Where $(h - \sigma)$ is the mill stretch (or displacement) and M is the spring constant of the mill.

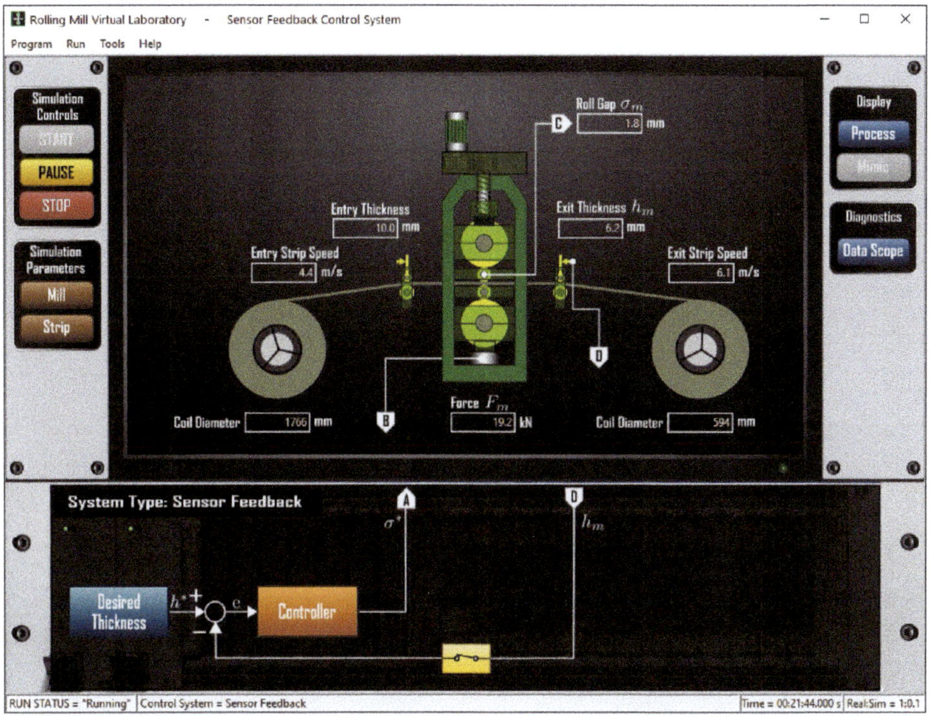

Figure 13.3 Schematic of rolling mill laboratory.

13.4.3 Composite Model

In equilibrium, F_1 and F_2 must be equal. Hence, equating Eqs. (13.1) and (13.2) leads to

$$h(M + W) = M\sigma + WH$$

or

$$h = \frac{M}{M + W}\sigma + \frac{W}{M + W}H.$$

Typical values for the parameters in the above relationship are of the following order:

$$\frac{M}{M + W} = \frac{1}{3} \quad \text{and} \quad \frac{W}{M + W} = \frac{2}{3}.$$

Traditionally, the exit thickness, h, of the strip is controlled by varying σ, the unloaded roll gap. The actuator also has dynamics, which can be approximated by a simple lag. A typical time constant is of the order of 10 ms. In a physical rolling mill, an X-ray gauge is typically used to measure h. It cannot be located at the exit of the roll gap but is placed some distance away. This introduces a measurement delay into the system. Obviously, the delay depends on the exit strip's speed. In this laboratory, it is assumed to be a constant of 0.5 seconds. In practice, the delay is a function of mill speed but can be calculated.

Putting all of this together leads to the open-loop model shown in Figure 13.4. The disturbance variable e will be studied in Chapter 24. For the moment, one can take $e = 0$. Hence, the plant model is given by

$$G(s) = \frac{Ke^{-sT_d}}{\tau s + 1}.$$

where $K = \frac{1}{3}$, $T_d = 0.5$ and $\tau = 0.01$.

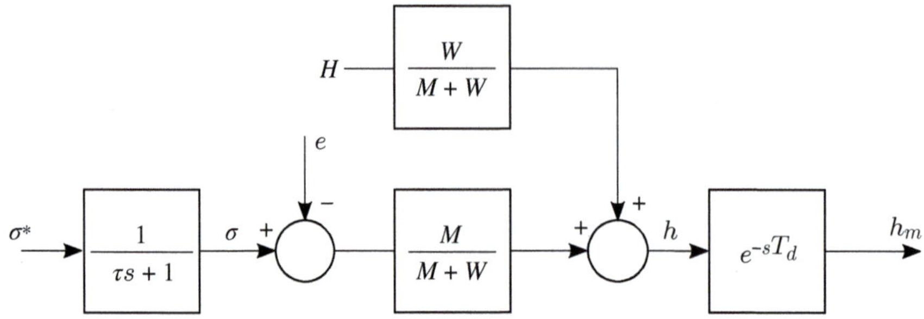

Figure 13.4 Open-loop rolling mill model.

13.4.4 Open-Loop Steady-State Performance

The starting point is the model for the output thickness

$$h = \frac{M}{M+W}\sigma + \frac{W}{M+W}H. \tag{13.3}$$

If σ is constant, then a relevant question is, 'what would be the effect on the output thickness h, if there was a 1 unit increase in the input thickness H?' For the purposes of this question, it is assumed that $\frac{M}{M+W} = \frac{1}{3}$.

With $\frac{M}{M+W} = \frac{1}{3}$ then $\frac{W}{M+W} = \frac{2}{3}$. Since

$$h = \frac{M}{M+W}\sigma + \frac{W}{M+W}H,$$

then initially

$$h_{initial} = \frac{1}{3}\sigma + \frac{2}{3}H. \tag{13.4}$$

After the unit step in H,

$$h_{step} = \frac{1}{3}\sigma + \frac{2}{3}(H+1).$$

Subtracting Eq. (13.4) gives

$$h_{step} = h_{initial} + \frac{2}{3}.$$

Thus, for the open-loop system a unit disturbance in H results in a steady-state thickness error of $\frac{2}{3}$ units.

13.5 Basic Control System Design

From Section 13.4.4, it follows that changes to the input thickness result in significant steady-state errors in the output thickness. To remove these errors, one can, in principle, use a single degree-of-freedom control architecture as illustrated in Figure 13.5, with the controller, $C(s)$, having integral action.

In Sections 13.6.1 and 13.7.1, the performance that can be achieved using this architecture will be examined.

The control system is designed using a *nominal* model of the system.

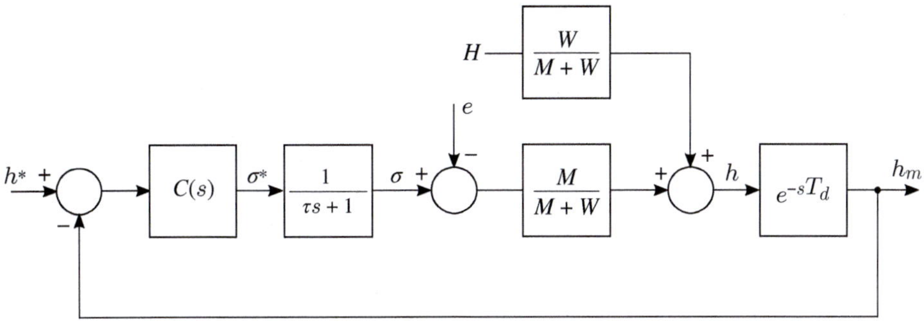

Figure 13.5 Single degree-of-freedom control loop using measured exit thickness.

Based on the previous discussion, the plant is approximately described by the following model relating changes in the passive roll gap to the exit thickness,

$$G_o(s) = \frac{\widehat{K}e^{-s\widehat{T}_d}}{\hat{\tau}s+1}.$$

For simplicity it is assumed that the parameters take the following values $\widehat{K} = K = \frac{1}{3}$, $\widehat{T}_d = T_d = 0.5$ and $\hat{\tau} = \tau = 0.01$.

Since $\widehat{T}_d > \hat{\tau}$, the plant is said to be time-delay dominated. In Sections 13.6 and 13.7, various approximations to the above model will be utilised for design purposes.

13.6 Linear Control Ignoring the Time Delay

As a first approximation, the time delay will be ignored leading to the following much simplified model:

$$G_o^*(s) = \frac{\widehat{K}}{\hat{\tau}s+1}.$$

Assuming that a PI controller will be used of the form:

$$C(s) = K_p + \frac{K_i}{s}, \tag{13.5}$$

$$= \frac{K_i\left(\frac{K_p}{K_i}s+1\right)}{s}.$$

As a simple design strategy, the term $\left(s + \frac{K_i}{s}\right)$ will be used, to cancel the nominal plant pole, i.e.

$$\frac{K_p}{K_i} = \hat{\tau}. \tag{13.6}$$

After cancellations, this leads to

$$G_o^*(s)C(s) = \frac{\widehat{K}K_i}{s}. \tag{13.7}$$

Next, it is noted that the closed-loop characteristic polynomial is

$$A_{cl}(s) = \left(s + \widehat{K}K_i\right). \tag{13.8}$$

Thus, if one wishes to have a closed-loop time constant of β (seconds), then it suffices to choose

$$\beta = \frac{1}{\widehat{K}K_i}. \tag{13.9}$$

Table 13.1 Laboratory tests.

Test	β
1	$3T_d$
2	$1.2T_d$
3	$0.7T_d$
4	$0.6T_d$

Solving (13.6) and (13.9) for K_p and K_i, gives

$$K_p = \frac{\hat{\tau}}{\hat{K}\beta} \quad \text{and} \quad K_i = \frac{1}{\hat{K}\beta}, \tag{13.10}$$

where β is the desired closed-loop time constant (in seconds). The parameter β can also be interpreted as being related to the desired closed-loop bandwidth ω_{cl} where $\omega_{cl} = \frac{1}{\beta}$ rad/s or $f_{cl} = \frac{1}{2\pi\beta}$ Hz. Thus, this parameter acts as a single-tuning parameter for controller synthesis.

13.6.1 Experiment: Testing a PI Controller

Using the rolling mill laboratory, assume that the desired product specification is a constant exit thickness of $h^* = 8$ mm with a tolerance of ± 0.02 mm, i.e. a tolerance of approximately of 0.25% for $h^* = 8$ mm. To enable direct comparison of the different architectures, the same test conditions will be used. Set the input thickness Hin to 9 mm.

To initially stabilise the system set $K_i = 0$ and K_p to a small value (say 1.0).

1. Generate controller values for each β chosen as a function of the process time delay T_d as in Table 13.1.
2. Select the sensor feedback control system in the virtual lab.
3. For each controller, perform a step change to the Input Thickness Hin (say from 9 to 10 mm). (You can use the square wave signal by setting with a long period.) Observe the measured output thickness, hm and record the settling time (if the system is stable). This is the disturbance response.
4. Tabulate your observations for each test listed in Table 13.1 and record the values of β in seconds, K_p, K_i, stable (yes/no) and if stable the disturbance response settling time in seconds.
5. Using the controller with the minimum settling time, perform a step change to the desired thickness, h^*. Observe the measured output thickness, hm (set point step response).
6. Record the disturbance and step response of this controller.
7. Note that, whilst instability is not important in the virtual laboratory, in a real rolling mill instability could have catastrophic consequences.

Reflect on your observations and answer *briefly*:

1. When stable, does this configuration remove steady-state errors?
2. What is the key difference between the reference and disturbance responses?
3. What happens as β is reduced? Why does this happen?

13.7 Linear Control Based on Rational Approximation to the Time Delay

It has been shown in Section 13.6 that robustness issues set severe limits on what the linear control system can achieve when a delay is ignored. In fact, the larger the time delay, the 'slower' the final system must be to remain stable.

A reasonable conjecture is that a better linear controller may be possible if the time delay is explicitly considered in the formulation of the controller, i.e. reducing the modelling error. Unfortunately, the time delay is irrational. Thus, simple linear controller design requires that the time delay be approximated by a rational function. A commonly used approximation is known as the first-order Padé approximation described in Section 12.4:

$$e^{-sT_d} \approx \frac{1 - \frac{T_d}{2}s}{1 + \frac{T_d}{2}s}. \tag{13.11}$$

This approximates the time delay's magnitude correctly at all frequencies and matches the phase reasonably well at low frequencies, i.e. $\omega \ll \frac{2}{T_d}$.

Utilising the Padé approximation leads to the following revised nominal model:

$$G(s) = \frac{\hat{K}(1 - \frac{\hat{T}_d}{2}s)}{(\frac{\hat{T}_d s}{2} + 1)(\hat{\tau}s + 1)}. \tag{13.12}$$

A PID design will now be carried out using the ideas described in Section 7.5. In particular, the following choice is made for $Q(s)$:

$$Q(s) = \hat{G}(s)^{-1}F(s), \tag{13.13}$$

where $\hat{G}(s)^{-1}$ is a proper stable approximation to the inverse of the nominal model $G_o(s)$ and $F(s)$ is an extra filter which is used to shape the closed-loop response. Since $G_o(s)$ has a zero outside the stability boundary, it cannot be included in the approximate inverse. This leads to the choice

$$\hat{G}(s)^{-1} = \frac{\left(1 + \frac{T_d s}{2}\right)(\tau s + 1)}{\hat{K}}. \tag{13.14}$$

This filter $F(s)$ is chosen as

$$F(s) = \frac{1}{\beta^2 s^2 + 2\beta s + 1}. \tag{13.15}$$

Finally, the control law is expressed in the usual form by calculating

$$C(s) = \frac{Q(s)}{1 - G_o(s)Q(s)}. \tag{13.16}$$

This leads to a PID control law of the form:

$$C_{PID}(s) = K_P + \frac{K_I}{s} + \frac{K_D s}{\gamma s + 1}. \tag{13.17}$$

The details of the algebra to go from (13.16) to (13.17) are rather messy and hence they are simply quoted below:

$$K_P = \frac{4\hat{T}_d\beta + 8\hat{\tau}\beta + \hat{T}_d^2 + 2\hat{\tau}\hat{T}_d - 4\beta^2}{16\hat{K}\beta^2 + 8\hat{K}\hat{T}_d\beta + \hat{T}_d^2\hat{K}},$$

Table 13.2 Laboratory tests.

Test	β in terms
1	$0.8T_d$
2	$0.6T_d$
3	$0.15T_d$
4	$0.1T_d$

$$K_I = \frac{2}{4\hat{K}\beta + \hat{T}_d\hat{K}},$$

$$K_D = \frac{(4\hat{K}\beta + \hat{T}_d\hat{K})^2\hat{T}_d\hat{\tau} - 2\hat{K}\beta^2(4\hat{K}\beta + \hat{T}_d\hat{K})(\hat{T}_d + 2\hat{\tau}) + 8\hat{K}^2\beta^4}{(4\hat{K}\beta + \hat{T}_d\hat{K})^3},$$

$$\gamma = \frac{2\beta^2}{4\beta + \hat{T}_d}.$$

13.7.1 Experiment: Testing PID Design

1. Using the above equations, calculate PIDγ controllers for the test values of β given in Table 13.2.
2. Use the controllers found above inside the sensor feedback architecture in the virtual lab. Perform an input disturbance test for each controller.
3. Tabulate your results for each test specified in Table 13.2 and record the values of β in seconds, K_p, K_i, K_d, γ, stable (yes/no) and if stable the disturbance response settling time in seconds.
4. Include a plot of the disturbance response with the shortest settling time.
5. What has been the effect of including an approximation of the time delay into your controller design?

13.8 Control System Design Based on Smith Predictor

So far, the controllers have been constructed either by ignoring the delay or by using a rational approximation to the time delay. An alternative approach uses a Smith predictor. The Smith predictor uses a parallel model of the plant to deal with the time delay. The controller design can then be carried out on the rational (undelayed) part of the plant model. It is strongly suggested that the reader reviews Sections 12.5 and 12.6 before proceeding.

The Smith predictor architecture is shown in Figure 12.2, where $G_o = e^{-s\hat{T}_d}G_o^*(s)$ is the true plant and $\overline{C}$ is the controller based on the rational (non-time delayed) component of the nominal model, G_o^*. For the purpose of design, use $\hat{K}, \hat{\tau}$ as approximations to the true values K, τ.

For the current problem

$$G(s) = \frac{Ke^{-sT_d}}{\tau s + 1},$$

$$G_o(s) = \frac{\hat{K}e^{-s\hat{T}_d}}{\hat{\tau}s + 1},$$

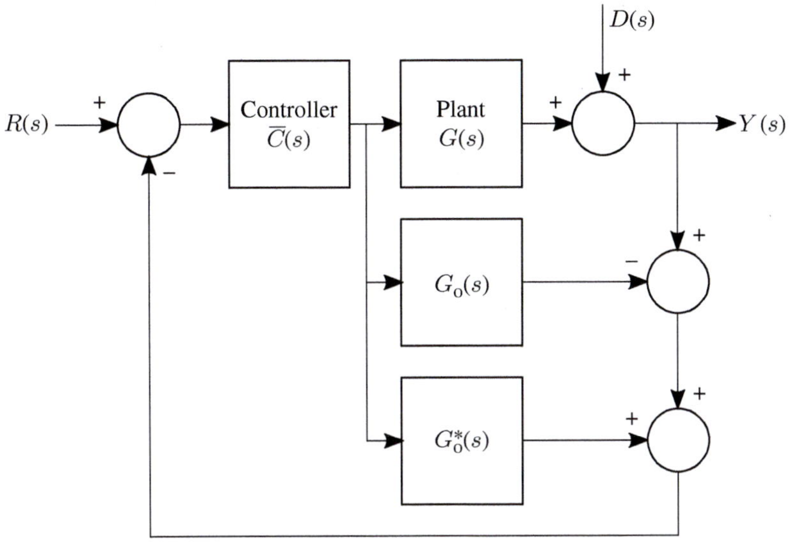

Figure 13.6 Smith predictor architecture.

$$G_o^*(s) = \frac{\widehat{K}}{\hat{\tau}s + 1}.$$

The Smith predictor can then be implemented as in Figure 13.6.

13.8.1 Experiment: Testing Smith Predictor

Design the controller (based on the Smith predictor) using nominal values $\widehat{K}$, $\hat{\tau}$ and $\widehat{T}_d$.

1. Calculate a controller $\overline{C}$, corresponding to each value of β shown in Table 13.3. Remember, $\overline{C}$ is based on the delay-free nominal model, G_o^*.
2. Select the Smith predictor control system in the virtual laboratory. (Go to 'Program' and then 'Load controller'.)
3. For each set of parameters in Table 13.3, perform disturbance tests. Note, when designing the Smith predictor based on model parameters that are different from those in the plant, then, it may help to first use a lower controller bandwidth to stabilise the system, then change to a faster bandwidth and complete the experiment.
4. Tabulate your results for each test specified in Table 13.3 and record the values of K_p, K_i, stable (yes/no) and if stable the disturbance response settling time in seconds.
5. Examine plots of the step and disturbance responses for the case where $\beta = 0.02T_d$ and $\widehat{T}_d = T_d$.

Considering both your current results and those of the previous experiments, answer the following questions:

(a) System performance is most sensitive to errors in which model parameter?
(b) How does this architecture perform compared to the previous architectures? Why?
(c) Assume a correct nominal model. What value does the settling time approach as the closed-loop bandwidth is increased?
(d) In practice, what would limit the closed-loop bandwidth chosen?

Table 13.3 Laboratory tests.

Test	β	β (seconds)	$\hat{T}_d$ (seconds)
Using $\hat{K} = K$, $\hat{\tau} = \tau$			
1	$0.1T_d$	0.05	0.5
2	$0.1T_d$	0.05	0.52
3	$0.02T_d$	0.01	0.5
4	$0.02T_d$	0.01	0.51
5	$0.02T_d$	0.01	0.52
Using $\hat{K} = 0.3$ ($\approx$10% error), $\hat{\tau} = \tau$			
6	$0.02T_d$	0.01	0.5
Using $\hat{K} = K$, $\hat{\tau} = 0.011$ (10% error)			
7	$0.02T_d$	0.01	0.5

13.9 Use of a Soft Sensor

It was seen in Section 13.8 that fundamental design difficulties arise from the time delay in the exit thickness measurement. Robustness issues have precluded the use of excessively high bandwidth for the controller. Moreover one can never compensate for disturbances inside the measurement delay time.

The ideal solution to this architectural problem would be to remove the measurement delay! Alas, the physical sensor cannot be located inside the roll gap. However, an instantaneous thickness value can be inferred from other measurements using a, so-called, *soft sensor* or (*virtual sensor*). There are many techniques in common use which achieve this functionality. One such device (known as a BISRA[7] gauge) infers the exit thickness from the roll force and the unloaded roll gap, both of which can be easily measured.

13.9.1 The BISRA Gauge

Recall, that the following relationship described the force from stretching the mill (see Eq. (13.1)):

$$F = M(h - \sigma). \tag{13.18}$$

Say that, in addition to exit thickness, it was possible to also measure the roll Force (F). Then rearranging (13.18) and substituting for the available measurement results in the following 'soft sensor' for exit thickness,

$$\hat{h} = \frac{F}{M} + \sigma, \tag{13.19}$$

where $\hat{h}$ is an estimate of the exit thickness, M is a constant (which is typically known for a given mill), F is the measured roll force and σ is the known unloaded roll gap.

7 British Institute of Steel Research Association.

The estimate given in (13.19) is called a 'soft sensor' since it uses a model (i.e. software) together with other variables (here F and σ) to infer another quantity (here h).

13.9.2 Experiment: Testing the BISRA Gauge

1. Select the 'BISRA controller' in the virtual laboratory.
2. Ensure $M = 3.5$ inside the BISRA gauge.
3. Close the feedback loop.
4. Perform disturbance tests by varying the incoming strip thickness for each set of parameters (see Section 13.8.1 for details). Use the estimated thickness as your output measurement.

13.10 Robustness of BISRA Gauge

Both the Smith predictor and BISRA gauge are sensitive to the fidelity of the models used. Hence, one needs to proceed with due caution as was previously demonstrated for the Smith predictor.

In Section 12.9, it was shown that the Smith predictor was sensitive to the value of the delay. This delay is not present in the BISRA gauge. However, sensitivity to the value of the mill modulus, M, remains an issue. Thus, the next step is to investigate the impact of having an error in, M, when implementing the BISRA gauge.

13.10.1 Experiment: Testing Sensitivity to Mill Modulus

Test the control law using an incorrect value for M in the Smith predictor. For example, change the true (but unknown) value of 3.5 to say 3.

Note that a steady-state error appears. (see Question 8 below).

13.10.2 Experiment: Alternative Solution to Achieve Steady-State Tracking

Can you think of a way that the above problem can be corrected using the true (but delayed) measurement of exit thickness? (See Question 9 of the Revision Questions.)

Test your suggested solution in the virtual laboratory.

13.11 Revision Questions

1. If the input thickness H is constant what would be the effect on output thickness, h, if there was a 1 unit increase in the passive roll gap σ? For the purposes of the question let $\frac{M}{M+W} = \frac{1}{3}$.
2. Consider the design tested in Section 13.6.1. Derive an algebraic expression in terms of T_d defining the range of β where the true feedback system is unstable. (Assume the gain and time constant for the nominal model match those of the true plant.)
(Hint: Consider the open-loop transfer function when the controller is applied to the true plant, $L(s) = C(s)G(s)$, where $G(s)$ includes the delay. Using the Nyquist Stability Criterion, the system becomes unstable when the phase of $L(s)$ crosses $-\pi$ radians and $|L(j\omega)| \geq 1$. Hence calculate ω such that the phase of $L(j\omega)$ is $-\pi$, then using this frequency and $|L(j\omega)| \geq 1$ solve for β).

3. Compare the answer obtained in Question 2 with the sufficient condition for robust stability studied in Question 3 of Chapter 12.
4. Given your answers to Question 2 above and Question 3 of Chapter 12, why is a Smith predictor preferable to simply ignoring the delay.
5. Derive the expression for K_P, K_I, K_D, and σ given in Section 13.7.
6. It was found in Section 13.7 that the system becomes unstable for small values of β. Why do you
7. think this happens?
   (a) The Smith predictor seems to gives very good performance. However, if a disturbance occurs in the input strip at time t_0. How much of the exit strip is impacted prior to the control law acting.
   (b) If the input thickness were to be measured up-stream of the mill how could feedforward improve the answer given in part (a).
8. Say that $\widehat{M} \neq M$ in the BISRA gauge. What effect will this have on steady-state errors?
9. How could you use the measured value of h to correct the steady-state error found in Question 8?

Further Reading

A great deal is available on rolling mill control. Readers may be interested in the following sources There are also several courses available on Rolling Technology.

1 G.F. Bryant (Ed.), *Automation of tandem mills*, London Iron and Steel Institute, 1973.
2 A.J. Carlton, W.J. Edwards and P.J. James, "Formula for Cold Rolling Analysis", *Proceedings AIME Symposium on Automation in the Iron and Steel Industry*, p. 223, Atlanta, Georgia, USA, 1977.
3 J. Gerry, "How to Control a Process with Long Dead Time", *Control Engineering*, 1998.
4 Y. Hosoya, E. Roller, H. Yoshimi, Y. Endou, K. Takahashi and S. Hiruta, "State-of-the-Art Noncontact Gauge Meter for Flat Rolling and Cold-Rolled Precision Metals Expected for Microprocessor Application", *Sensors and Materials*, 31(10):3075–3086, 2019.

14

Control System Design for Open-Loop Unstable Systems

14.1 Introduction

Up to this point in the book, it has been implicitly assumed that the system under study is open-loop stable. Whilst most systems encountered in practice are open-loop stable, there do exist important systems that are definitely not open-loop stable.

This chapter examines the control of such systems. Chapter 15 will illustrate the ideas using the control of a rocket in two-dimensional flight. This is a quintessential example of an open-loop unstable system.

14.2 Some Simple Examples of Open-Loop Unstable Systems

The first difficulty one faces in controlling an unstable system is that one needs to have a minimum gain to make the system stable.

To illustrate consider the following simple first-order unstable system:

$$\dot{x} = x + u. \tag{14.1}$$

Note that if one sets $u = 0$, then the response satisfies

$$x(t) = e^{t}x(0), \tag{14.2}$$

i.e. the response diverges exponentially.

Next consider a closed-loop, in which a simple proportional feedback control law of the following form is used:

$$u(t) = -kx(t). \tag{14.3}$$

Then the closed loop system satisfies

$$\dot{x} = x - kx. \tag{14.4}$$

Clearly, we need $k > 1$ to achieve closed-loop stability.

Chapter 15 will be devoted to the study of a practical system, namely, a rocket. Before tackling this problem, we first consider a much simpler problem but one in which the dynamics are an

Practical Control System Design: Real World Designs implemented on Emulated Industrial Systems,
First Edition. Adrian Medioli and Graham Goodwin.

Companion website: www.wiley.com/go/medioli/practicalcontrolsystemdesign

approximation to that of a rocket. Thus, consider the following linearised transfer function linking the angle of the motor gimbal, δu, to the angle of rocket, $\delta\theta$, where the thrust, T, is constant:

$$\frac{\delta\theta(s)}{\delta u(s)} = \frac{-lT}{Js^2 - \frac{1}{2}\rho\mu v_2^2}. \tag{14.5}$$

The following coefficient values are used:

$$T = 35 \times 10^6, \quad l = 15, \quad J = 2.67 \times 10^8, \quad \rho = 20, \quad \mu = 5, \quad v_2 = 0.7123 \times 10^3, \tag{14.6}$$

giving

$$G_{sr}(s) = \frac{\delta\theta(s)}{\delta u(s)} = \frac{-1.966}{s^2 - 0.09501} = \frac{-1.966}{(s - 0.3082)(s + 0.3082)}. \tag{14.7}$$

There are some interesting things about this simple rocket transfer function, $G_{sr}(s)$, which exemplify some of the issues in rocket control (in particular), and unstable systems (in general).

First, the gain is apparently negative. However, if one sets $s = j\omega$ and considers small frequencies, then the gain is positive. This is a consequence of the unstable pole at $s = 0.3082$.[1]

Second, the system is open-loop unstable so it is necessary to actively control the system. (Indeed, there is a minimum feedback gain necessary to ensure that the closed-loop system is stable.) In the sequel a simple control law will be studied that ensures stability.

Assume that one uses a prototype controller $C(s) = -1$. (The negative sign cancels the negative sign in the numerator of (14.7).) The reader is asked to sketch the Nyquist and Bode Diagrams for the system.

$$C(s)G_{sr}(s) = \frac{1.966}{(s - 0.3082)(s + 0.3082)}. \tag{14.8}$$

The design objective is to design a controller which ensures that the closed-loop system has no unstable poles. It is further required that the close-loop system have acceptable gain and phase margin.

The Nyquist stability theorem tells us that, if there are N clockwise encirclements of the point (−1,0) and there are P open-loop poles in the right-hand plane (RHP), then the number of unstable *closed-loop* poles, Z, is given by

$$Z = P + N.$$

For $G_{sr}(s)$ the number of unstable open-loop poles is $P = 1$.

Thus for the simple system, $G_{sr}(s)$ one needs to modify $C(s)$ to ensure that $N = -1$, e.g. one anti-clockwise encirclement of the origin.

With the open-loop Nyquist diagram of $C(s)G_{sr}(s)$ lying on the negative real axis, one can deduce that a phase advance compensator would shift the Nyquist diagram in such a way that would ensure $N = -1$.

This suggests a compensator of the form

$$C(s) = -g\frac{1 + \tau_1 s}{1 + \tau_2 s}. \tag{14.9}$$

The reader should try values of τ_1, τ_2 and verify that closed-loop stability is indeed achieved. Also, the reader should choose a value of g which gives reasonable gain and phase margins.

1 Actually because of the symmetry of the singularities about the $j\omega$ axis, this is a zero-phase system.

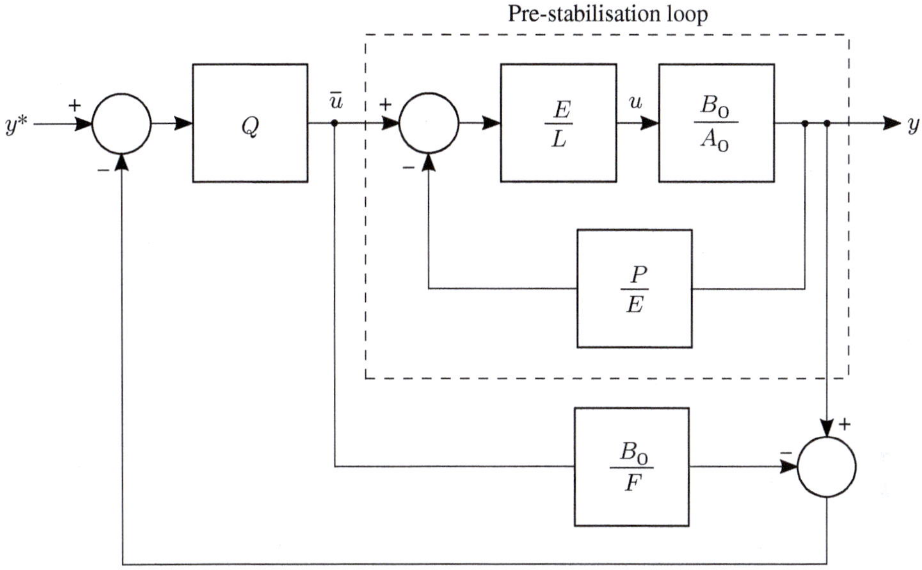

Figure 14.1 Approximate Youla–Kucera parameterisation for unstable plant.

14.3 All Stabilising Control Laws for Systems Having Undesirable Open-Loop Poles

Readers will recall that, when the Youla–Kucera parameterisation of all stabilising control laws was studied in Section 7.4, a restriction was made that the open-loop system should be stable. This clearly begs the question, 'Is there an equivalent characterisation of all stabilising control laws for open-loop unstable systems?' First, the idea of 'unstable' open-loop poles is expanded to include 'undesirable' poles. An example of an 'undesirable' pole could be a pole having an excessively long transient time.

The basic idea inherent in extending the Youla–Kucera parameterisation from open-loop stable systems to systems having 'undesirable' open-loop poles is to apply 'pre-stabilizing' feedback so as to create a new system having 'desirable' poles.

To show how this can be done consider an open-loop plant having transfer function

$$G_o(s) = \frac{B_o(s)}{A_o(s)}, \tag{14.10}$$

where $A_o(s)$ may have 'undesirable' roots.

First a design is carried out for a pre-stabilising controller

$$C(s) = \frac{P(s)}{L(s)} = \frac{P(s)E(s)}{L(s)E(s)}, \tag{14.11}$$

via the pole-assignment identity

$$A_o(s)L(s) + B_o(s)P(s) = E(s)F(s), \tag{14.12}$$

where both $E(s)$ and $F(s)$ have 'desirable' roots.

Then the control law is implemented as in the dotted subsection Figure 14.1 so that the estimation polynomial $E(s)$ cancels in the closed loop.

The closed-loop transfer function from $\bar{u}$ to y is readily seen to be

$$G_o'(s) = \frac{\dfrac{E(s)B_o(s)}{L(s)A_o(s)}}{1 + \dfrac{E(s)}{L(s)}\dfrac{B_o(s)}{Ao(s)}\dfrac{P(s)}{E(s)}}, \tag{14.13}$$
$$= \frac{E(s)B_o(s)}{L(s)A_o(s) + B_o(s)P(s)},$$
$$= \frac{B_o(s)}{F(s)},$$

where Eq. (14.12) has been used.

Finally, the Youla–Kucera parameterisation of all stabilising control can be used for the new transfer function $G_o'(s)$. The result is shown in Figure 14.1 where $Q(s)$ is now required to be proper and to have 'desirable' poles. As usual, $Q(s)$ should be an approximate inverse to $G_o'(s)$.

It is trivially seen that the architecture of Figure 14.1 yields a closed-loop (nominal) system having 'desirable' poles.

Actually, there do exist some feedback systems having 'desirable' closed-loop poles that cannot quite be drawn as in Figure 14.1. However, these are exceptional cases and would not normally be met in practice. For further details the reader is referred to the reference [1] given in Further Reading. The topic of all stabilising control laws for open-loop unstable systems will be revisited in Section 22.13 where state-space ideas will be used.

14.4 Revision Questions

1. Consider a system having transfer function $G(s)$ where

$$G(s) = \frac{K(s+b)}{(s+a)(s-a)}, \quad 0 < b < a. \tag{14.14}$$

   (a) If b is positive, draw a root locus diagram of the closed-loop poles as a function of gain, K_p, in a proportional control law.
   (b) If b is negative, what happens to the closed-loop poles as K_p is made large?
2. Consider transfer function $G(s) = \frac{K(s+b)}{(s+a)(s-a)}$, as in Question 1, where b is negative and $0 < |b| < |a|$.
   (a) Draw the root locus.
   (b) For what values of gain is the closed-loop system stable?
3. For the system in Question 2, let $K = 1, b = -\frac{1}{2}, a = 1$. Use pole assignment to find a control law of the form:

$$C(s) = \frac{p_1 s + p_0}{s + \ell_0}, \tag{14.15}$$

   which places the closed-loop poles at -1.
4. Draw the root locus diagram for the system of Eq. (14.15) as the gain of the control law is varied.
5. Say the plant gain changes to $(1 + \epsilon)$. What value of ϵ would make the closed-loop unstable?
6. What do you see as a potential problem with this system and the proposed control law?

Further Reading

1 G.C. Goodwin, S.F. Graebe and M.E. Salgado, *Control system design*, Prentice Hall, 2000.

2 G. Stein, "The practical (and sometimes dangerous) consequences of control must be respected, and the underlying principles must be clearly and well taught", *IEEE Control Systems Magazine*, 23(4):12–25, 2003. "Respect the unstable".

15

Laboratory 7: Control of a Rocket

15.1 Introduction

This chapter deals with the modelling and control of a rocket operating in two dimensions. A suitable nonlinear model will be developed in Section 15.2. Section 15.3 shows how equilibrium states can be evaluated. Section 15.4 linearises the model about an equilibrium state. Section 15.5 studies the rocket in open-loop flight. Section 15.6 outlines the design of the control system, and Section 15.6 tests the design using the virtual laboratory. A model for a rocket in 3D flight will be developed later in Chapter 28.

A screen shot of the virtual laboratory is shown in Figure 15.1. This laboratory is based on an original design by Jim Kusters[1] with advice from Professor Robert Skelton,[2] and Professor Peter Wellstead.[3]

In this laboratory, you will design a controller for the rocket and learn some limitations regarding the design of a controller to achieve good performance. The design will be validated using linear and nonlinear models of the rocket.

15.2 Dynamics of a Rocket in 2D Flight

This section describes a coordinate system and derives all of the dynamic elements of a rocket in two-dimensional space. Figure 15.2 shows the key variables that will be used throughout this chapter when developing a two-dimensional rocket model.

15.2.1 Coordinate Systems

It is convenient to use two coordinate systems when describing the operation of a rocket, namely, a coordinate system orientated with the rocket with basis vectors $\underline{B} = (\underline{b}_1, \underline{b}_2)$ and a stationary reference frame with basis vectors $\underline{E} = (\underline{e}_1, \underline{e}_2)$ (both of these are shown in Figure 15.2).

A vector $\underline{r}$ can be expressed in terms of $\underline{B}$ or $\underline{E}$ as follows:

$$\underline{r} = \underline{E}\mathbf{r}^e = \underline{B}\mathbf{r}^b, \tag{15.1}$$

1 A masters, student visiting from the Technical University of Eindhoven, Netherlands.

2 Bob, at the time of developing this laboratory, was in the Mechanical and Aerospace Engineering. Department at University of California, San Diego. He was a visiting academic at the University of Newcastle and gave advice to Jim Kusters during the preparation of this laboratory.

3 Peter was working at the Hamilton Institute of the National University of Ireland at the time this laboratory was developed.

Practical Control System Design: Real World Designs implemented on Emulated Industrial Systems,
First Edition. Adrian Medioli and Graham Goodwin.

Companion website: www.wiley.com/go/medioli/practicalcontrolsystemdesign

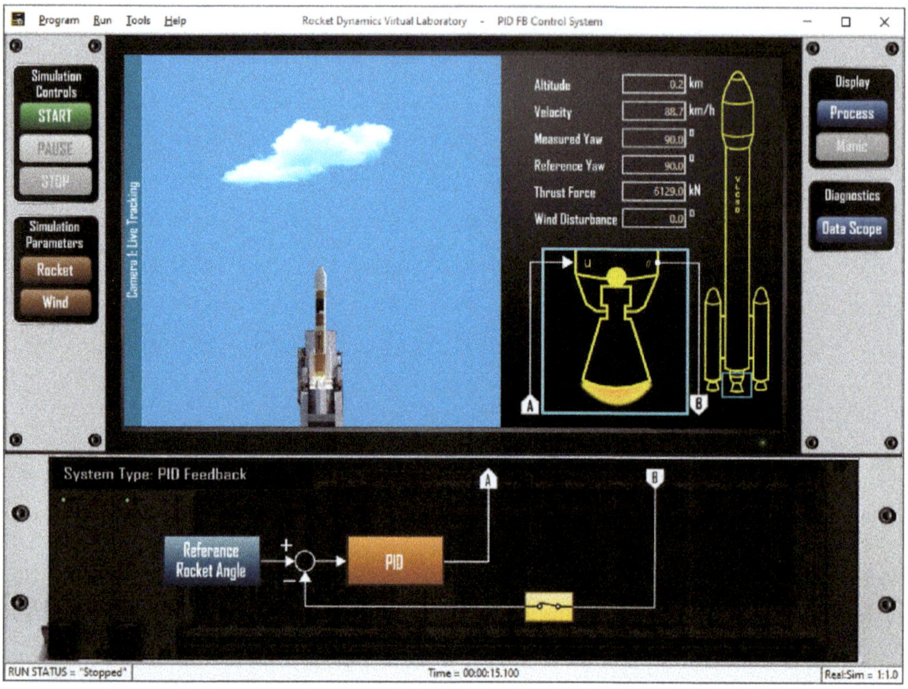

Figure 15.1 Screen shot.

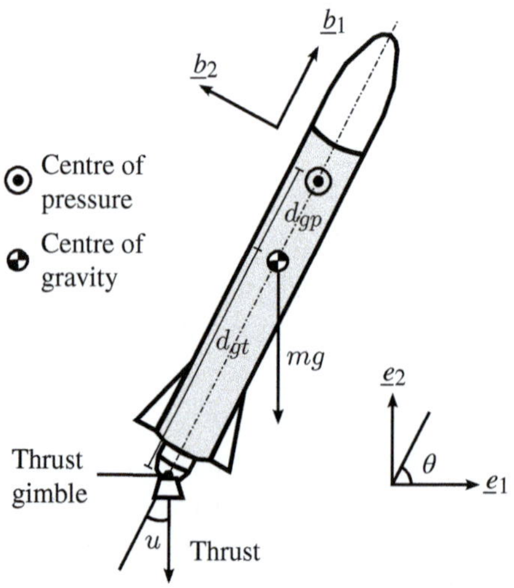

Figure 15.2 2D rocket model (d_{gp} is positive if the centre of pressure is above the centre of gravity).

The relationship between $\underline{E}$ and $\underline{B}$ is given by a rotational transformation. In the 2D case, this is simple, i.e.

$$\underline{B} = \underline{E}R, \tag{15.2}$$

where

$$R = \begin{bmatrix} \cos\theta & -\sin\theta \\ \sin\theta & \cos\theta \end{bmatrix}. \tag{15.3}$$

15.2.2 Forces

As shown in Figure 15.2, there are three forces acting on the rocket, namely,

i) Gravity
   This force is described in the $\underline{E}$ coordinate systems as:

$$\underline{f}_g = \underline{E}\begin{bmatrix} 0 \\ -mg \end{bmatrix} = \underline{E}\mathbf{f}_g^e. \tag{15.4}$$

ii) Thrust
   As shown in Figure 15.2, the thrust is orientated at angle u to the body of the rocket. Thus, in the body frame:

$$\underline{f}_t = \underline{B}\begin{bmatrix} -\cos u \\ -\sin u \end{bmatrix} T = \underline{B}\mathbf{f}_t^b, \tag{15.5}$$

   where T denotes the magnitude of the thrust produced by the rocket engine.

iii) Drag
   The drag force opposes the motion of the rocket. It is assumed to operate through the centre of pressure of the rocket as shown in Figure 15.2. Describing drag forces is a complex issue. However, as a reasonable approximation, the magnitude of the drag force can be modelled as being proportional to the square of the speed of the rocket, i.e.

$$|f_d| = \frac{1}{2}\rho C_d A s^2 = \frac{1}{2}\sigma s^2, \tag{15.6}$$

   where ρ is the density of the fluid in which the rocket is operating, C_d is the drag coefficient, A denotes an effective area for the rocket, and s is the speed of the rocket. For simplicity, we combine the constants into an aerodynamic constant denoted by σ.

Let the coordinates of the centre of mass be x, y in the $\underline{E}$ coordinate system. The corresponding velocities are denoted as v_1 and v_2. Then the velocity of the rocket can be expressed as

$$\underline{v} = \underline{E}\mathbf{v}^e = \underline{E}\begin{bmatrix} v_1 \\ v_2 \end{bmatrix}, \tag{15.7}$$

and the speed is

$$s = \sqrt{v_1^2 + v_2^2}, \tag{15.8}$$

where

$$v_1 = \frac{dx}{dt}, \tag{15.9}$$

$$v_2 = \frac{dy}{dt}. \tag{15.10}$$

The drag force, $\underline{f}_d$, opposes the velocity. Hence in the $\underline{E}$ coordinate system:

$$\begin{aligned} \underline{f}_d &= \underline{E}\begin{bmatrix} -v_1 \\ -v_2 \end{bmatrix}\left(\frac{1}{2}\sigma\right)(v_1^2 + v_2^2)^{\frac{1}{2}}, \\ &= \underline{E}\mathbf{f}_d^e. \end{aligned} \tag{15.11}$$

15.2.3 Translational Dynamics

The motion of the centre of mass will be described in the earth-centred frame $\underline{E}$. Newton's Law relates acceleration to force. (For simplicity, the rocket mass is taken to be constant.[4]) Hence,

$$\underline{E}\begin{bmatrix} m\frac{dv_1}{dt} \\ m\frac{dv_2}{dt} \end{bmatrix} = \underline{E}\mathbf{f}_d^e + \underline{E}\mathbf{f}_g^e + \underline{B}\mathbf{f}_t^b, \tag{15.12}$$

$$= \underline{E}\mathbf{f}_d^e + \underline{E}\mathbf{f}_g^e + \underline{E}R\mathbf{f}_t^b.$$

Using (15.4), (15.5), (15.11) and (15.3), then in the $\underline{E}$ coordinate system:

$$\begin{bmatrix} m\frac{dv_1}{dt} \\ m\frac{dv_2}{dt} \end{bmatrix} = \left(\frac{1}{2}\sigma\right)\left(v_1^2 + v_2^2\right)^{\frac{1}{2}}\begin{bmatrix} -v_1 \\ -v_2 \end{bmatrix} + \begin{bmatrix} 0 \\ -mg \end{bmatrix} + T\begin{bmatrix} \cos\theta & -\sin\theta \\ \sin\theta & \cos\theta \end{bmatrix}\begin{bmatrix} \cos u \\ \sin u \end{bmatrix}, \tag{15.13}$$

or

$$m\frac{dv_1}{dt} = -\left(\frac{1}{2}\sigma\right)\left(v_1^2 + v_2^2\right)^{\frac{1}{2}} v_1 + T\cos\theta\cos u - T\sin\theta\sin u \tag{15.14}$$

$$m\frac{dv_2}{dt} = -\left(\frac{1}{2}\sigma\right)\left(v_1^2 + v_2^2\right)^{\frac{1}{2}} v_2 - mg + T\sin\theta\cos u + T\cos\theta\sin u. \tag{15.15}$$

15.2.4 Rotational Dynamics

Rotation about the centre of gravity will be considered. The moment of inertia is assumed to be J. Newton's Law gives

$$J\frac{d^2\theta}{dt^2} = \text{torque}. \tag{15.16}$$

There are two torques of relevance; one due to drag, the other due to thrust. The distance from the centre of gravity to the centre of pressure is denoted as d_{gp}, and the distance from the centre of gravity to the thrust gimble is denoted as d_{gt}. The convention adopted is that any point further toward the tip of the rocket than the centre of gravity is considered positive while any point further toward the thruster is negative.[5] The angle θ is considered positive as shown in Figure 15.2. Hence, by these definitions, a counter-clockwise rotation about the centre of gravity is positive.

Thus, the torque due to thrust is

$$t_T = d_{gt} T \sin u, \tag{15.17}$$

and the torque due to drag is

$$t_d = d_{gp}\frac{1}{2}\sigma\left(v_1\sin\theta - v_2\cos\theta\right)\left(v_1^2 + v_2^2\right)^{\frac{1}{2}}. \tag{15.18}$$

Finally,

$$\text{total torque} = t_T + t_d. \tag{15.19}$$

---

4 In practice, the mass will change as fuel is used.

5 This implies that $d_{gt} \leq 0$.

15.2.5 Composite Model

The full model has states x, y, v_1, v_2, θ, and $\frac{d\theta}{dt}$. The model equations are (15.9), (15.10), (15.14), (15.15), (15.16), (15.19), (15.18) and (15.17).

In summary, the model is as follows:

$$m\dot{v}_1 = -\frac{1}{2}\sigma v_1 \sqrt{v_1^2 + v_2^2} + T(\cos\theta\cos u - \sin\theta\sin u), \tag{15.20}$$

$$m\dot{v}_2 = -\frac{1}{2}\sigma v_2 \sqrt{v_1^2 + v_2^2} + T(\cos\theta\sin u + \sin\theta\cos u) - mg, \tag{15.21}$$

$$J\ddot{\theta} = \frac{1}{2}\sigma d_{gp}\sqrt{v_1^2 + v_2^2}\left(v_1\sin\theta - v_2\cos\theta\right) + d_{gt}T\sin u. \tag{15.22}$$

A simplified model when rocket points upwards is discussed in Revision Question 3.

15.3 Equilibrium

A symbolic solver such as those found in MATLAB® or GNU Octave can be used to calculate the equilibrium values.

For the rocket pointing straight up, $\overline{\theta} = 0.5\pi$, the answer should be:

$$\overline{v}_1 = 0.0, \quad \overline{v}_2 = 1.2402 \times 10^3, \quad \overline{\theta} = 0.5\pi, \quad \overline{u} = 0.0$$

15.4 Linearised Model

Next, the model dynamics will be linearised about an operating point leading to

$$\dot{\tilde{x}} = A\tilde{x} + B\tilde{u}, \tag{15.23}$$

where $\tilde{x} = x - \overline{x}$, $\tilde{u} = u - \overline{u}$ and where

$$\mathbf{A} = \left.\frac{\partial f}{\partial x}\right|_{\overline{x},\overline{u}}, \quad \mathbf{B} = \left.\frac{\partial f}{\partial u}\right|_{\overline{x},\overline{u}}. \tag{15.24}$$

The details for $\mathbf{A}$ and $\mathbf{B}$ are as follows:

$$\mathbf{A} = \begin{bmatrix} -\frac{\sigma\left(2\overline{v}_1^2+\overline{v}_2^2\right)}{2m\sqrt{\overline{v}_1^2+\overline{v}_2^2}} & -\frac{\sigma\overline{v}_2\overline{v}_1}{2m\sqrt{\overline{v}_1^2+\overline{v}_2^2}} & -\frac{T\sin(\overline{\theta}+\overline{u})}{m} & 0 \\ -\frac{\sigma\overline{v}_2\overline{v}_1}{2m\sqrt{\overline{v}_1^2+\overline{v}_2^2}} & -\frac{\sigma\left(\overline{v}_1^2+2\overline{v}_2^2\right)}{2m\sqrt{\overline{v}_1^2+\overline{v}_2^2}} & \frac{T\cos(\overline{\theta}+\overline{u})}{m} & 0 \\ 0 & 0 & 0 & 1 \\ \frac{\sigma d_{gp}\left(2\overline{v}_1^2\sin(\overline{\theta})-\overline{v}_1\overline{v}_2\cos(\overline{\theta})+\sin(\overline{\theta})\overline{v}_2^2\right)}{2J\sqrt{\overline{v}_1^2+\overline{v}_2^2}} & -\frac{\sigma d_{gp}\left(2\overline{v}_2^2\cos(\overline{\theta})-\overline{v}_2\overline{v}_1\sin(\overline{\theta})+\cos(\overline{\theta})\overline{v}_1^2\right)}{2J\sqrt{\overline{v}_1^2+\overline{v}_2^2}} & \frac{\sigma d_{gp}\sqrt{\overline{v}_1^2+\overline{v}_2^2}\left(\overline{v}_1\cos(\overline{\theta})+\overline{v}_2\sin(\overline{\theta})\right)}{2J} & 0 \end{bmatrix}$$

$$\mathbf{B} = \begin{bmatrix} \frac{-T\sin(\overline{\theta}+\overline{u})}{m} \\ \frac{T\cos(\overline{\theta}+\overline{u})}{m} \\ 0 \\ \frac{d_{gt}T\cos(\overline{u})}{J} \end{bmatrix}. \tag{15.25}$$

Substituting for the operating condition found in Section 15.3 into the above linearised-state equations gives the following matrices for the linearised state-space model (15.23).

$$\mathbf{A} = \begin{bmatrix} -0.0062 & 0.0 & -17.5 & 0 \\ 0.0 & -0.0124 & 0.0 & 0 \\ 0 & 0 & 0 & 1 \\ 0.0002 & 0.0 & 0.2883 & 0 \end{bmatrix}; \quad \mathbf{B} = \begin{bmatrix} -17.5 \\ 0 \\ 0 \\ -1.9688 \end{bmatrix}, \tag{15.26}$$

$$\mathbf{C} = \begin{bmatrix} 0 & 0 & 1 & 0 \end{bmatrix}. \tag{15.27}$$

The **C** vector is selected in this way since for rotational control, the rocket angle θ is the only signal of relevance.

The vector of eigenvalues of **A** is easily seen to be:

$$\boldsymbol{\lambda} = \begin{bmatrix} -0.5437 \\ 0.5296 \\ 0.0079 \\ -0.0124 \end{bmatrix}, \tag{15.28}$$

Thus, the rocket, around the equilibrium point, has two unstable poles, one at 0.5296 and one at 0.0079. The latter pole is essentially at the origin implying that the system contains a (near) integrator plus one unstable pole.

15.5 Open-Loop Flight

The nature of the dynamics depends heavily on d_{gp} (the relative location of the centre of gravity and the centre of pressure).

The eigenvalues λ for two possible values of d_{gp} are given below:

- For $d_{gp} = 5: \quad \lambda = [-0.5437, \quad 0.5296, \quad 0.0079, \quad -0.0124]$
- For $d_{gp} = -5: \lambda = [-0.0124, \quad -0.0071 + 0.5370i, \quad -0.0071 - 0.5370i, \quad 0.0079]$

As can be seen both configurations are unstable. However, there is a significant qualitative difference. For d_{gp} negative (i.e. when the centre of pressure is behind the centre of gravity), the unstable pole is almost an integrator whereas for d_{gp} positive (i.e. when the centre of pressure is above the centre of gravity) the unstable pole is in the far right-half plane, i.e. the system is significantly unstable.

To increase the stability of a rocket in open-loop one could add weight to the nose (thus moving the centre of gravity up the rocket) or increase the area of the fins (thus moving the centre of pressure down the rocket).

15.6 Controller Design for the Rocket

15.6.1 Simplified Design of PID

Section 15.4 shows that an approximate model for the rocket in vertical flight is of the form:

$$\frac{\tilde{\theta}}{\tilde{u}} = \left[\frac{-K}{s^2 - a^2}\right] = G(s). \tag{15.29}$$

This model has poles at $\pm a$. For the rocket under study $a \simeq 0.6$.

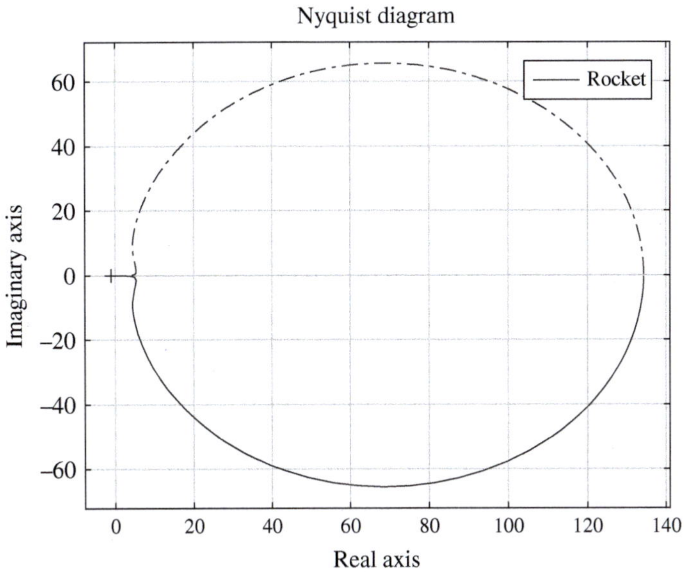

Figure 15.3 Nyquist plot.

Say that one wishes to have an integrator in the control law, then since an integrator is potentially destabilising, it is prudent to place a zero at a relatively low frequency. A second zero can be used to cancel the stable pole at 0.6. This line of reasoning leads to a control law.

$$C(s) = \frac{\overline{K}(s+0.6)(s+0.2)}{s}, \qquad (15.30)$$

$$= \overline{K}\left(s + 0.8 + \frac{0.12}{s}\right).$$

Noting that there is a negative sign in the plant model, then the approximate closed-loop polynomial is

$$P_{cl} = \left(s + K\overline{K}\right). \qquad (15.31)$$

15.6.2 Frequency Domain Design

The number of unstable poles is **2** and so, for a controller to stabilise the system, the Nyquist plot should make two counter-clockwise encirclements of the -1 point.

The Nyquist plot corresponding to the open-loop transfer function with a negative unity gain is shown in Figure 15.3. It is suggested that the reader use this plot to suggest a controller structure that will give the necessary two counter-clockwise encirclements needed for closed-loop stability.

By looking closely at the plot around the $(-1, 0)$ point, it can be seen that there are no net encirclements. However, by including a single stage of phase advance, then a Nyquist plot with two counter-clockwise encirclements can be obtained.

Finally, an integrator can be added to the controller to eliminate steady-state errors.

With the inclusion of all of the above components, the full controller takes the form

$$C(s) = -P\left(\frac{\tau_i s + 1}{\tau_i s}\right)\left(\frac{\tau_1 s + 1}{\tau_2 s + 1}\right). \qquad (15.32)$$

In other words, a controller with negative gain, a 'lead' filter and an integrator has been designed. Testing $\theta = 0.5\pi$ and $\theta = 0.3\pi$ leads to the conclusion that the following is a suitable controller:

$$C(s) = -P\frac{\tau_i s + 1}{\tau_i s}\frac{\tau_1 s + 1}{\tau_2 s + 1},$$

with,

$$P = 0.1, \tag{15.33}$$

$$\tau_i = \frac{1}{0.05 \cdot 2\pi} = 3.18, \tag{15.34}$$

$$\tau_1 = \frac{1}{0.01 \cdot 2\pi} = 15.92, \tag{15.35}$$

$$\tau_2 = \frac{1}{1 \cdot 2\pi} = 0.27. \tag{15.36}$$

The above control law defined by (15.32)–(15.36) can be written as a PIDγ control law.

$$C(s) = -\left(K_P + \frac{K_I}{s} + \frac{K_D s}{\gamma s + 1}\right), \tag{15.37}$$

where $K_P = 0.592$, $K_I = 0.0314$, $K_D = 1.432$ and $\gamma = 0.27$.

Note that the derivative term needs a filter to make it proper hence the addition of γ to the PID.

15.7 Experiment: Testing the Control Law

Only angle control of the rocket will be considered. Provision has been made for two control architectures, namely, (i) PID and (ii) a sampled-data control law (denoted APTF) which is a second-order transfer function expressed in terms of the z-transform variable z. Only the PID design will be tested. The sampled-data control law will be left to the reader to explore after reading Chapter 17 (Principles of Sampled-Data Control System Design).

The reader is encouraged to run the following tests and comment on the rocket dynamics and steady-state results.

In the Program box, load the PID control law. Initial set the reference angle as 90° (i.e. vertical motion).

15.7.1 Testing the Design Mode in Section 15.6.1

Click on the PID box and enter the parameters found in Section 15.6.1. (Use the gain $\overline{K} = 1$.)

(a) Use a sinusoidal wind disturbance of peak amplitude 0.1 and frequency 1 Hz. Measure the angle response.
(b) Increase $\overline{K}$ to 2 and repeat part (a).
(c) Increase $\overline{K}$ to 4 and repeat part (a).
(d) Try $\overline{K} = 0.5$.
(e) Try $\overline{K} = 0.1$.

Note that the initial conditions for the rocket may play an important role.

15.7.2 Testing the Design Made in Section 15.6.2

(a) Use the parameters found in Section 15.6.1 with $\overline{K} = 1$.
With the sinusoidal wind disturbance used in Section 15.7.1 (a).
(b) Try a sequence of sudden wind gusts for the wind. (Use a square wave for the wind Peak Amplitude 0.5, Frequency 0.1 Hz, Duty Cycle 10%.)

15.8 Revision Questions

1. Evaluate the poles and zeros of the system described in (15.26) and (15.27).
2. This question studies the dynamics of an inverted pendulum in two-dimensions.
   Consider the inverted pendulum shown in Chapter 3 Figure 3.4. Derive the equations of motion.
3. Consider the part of the model describing the angle dynamics given in (15.22). Assume that the rocket is rising at an angle $\theta = 0.3\pi$. Linearise the model about this operating point.

Further Reading

1 J.D. Anderson, *Introduction to flight*, McGraw Hill, 2004.
2 A.B. Kisabo, A.F. Adebimpe, O.C. Okwo and S.O. Samuel, "State-space modelling of a rocket for optimal control system design", *Journal of Aircraft and Spacecraft Technology*, 3:128–137, 2019.
3 M.J. Sidi, *Spacecraft dynamics and control*, Cambridge University Press, 1997.

16

Bode Sensitivity Trade-Offs

16.1 Introduction

It has been seen in the laboratories presented so far that one can shape the sensitivity and complementary sensitivity functions by the use of carefully designed feedback.

However, there are limitations associated with what can, and cannot, be achieved. The reader has already seen a first instance of these limitations in Chapter 2 where it was shown that, for a non-dynamical system, the sum of sensitivity and complementary sensitivity was one. Similarly, it was shown in Section 6.3 of Chapter 6 that the sum of sensitivity and complementary sensitivity was 1 at each frequency. In the current chapter, several other constraints will be explored. Namely, when all frequencies are considered simultaneously, then there are constraints on the trade-offs between different frequencies.

The key conclusion presented here is that one can think of disturbances as a form of 'dirt' in the frequency domain.[1] A linear control law can shift the 'dirt' around. (e.g. from low frequencies to high frequencies) but the 'dirt' remains somewhere.

This chapter will give a brief overview of these ideas. Most of the results will be presented without proof since to give the proofs would be too distracting for a book of this level. The interested reader can find full proofs in the literature quoted at the end of the chapter.

16.2 System Properties

Before describing the results, it is helpful to recall several properties of transfer functions.

Stable:	No poles in right half plane (RHP)
Minimum phase:	No zeros in RHP
Relative degree:	The difference between the numerator and denominator degree

Loop transfer function:

$$L(s) = G_0(s)C(s), \tag{16.1}$$

where $G_0(s)$ is the plant transfer function and $C(s)$ is the controller transfer function.

1 This terminology is due to Gunter Stein who used it in his Bode lecture in 1989 – see reference [1] given in Further Reading.

Practical Control System Design: Real World Designs implemented on Emulated Industrial Systems,
First Edition. Adrian Medioli and Graham Goodwin.

Companion website: www.wiley.com/go/medioli/practicalcontrolsystemdesign

Loop constant:

$$K \triangleq \lim_{s\to\infty} sL(s). \tag{16.2}$$

(Note, $K = 0$ if the Loop Transfer Function has relative degree greater than 1.)

16.3 Bode Integral Constraints

16.3.1 Open-Loop Stable Systems

To explain the Bode integral, consider the simple continuous-time feedback circuit shown in Figure 16.1. The transfer function $L(s) = G_o(s)C(s)$ is the loop-transfer function.

The transfer function from d to y is the 'Sensitivity Function' and is given by:

$$S_o(s) = \frac{1}{1 + L(s)}. \tag{16.3}$$

If one thinks of d as a disturbance source, then it would seem to be highly desirable to make $|S_o(j\omega)|$ small at *all* frequencies. However, a key requirement is that $S_o(s)$ remains stable, i.e. have all poles in the open left half plane. Surprisingly, the requirement of stability means that not all functions $S_o(s)$ are achievable. Indeed, assuming that $L(s)$ is open-loop stable with relative degree greater than 0, then analytic function theory can be used to show that $|S_o(j\omega)|$ must satisfy an integral of the form:

$$\int_0^\infty \ln |S_o(j\omega)| d\omega = -K\frac{\pi}{2}, \tag{16.4}$$

where K is as in (16.2).

When $L(s)$ has a relative degree greater than or equal to 1, then

$$\int_0^\infty \ln |S_o(j\omega)| d\omega = 0. \tag{16.5}$$

The interpretation of Eq. (16.5) is as follows: since $\ln(1) = 0$, then Eq. (16.5) says that $\ln |S(j\omega)|$ must be 'as much above 1 as it is below 1'. This is sometimes called the 'Water Bed Effect', i.e. if one 'pushes the sensitivity down' at some frequencies then it necessarily 'pops up' at other frequencies such that (16.5) is always satisfied. This is illustrated in Figure 16.2.

16.4 Examples of Bode Sensitivity Trade-Offs

(i) Consider

$$L(s) = G_o(s)C(s) = \frac{9}{s(s+4)}. \tag{16.6}$$

Then Eq. (16.5) predicts

$$\int_0^\infty \ln |S_o(j\omega)| d\omega = 0. \tag{16.7}$$

(ii) Consider

$$L(s) = G_oC(s) = \frac{17s + 100}{s(s+4)}. \tag{16.8}$$

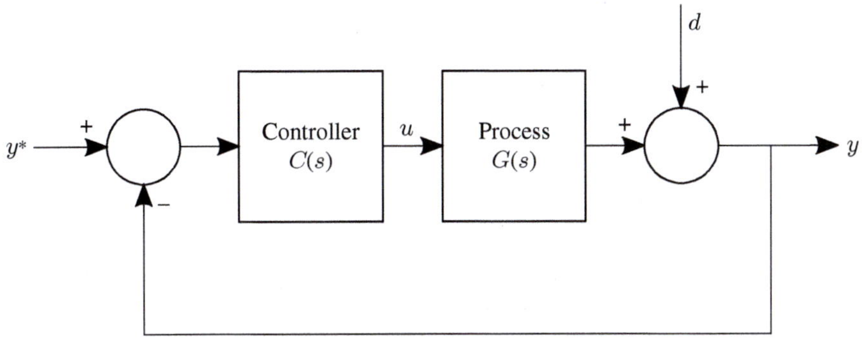

Figure 16.1 Simple continuous-time feedback circuit.

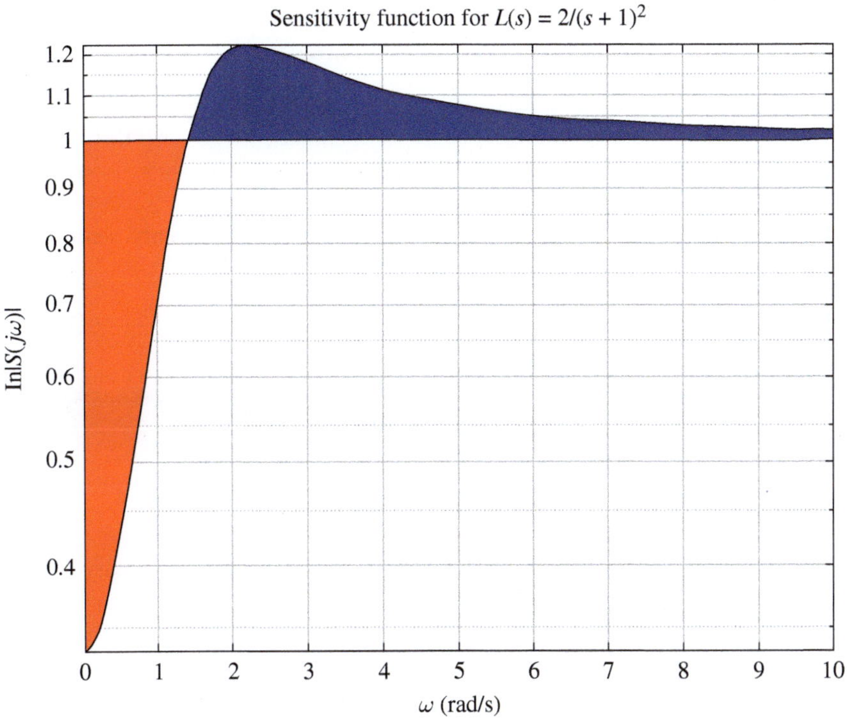

Figure 16.2 Graphical interpretation of the Bode integral.

Then

$$S_o(s) = \frac{s(s+4)}{s^2 + 21s + 100}. \tag{16.9}$$

Note that $K = \lim_{s\to 0} sG_o(s)C(s) = 17$.
Hence, (16.11) gives

$$\int_0^{\infty} \ln |S_o(j\omega)| d\omega = -8.5\pi. \tag{16.10}$$

16.4.1 Open-Loop Unstable Systems

If the loop transfer function $L(s)$ is open-loop *unstable*, then the result in (16.4) becomes

$$\int_0^\infty \ln |S_o(j\omega)| d\omega = \pi \sum_{i=1}^{Np} \Re(p_i) - K\frac{\pi}{2} \tag{16.11}$$

where $\{p_i, i = 1, \dots, N_p\}$ denote the set of open-loop unstable poles of $L(s)$ and $\Re(\cdot)$ is a function that extracts the real part of a number.

Note that (16.11) implies that the sensitivity trade-off is harder for open-loop unstable systems than it is for open-loop stable systems since the right-hand side of (16.11) is positive and hence $\ln |S_o(j\omega)|$ must be above 1 'more often' than it is below. The reader will recall from Chapters 14 to 15 that it is indeed true that open-loop unstable systems present more difficult control system design challenges than open-loop stable systems. (See also reference [1] given in Further Reading.) Indeed, a useful rule of thumb is that the closed-loop bandwidth should be greater than any unstable pole.

16.5 Bode Complementary Sensitivity Integrals

16.5.1 Minimum Phase Plants

For a minimum phase plant, the following integral holds for the complementary sensitivity function:

$$\int_0^\infty \frac{1}{\omega^2} \ln |T_o(j\omega)| d\omega = \frac{-\pi}{2k_v}, \tag{16.12}$$

where $\frac{1}{k_v} = -\lim_{s\to 0} \left[\frac{1}{sL(s)}\right]$. The reader will be asked in Question 4 of the Revision Questions section to derive (16.12) from (16.4) by considering a pseudo sensitivity function.

16.5.2 Non-minimum Phase Plants

If the system has RHP zeros (i.e. it is non-minimum phase) then the Bode Complementary Sensitivity Integral becomes

$$\int_0^\infty \frac{1}{\omega^2} \ln |T_o(j\omega)| d\omega = \frac{-\pi}{2k_v} + \pi \sum_{i=1}^{M} \frac{1}{z_i}, \tag{16.13}$$

where $\{z_i,\ i = 1, \dots, M\}$ denotes the set of RHP zeros.

Obviously from (16.13), small RHP zeros make the trade-off more difficult. Indeed, a useful rule of thumb is that the closed-loop bandwidth needs to be less than the location of the smallest RHP zero.

16.6 Bode Sensitivity for Time-Delay Systems

The reader is asked to revisit Section 12.8. There it was shown that time delays give additional phase shift which can lead to difficulties in assigning desirable sensitivity properties. These results can also be further embellished from the perspective of Bode integral constraints.

When the loop-transfer function has a pure time delay, τ, then the Bode integral for complementary sensitivity becomes

$$\int_0^\infty \frac{1}{\omega^2}|T_o(j\omega)|d\omega = \pi\tau - \frac{\pi}{2k_v} + \pi\sum_{i=1}^{M}\frac{1}{z_i}, \tag{16.14}$$

where $\{z_i,\ i = 1, \dots, M\}$ are the set of RHP zeros.

It thus follows that not only do RHP zeros make the complementary sensitivity more difficult to reduce but time delays also make sensitivity allocation difficult.

16.7 Revision Questions

1. Consider the following system:

$$G(s) = \frac{s-10}{(s-1)(s+1)}. \tag{16.15}$$

Assume that the control law is stable and minimum phase and of relative degree 1.
What is

$$\int_0^\infty \ln|S(j\omega)|d\omega, \tag{16.16}$$

where S is the closed-loop sensitivity function?

2. Consider the same system as in Question 1. What is

$$\int_0^\infty \frac{1}{\omega^2}\ln|T(j\omega)|d\omega. \tag{16.17}$$

3. How do the Bode integrals change when the sensitivities are expressed on a dB scale, where

$$|S| \text{ in dB} = 20\log_{10}|S|? \tag{16.18}$$

4. This question derives (16.12) from (16.4). Consider a loop transfer function $L(s)$. Define

$$T(s) = \frac{L(s)}{1+Ls}. \tag{16.19}$$

Let $\overline{S}(s)$ be a 'pseudo sensitivity function' defined by

$$\overline{S}(s) = T\left(\frac{1}{s}\right). \tag{16.20}$$

   (a) What is the Bode Sensitivity Integral for $\overline{S}(s)$?
   (b) What are the zeros of $\overline{S}(s)$ as a function of the zeros of $T(s)$?
   (c) Use the above connection to derive the Bode Integral for $T(s)$ using the Bode Integral for $\overline{S}(s)$.

5. Directly compute $\int_0^\infty \ln|S_o(j\omega)|d\omega$ for the system given in Eq. (16.6).
6. Repeat Question 5 for the system given in Eq. (16.8).

Further Reading

Bode's integral is covered in many books. The original result appears in

1 G. Stein, "Respect the unstable", *IEEE Control Systems Magazine*, 23(4):12–25, 2003.

Bode's integral is covered in many books. The original result appears in

2 H. Bode, *Newton's analysis and feedback amplifier design*, D. van Nostrand, New York, 1945.

Proofs of the Bode integral can be found in

3 G.C. Goodwin, S.F. Graebe and M.E. Salgado, *Control system design*, Prentice Hall, 2000.
4 M.M. Seron, J.H. Braslavsky and G.C. Goodwin, *Fundamental limitation in filtering and control*, Springer-Verlag, London, 1997.

Part IV

Sampled Data Control Systems

The material presented so far in the book has assumed that the control system will be implemented in continuous time. This assumption was valid in the past when control laws were implemented via hydraulic, pneumatic or analogue electronics. Whilst such system still exist, it is much more common these days to have controllers implemented in digital form using either an embedded digital controller, programmable logic controller or computer control. With this as background, Chapter 17 describes control system design when the control law is implemented in digital form. Chapter 18 presents a real-world example namely the design of a noise shaping quantiser as used in audio-signal processing.

Practical Control System Design: Real World Designs implemented on Emulated Industrial Systems,
First Edition. Adrian Medioli and Graham Goodwin.

Companion website: www.wiley.com/go/medioli/practicalcontrolsystemdesign

17

Principles of Sampled-Data Control System Design

17.1 Introduction

The chapter describes specific tools that apply to digital (sampled data) control systems. Digital (or sampled data) control laws take in sampled values and output sample values. The output samples, from the controller, are converted into a form where they can be applied to drive the input of the system by using a digital to analogue converter. Thus, two key elements of a sampled-data control system, are (i) the sampler (on the output of the system) called an A/D converter, and (ii) the device that converts sampled control signals (emanating from the controller) into an analogue form so they can be applied to the input of a process (called a D/A converter).

17.2 A/D Conversion

On the input side of the controller, samples are taken by an A/D converter. To avoid the impact of high-frequency noise on the sampled output, it is usual to use some form of low-pass filter prior to the sampler. This is often termed an anti-aliasing filter. Various options exist, but a simple option is to average the analogue signal over the sample period. (Alternatively, one can use a simple low-pass filter.) In the sequel, continuous time is denoted $t = k\Delta$ where Δ is the sample period. To simplify notation, a subscript can be used to denote sampled values, i.e.

$$y_k \triangleq y(k\Delta). \tag{17.1}$$

As explained above, a simple anti-aliasing filter integrates the analogue signal over the sample point and then divides it by the period. This can be represented in transfer function form as shown in Figure 17.1.

When the sample period is small compared to the rate of change of $y(t)$ the anti-aliasing filter can be ignored. The anti-aliasing filter should, however, not be ignored when the closed-loop response time approaches the sampling period.

17.3 Sampled Output Noise

One case where it is necessary to explicitly account for the anti-aliasing filter is when one wants to treat the impact of continuous-time noise. For example, if $y(t)$ is measured in the presence of wide-band noise, of spectral density R, then this can be approximated as a signal having constant spectral density, R, over a wide (theoretically infinite) bandwidth. This looks very simple.

Practical Control System Design: Real World Designs implemented on Emulated Industrial Systems,
First Edition. Adrian Medioli and Graham Goodwin.

Companion website: www.wiley.com/go/medioli/practicalcontrolsystemdesign

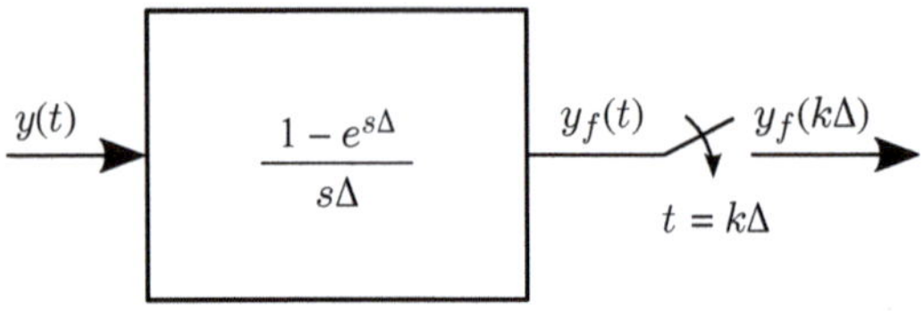

Figure 17.1 A/D converter representation including anti-aliasing filter.

However, often variances are used to describe wide-band *sampled-data* noise. The variance contains a surprising result.[1] Specifically, the sampled signal will have variance given by the integral of the spectral density over the Nyquist bandwidth[2] from $-\frac{1}{\Delta}$ to $\frac{1}{\Delta}$. If an anti-aliasing filter were not used, then the full noise spectrum would be folded into the range $\left(-\frac{1}{\Delta}, \frac{1}{\Delta}\right)$. In this case, the variance of the sampled noise would be very large (theoretically infinite). This is clearly unsatisfactory. However, if an anti-aliasing filter is used, then the impact of folding is significantly reduced. The sampled-data noise will be R as for the underlying continuous-time noise. However, the sampled noise variance will be approximately $\frac{2R}{\Delta}$. It can thus be seen that the sampled-data variance will be inversely proportional to the sample period provided an anti-aliasing filter is used. Moreover, the discrete spectral density is R, as for the continuous case, but is now defined on the discrete frequency interval $\left(-\frac{1}{\Delta}, \frac{1}{\Delta}\right)$.

17.4 D/A Conversion

The connection between the controller output and the plant input is termed a D/A converter.

Say one denotes the sampled output from the controller as u_k at time $t = k\Delta$. Obviously, a single sample has no energy and hence it cannot 'drive' a system.

Hence the sampled signal u_k is typically converted into a continuous-time signal via a hold device. This is illustrated in Figure 17.2 for a, so-called, 'zero-order hold'.

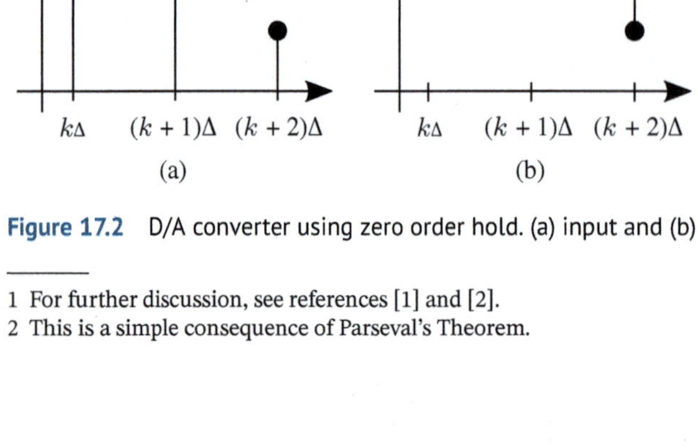

Figure 17.2 D/A converter using zero order hold. (a) input and (b) output.

1 For further discussion, see references [1] and [2].

2 This is a simple consequence of Parseval's Theorem.

17.5 Sampled-Data Models

The next topic to be explored is the representation of systems in the sampled-data domain. This is most conveniently explained using state-space ideas.

Thus, assume that the plant is modelled by continuous-time state-space model of the form:

$$\dot{x} = Ax + Bu. \tag{17.2}$$

The associated sampled-data model links the sampled output to the sampled input. The key idea used in developing a sampled-data model is to solve the continuous-time model over the sample period Δ. Assuming that the input is implemented via a zero-order hold, then the solution to (17.2) over the interval $k\Delta$ to $(k+1)\Delta$ is,

$$x((k+1)\Delta) = e^{A\Delta}x(k\Delta) + \int_0^{\Delta} e^{A(t-\tau)}Bu_k\, d\tau. \tag{17.3}$$

The above equation can be written more compactly as a discrete-time model of the form:

$$x_{k+1} = A'x_k + B'u_k, \tag{17.4}$$

where x_k is the sampled value of $x(t)$, $t = k\Delta$, and u_k is the sampled input at the kth sample. Comparing (17.4) and (17.3) shows that

$$A' = e^{A\Delta}, \tag{17.5}$$

$$B' = \int_0^{\Delta} e^{A(\Delta-\tau)}B\, d\tau. \tag{17.6}$$

17.6 Shift Operator Models

It is common in the literature (see [3] and [5]) relating to the analysis of sampled-data systems to utilise a 'forward shift operator', q, having the property that

$$qx_k = x_{k+1}. \tag{17.7}$$

Using the above notation, the model (17.4) can be expressed more compactly as

$$qx = A'x + B'u. \tag{17.8}$$

17.7 Divided Difference Models

Shift operator models are very widely used in the design and analysis of sampled-data systems. However, they have several associated difficulties. These difficulties are most pronounced when the sample period, Δ, is small compared to the dynamics of the system. Thus, whilst the model (17.8) looks reasonable at first glance, it should be noted that $A' \to I$ for all systems as $\Delta \to 0$ and $B' \to 0$ for all systems as $\Delta \to 0$. Thus, the connection between the sample data model (17.8) and the continuous-time model (17.2) is lost. Another issue with shift operator models is that the discrete-time output noise (introduced in Section 17.3) has the property that the variance, i.e. $\frac{2R}{\Delta}$, tends to ∞ as $\Delta \to 0$. A way around the above difficulties is as follows: Instead of using the model (17.8), one can replace the shift operator, q, by a 'divided difference operator', δ, defined[3] as follows:

$$\delta = \frac{q-1}{\Delta}. \tag{17.9}$$

Utilising this operator in (17.8) leads to an alternative form for the sampled-data model, namely,

$$\delta x = \left[\frac{q-1}{\Delta}\right] x = \frac{[A' - I]}{\Delta} x + \frac{B'}{\Delta} u. \tag{17.10}$$

This leads to an alternative (exact) discrete model:

$$\delta x = A'' x + B'' u, \tag{17.11}$$

where

$$A'' = \frac{A' - I}{\Delta} = \frac{e^{A\Delta} - I}{\Delta}, \tag{17.12}$$

$$B'' = \frac{1}{\Delta} \int_0^{\Delta} e^{A(t-\tau)} B \, d\tau. \tag{17.13}$$

It is important to note that the model (17.11)–(17.13) is exact. A useful approximation to both the shift and delta-domain models (17.8) and (17.11) will be described in the sequel.

17.8 Euler Approximate Model

In many practical problems, the sample period, Δ, is small compared to the time constant of the system. In this case, one can make the following useful approximations for the shift-operator model.

$$A' = e^{A\Delta} \simeq I + A\Delta = \overline{A}', \tag{17.14}$$

$$B' = \int_0^{\Delta} B \, d\tau \simeq B\Delta = \overline{B}'. \tag{17.15}$$

This approximation is commonly termed an 'Euler approximation'. Using this approximation, the discrete shift-operator model can be rewritten in *approximate* form as

$$qx \simeq (I + A\Delta)x + B\Delta u_k. \tag{17.16}$$

17.9 Euler Approximate Model in Delta Domain

The approximate model (17.16) is very useful and intuitive. It becomes even more intuitive when using the divided difference operator. Specifically, it is easily seen from (17.11)–(17.14) and (17.16) that the corresponding Euler model in the delta domain is

$$\delta x \simeq Ax + Bx. \tag{17.17}$$

In other words, provided one works with divided differences, then the discrete-time dynamics are *approximately* described by the same A and B matrices as in continuous time.[4] This is not surprising since the delta operator acts like a derivative, i.e.

$$\delta x = \frac{x_{k+1} - x_k}{\Delta} = \frac{x((k+1\Delta) - x(k\Delta))}{\Delta} \simeq \frac{dx}{dt}. \tag{17.18}$$

3 Also known as the Delta operator – see reference [2].

4 Indeed, in many practical problems met in the industry it suffices to do the control system design in continuous time and then to implement the control law via the Euler approximate model using the Delta operator.

17.10 Delta Analysis

The ideas described in Sections 17.7 and 17.9 can be made rigorous using delta operators. A full exploration of this idea lies outside the scope of this introductory text. Suffice it to say here that, for sufficiently small Δ, *all known discrete-time results* can be interpreted as *continuous results* in the divided difference domain see [1, 2] and [4].

17.11 Historical Notes

For historical reasons, sampled-data results have traditionally been expressed in terms of the forward shift operator q. The corresponding transformation is called the z-transform. Indeed, most of the available text books use the shift operator and z-transforms when describing sampled-data results. An unfortunate consequence of this choice is that it highlights the *difference* between the sampled data and continuous cases rather than making it clear that, for small Δ, all sampled-data results converge to the continuous cases. Indeed, this convergence is so startling that, in many cases of practical interest, it suffices to use the continuous result.

The link between sampled and continuous time is clear when one uses the delta operator, i.e.

$$\delta x = \frac{x((k+1)\Delta) - x(k\Delta)}{\Delta} = \frac{(q-1)x}{\Delta}. \tag{17.19}$$

The corresponding transfer function variable is called the Delta transform, $\overline{\delta}$. The relationship between the operators and transforms is summarised in Table 17.1.

The connection between sampled-data and continuous-time control is explained in detail in Ref. [1, 2] and [4].

17.12 An Example of Shift and Delta Models

Say that one wishes to design a first-order low-pass filter with unity gain and time constant of 10 seconds using a sample period of $\Delta = 0.1$ seconds.

The corresponding continuous-time filter is described by

$$\frac{d}{dt}x = -\frac{1}{10}x + \frac{1}{10}u = ax + bu. \tag{17.20}$$

Say that one wishes to implement a sampled-data version of this filter. If one uses a zero-order hold for the input, then the *exact* sampled-data model using shift operators is given by (17.5) and (17.6), i.e.

$$x_{k+1} = e^{-\frac{\Delta}{10}}x_k + \frac{1}{\Delta}\left(1 - e^{-\frac{\Delta}{10}}\right)u_k \tag{17.21}$$

Table 17.1 Operator and transform relationships.

	Symbol	Delta	Reverse relationship
Shift operator	q	$\delta = \frac{q-1}{\Delta}$	$q = \Delta\delta + 1$
Transfer function variable	z	$\overline{\delta} = \frac{z-1}{\Delta}$	$z = \Delta\delta + 1$

or

$$qx_k = a'x_k + b'u_k. \tag{17.22}$$

The corresponding transfer function in terms of the z variable is

$$X(z) = \frac{b'}{z - a'}U(z). \tag{17.23}$$

Note that b', a' are very different to the continuous values a, b. Indeed

$$a' = e^{-\frac{\Delta}{10}} \to 1 \qquad \text{as} \qquad \Delta \to 0, \tag{17.24}$$

$$b' = \frac{1}{b}\left(1 - e^{-\frac{\Delta}{10}}\right) \to 0 \qquad \text{as} \qquad \Delta \to 0. \tag{17.25}$$

The results in (17.24) and (17.25) are exactly as predicted in Section 17.7.

Next, say that the model is converted to the delta domain. From Section 17.11 this can be achieved via the substitution $q = \Delta\delta + 1$. Then (17.21) can equivalently be written as follows:

$$\begin{aligned} \delta x &= \frac{x_{k+1} - x_k}{\Delta}, \\ &= \left[\frac{a' - 1}{\Delta}\right] x_k + \left[\frac{b'}{\Delta}\right] u_k, \\ &= a''x_k + b''u_k. \end{aligned} \tag{17.26}$$

We immediately see that

$$a'' = \frac{a' - 1}{\Delta} = \left(\frac{e^{-\frac{\Delta}{10}} - 1}{\Delta}\right) \to -\frac{1}{10} \qquad \text{as} \qquad \Delta \to 0, \tag{17.27}$$

$$b'' = \left(\frac{1 - e^{-\frac{\Delta}{10}}}{\Delta}\right)\left(\frac{1}{10}\right) \to \frac{1}{10} \qquad \text{as} \qquad \Delta \to 0. \tag{17.28}$$

Thus as $\Delta \to 0$, the delta-domain model converges to the continuous-time model. Again this is exactly as foreshadowed in Section 17.11.

17.13 Sampled-Data Stability

Given a discrete model (in the shift domain), stability is associated with the eigenvalues of the sampled data 'A matrix' having magnitude less than 1. Hence, the shift operator stability domain is as shown in Figure 17.3.

The reader will note that the above result is very different to the continuous case where the stability domain is the left-half plane. Hence, very different tests for stability are needed.

Next, consider the transformation to the delta domain using $\delta' = \frac{z-1}{\Delta}$. This can be interpreted as: shift the z-domain left by 1 and scale by $\frac{1}{\Delta}$. This transforms the stability domain to the one shown in Figure 17.4.

It now becomes clear that as $\Delta \to 0$, the sampled-data stability domain converges to the left-half plane, i.e. to the continuous-time stability domain.

17.14 Bode Sensitivity Integrals (Sampled Data Case)

Bode sensitivity integrals were presented in Chapter 16 for the case of continuous-time systems. Here, the corresponding results for the sampled-data case as quoted.

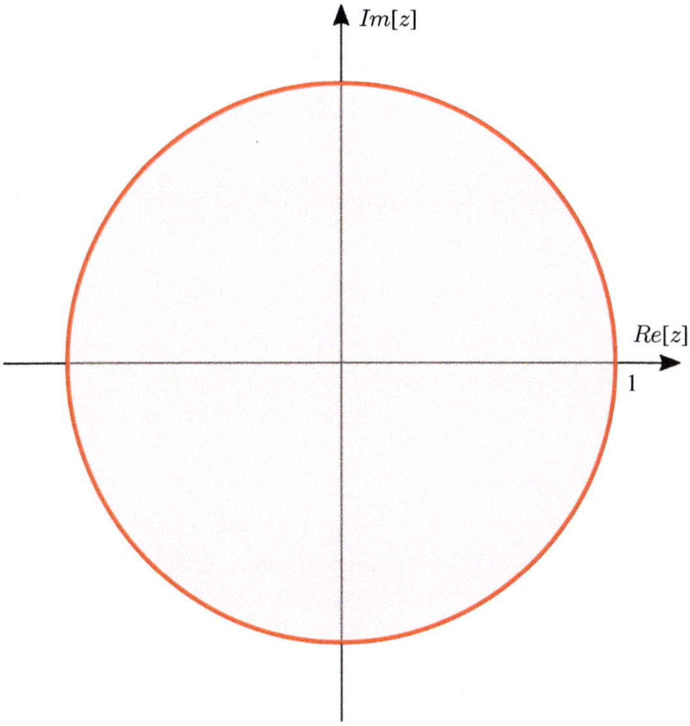

Figure 17.3 Stability domain for shift operator models.

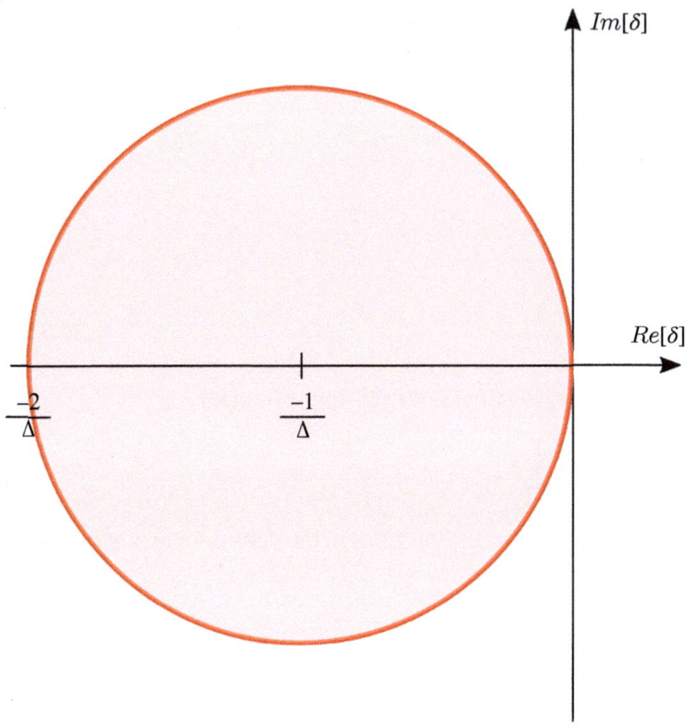

Figure 17.4 Stability domain for delta operator models.

The key difference, between the Bode sensitivity integrals for the discrete and continuous cases are the range of frequencies and the definitions of unstable open-loop poles and non-minimum phase open-loop zeros.

As an illustration we quote below the Bode sensitivity integral S expressed in both the Z-domain $as\ S_z$ and δ-domain as S_δ.

17.14.1 *Z*-Domain

Let the discrete-loop transfer function $L(z)$ has q open-loop unstable poles (outside the unit circle) at $p_1^z \dots p_q^z$ then

$$\int_0^\pi \ln\left|S_z(e^{j\omega})\right| d\omega_z = \pi \sum_{i=1}^{q} \ln\left|p_i^z\right|. \tag{17.29}$$

17.14.2 Delta Domain

The delta domain equivalent of the result in Section 17.14.1 is easily obtained by transforming (17.29) to the delta domain using $\delta = \frac{z-1}{\Delta}$. This can be done in several steps as explained below:

(i) Say that the delta-domain model has unstable poles at p_i^δ. Then, the equivalent Z-domain poles will be at

$$p_i^z = \Delta p_i^\delta + 1. \tag{17.30}$$

Hence, the right-hand side of (17.29) becomes

$$\text{RHS} = \pi \sum_{i=1}^{q} \ln\left|\Delta p_i^\delta + 1\right|. \tag{17.31}$$

(ii) The sensitivity function is evaluated such that

$$S_\delta\left(\frac{e^{j\omega'\Delta} - 1}{\Delta}\right) = S_z\left(e^{j\omega}\right); \quad \omega' = \frac{\omega}{\Delta}. \tag{17.32}$$

(iii) Hence, the left-hand side of (17.29) is

$$\Delta \int_0^{\frac{\pi}{\Delta}} \ln\left|S_\delta\left(\frac{e^{j\omega'\Delta} - 1}{\Delta}\right)\right| d\omega'. \tag{17.33}$$

(iv) Putting the above elements together gives the final Bode sensitivity integral:

$$\Delta \int_0^{\frac{\pi}{\Delta}} \ln\left|S_\delta\left(\frac{e^{j\omega'\Delta} - 1}{\Delta}\right)\right| d\omega' = \pi \sum_{i=1}^{q} \ln\left|\Delta p_i^\delta + 1\right|. \tag{17.34}$$

It is interesting to consider the case where all poles and zeros in the delta domain are small in comparison to $\frac{1}{\Delta}$. In this case,

$$\ln\left|\Delta p_i^\Delta + 1\right| \simeq \Delta p_i^\Delta, \tag{17.35}$$

$$\left[\frac{e^{j\omega'\Delta} - 1}{\Delta}\right] \simeq \omega' \text{ for } \omega' < \frac{\pi}{\Delta}. \tag{17.36}$$

Hence, provided $S_\delta\left(\frac{e^{j\omega'\Delta}-1}{\Delta}\right)$ is near one in the vicinity of $\omega' = \frac{\pi}{\Delta}$, then (17.34) is approximately

$$\int_0^\infty \ln\left|S_\delta\left(j\omega'\right)\right| d\omega' \simeq \pi \sum_{i=1}^{q} p_i^\delta. \tag{17.37}$$

This can be seen to be identical to the continuous case given in Chapter 16. Thus, the delta-domain result approaches the continuous-time case for Δ sufficiently small.

17.15 Sampling Zeros

One surprising fact about discrete models is that they are almost always of relative degree 1. This can be thought of as a consequence of the observation that discrete systems always show some response at the first sample, whereas if a continuous system has relative degree r, then the first $(r-1)$ derivations are zero when a step is applied.

The reader may wonder where the additional zeros come from. To better understand these zeros, consider an all-pole continuous system of the form:

$$G(s) = \prod_{i=1}^{n} \frac{1}{(s+p_i)}. \tag{17.38}$$

Then the corresponding delta-domain sample-data model turns out to be

$$G_o(\delta) = \prod_{i=1}^{n} \frac{B_{delta}(\delta)}{(\delta + p_i^\delta)}. \tag{17.39}$$

The extra polynomial $B_{delta}(\delta)$ is called the sampling-zero polynomial. Its properties have been extensively studied.

In particular, in the z-domain the zeros of the equivalent polynomial $B_z(z)$ are all located on the negative real axis to the left of -1., i.e. they are non-minimum phase. When transformed to the δ-domain, the roots lie to the left of $\frac{-2}{\Delta}$ and have magnitudes that go to ∞ as $\Delta \to 0$.

Thus as $\Delta \to 0$ the sampling zeros tend to $-\infty$ where they appear as the continuous-time relative degree.

The reader may wonder if one needs to worry about these rather odd sampling zeros. The answer is that it depends on the range of frequencies of interest.

For frequencies near $\frac{\pi}{\Delta}$, then the sampling zeros are certainly important.

For frequencies much less than $\frac{\pi}{\Delta}$, then the sampling zeros can be ignored.

In most cases, the sample period Δ will be chosen so that the target closed-loop bandwidth is much less than $\frac{\pi}{\Delta}$. In this case, one can (and indeed should) ignore the sampling zeros in sampled-data control system design. See also the comments made in Section 17.11.

17.16 Revision Questions

1. Consider the following continuous-time state-space model:

$$\dot{x} = \begin{bmatrix} 0 & +\omega_0 \\ -\omega_0 & 0 \end{bmatrix} x + \begin{bmatrix} 0 \\ 1 \end{bmatrix} u, \tag{17.40}$$

$$y = \begin{bmatrix} 1 & 0 \end{bmatrix} x. \tag{17.41}$$

What is the transfer function from u to y?

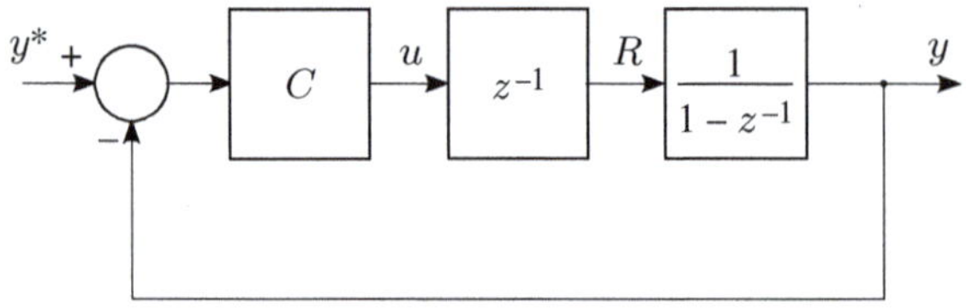

Figure 17.5 Question 10 diagram.

2. (a) What are the zeros and poles of the system in Question 1?
   (b) What is the relative degree?
3. Assuming a zero-hold input, what is the exact sampled data equivalent model for the system in Question 1?
4. What is the sampled-data transfer function of the system found in Question 3?
5. (a) What are the poles and zeros of the sampled-data system found in Question 4?
   (b) What is the relative degree?
6. What is the explanation for the extra zero found in the sampled-data model?
7. What is the exact delta-operator model?
8. What is the form of the exact delta-domain model when $\Delta \to 0$?
9. What is the limit, as $\Delta \to 0$, of the shift-operator model?
10. (This question is an actual problem discussed with Ericsson AB in relation to Power Control in 3G mobile telecommunications). Consider a simple feedback loop, as shown in Figure 17.5. The controller sends ± 1 (dB) correction from a Base Station to a mobile unit. Note there is one unit delay in the loop.

    Say that C, the controller in the base station, sends -1 if $y > y^*$ and $+1$ if $y \leq y^*$.

    Note that the mobile unit integrates the incoming communication, i.e. adds the received increment (in dB) to the transmitted power.

    Say that the system begins with $y = y^*$, $R = 1$. Calculate the commands u sent from the base station and the mobile response. (Do this by doing a simulation.)

    Hence show that the output power oscillates from $+2$ dB to -1 dB.
11. Consider the system described in question 17.10
    (a) What is the input sensitivity at $\omega = \pi$?
    (b) Link your result to the simulated results in Question 10.
12. Consider the following simple second-order model (a double integrator):

$$\ddot{y} = u. \tag{17.42}$$

    Assume that the input is implemented via a zero-order hold.
    (a) What is the exact shift-operator model?
    (b) What are the poles and zeros of the sampled-data model?
13. Repeat Question 12 for the delta-domain model.

Further Reading

1 G.C. Goodwin, J.C. Aguero, M.E. Garridos, M.E. Selgado and J.I. Yus, "Sampling and Sampled data models: The interface between the continuous world and digital algorithms", *IEEE Control Systems Magazine*, 33(5):34–53, 2013.

2 R.H. Middleton and G.C. Goodwin, *Digital control and estimation: a unified approach*, Prentice Hall, 1990.

3 G.F. Franklin, J.D. Powell and A. Ernamie-Naeini, *Digital control of dynamical systems*, 3rd Edition, Ellis-Kangle Press, 1988.
4 G.C. Goodwin, M.G. Cea and A. Feuer, "Reproachment between discrete and continuous nonlinear filtering", *Proceedings of the 2010 American Control Conference*, 2010.
5 P. Katz, *Digital control using microprocessors*, Prentice Hall, Englwood Cliffs, New Jersey, 1987.

18

Laboratory 8: Audio Signal Processing and Optimal Noise Shaping Quantisers

18.1 Introduction

This chapter bridges two areas, namely, audio signal processing and sampled-data feedback control. Readers need not have any prior knowledge of audio signal processing since the necessary ideas are developed as the laboratory evolves. The laboratory uses only relatively simple ideas in sampled data feedback control. However, the reader is encouraged to reread Chapter 17 before proceeding.

The authors' belief is that this laboratory is interesting since it demonstrates that relatively simple control ideas play a central role in modern high-technology equipment, e.g. compression of audio signals. The laboratory illustrates the 'Magic of Feedback' in a simple and easily understood setting. The users of this laboratory will actually 'hear' the difference that feedback makes. Also, the concept of 'Sensitivity Dirt' introduced in Chapter 16 will be illustrated via the perception of sounds.

18.2 The Physical Apparatus

In many applications continuous-time and continuous-valued signals have to be compressed by converting into sampled and quantised signals whose values are given only at specific times (i.e. at the sample times) and whose amplitudes belong to a finite set (i.e. at the available quantised levels). Both procedures form essential components of analogue to digital converters.

A related issue is that of re-quantisation which corresponds to lowering the resolution of a finely quantised signal in order to provide a coarser quantised signal. This can often be dealt with as if it were a quantisation problem.

Quantisation and sampling arise in many situations. A quintessential example occurs when data needs to be transmitted, or stored, with a small number of bits. An example occurred in the, now obsolete, mastering process of compact discs (see Figure 18.1). An equivalent problem arises when compressing audio files of any nature. In the CD mastering application, each sample was typically quantised to 16 bits, based upon master recordings of higher resolution (e.g. 24 bit) or of an analog nature. Although CDs are no longer produced the idea of audio quantisation provides a powerful way to demonstrate feedback trade-off principles.

Another area where quantisation plays a central role is in power electronics [15–18]. For example, a three-phase converter is often used to convert a DC source into three-phase AC source. Here, power electronic switches are used to produce an approximately three-phase sinusoidal waveform by switch the DC source in various ways. The resulting waveform is inherently quantised due to the finite levels available in the DC source. An example of a three-phase DC/AC converter will be given in Chapter 21 where the Electrical aspects of Wind Power generation will be studied.

Practical Control System Design: Real World Designs implemented on Emulated Industrial Systems,
First Edition. Adrian Medioli and Graham Goodwin.

Companion website: www.wiley.com/go/medioli/practicalcontrolsystemdesign

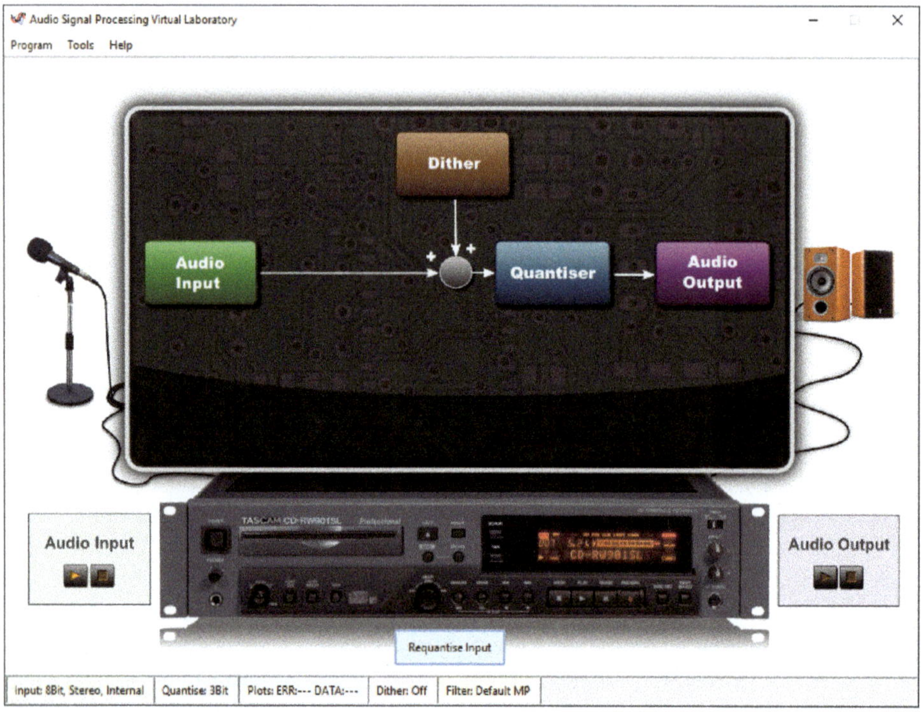

Figure 18.1 Screenshot of program.

Quantisation unavoidably introduces loss of information. In the case of audio signals, the challenge is to preserve the perceived sound quality as much as possible. The aim is thus to make the difference between the original and the quantised signal, as far as possible, inaudible to the listener.

In view of the importance of the perceived sound quality, significant research effort has been concentrated into the development of *psychoacoustically optimal* quantisers. One such scheme is an optimal *noise shaping quantiser* which will be the subject of this virtual laboratory. The same idea is also used extensively in the case of power electronics where the quantisation levels are fixed by the available switching 'states'.

The music used in this laboratory is an original piece composed by Andrew Goodwin (the second author's son) and played by him on guitar and electronic synthesiser.

18.3 Psychoacoustic Issues

It will be well known to users of this laboratory that humans do not hear all sounds with equal sensitivity. For example, few humans can hear anything above 20 kHz. A typical response of the human ear is shown in Figure 18.2.

In this experiment you will begin by measuring your own hearing gain.

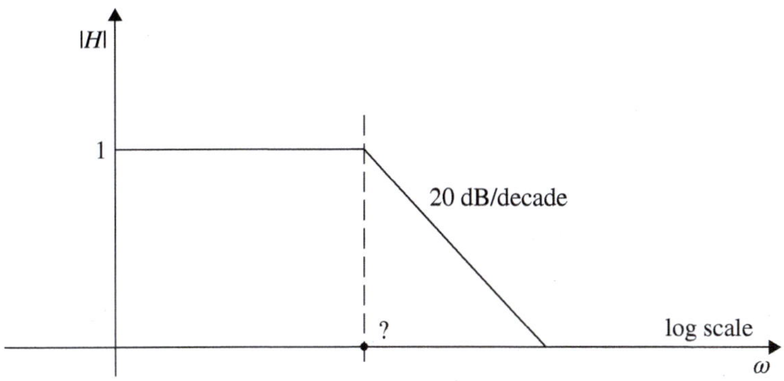

Figure 18.2 Developing a filter.

18.3.1 Experiment: Testing Your Hearing Sensitivity

There are several ways to perform this test. The first involves setting one of the frequencies to a volume and then matching all the others to this perceived level (see Section 18.3.1). The second is similar but involves setting each volume to be just within the threshold of hearing. You will notice that the setting of the volume controls will be different for each frequency.

Even though this test does not correct for the audio characteristics of your computer's sound card, speakers or headphones it can still be useful to give you some indication of your hearing sensitivity across the tested frequencies. One thing that you should correct for, though, is the interference of ambient sounds as these can have a large effect on distinguishing some of the tones. This is probably easiest to achieve through the use of headphones.

Firstly, run the laboratory application program. Once the main laboratory program is running select 'Test Hearing Sensitivity' in the Tools menu. This tool contains volume controls for a number of frequencies roughly covering the range of audible frequencies. The frequencies are evenly spaced logarithmically across the audible range. Each volume control also works on a logarithmic scale. The frequency band is toggled on or off by clicking the slider or the on/off button. The lowest setting of the control corresponds to the lowest possible volume for your sound card, though the sound will not be completely off.

You need to calibrate your computer's sound system so that you can test your hearing sensitivity and for the sake of your ears, do not start with too high a volume. Note that the sound you hear is affected by the characteristics of your speakers/headphones and the computer's sound card as well as your hearing.

To calibrate your sound system, you need to find which slider's frequency is the loudest for your sound card configuration, you may proceed as follows:

1. Set all the sliders to their minimum volume.
2. Identify which slider has the highest volume – even though the slider is set to the minimum: this is done by switching each slider on, one at a time. Once this has been identified this slider is the one you will use to calibrate the system volume.

3. Now with the 'loudest' slider activated, and still set to minimum, adjust the volume control of your computer (or speaker) till that slider's tone can just be heard, you are now ready to go on to the next step. If on checking each slider at the minimum setting none could be heard that is still okay as you are also ready to go on to the next step.

Strategy one: Matching All Frequencies to a Volume: Set your 'loudest frequency', as described above, with its slider at the lowest level and turn up your speaker/headphone volume until you can easily hear the tone but at a low volume. Now set all the other sliders so that all the tones sound equally loud.

Strategy two: Minimum Sound Threshold: Similarly with the 'loudest frequency' on, and set to the lowest level, turn the speaker/headphone volume up or down until the sound is just barely audible. Now adjust all the other bands until they are also just audible.

Once the volume controls are appropriately set you can produce a plot of your hearing sensitivity. In the Tools menu of the Hearing Sensitivity application select 'Plot Hearing Sensitivity From Slider Panel'. The plot is on a logarithmic frequency scale.

The plot window has a number of useful features. You can zoom in on the plot trace by dragging the mouse with left button down over the plot. Also, you can save or print the trace using the menu which appears when you right click on the plot window.

Plot result: Construct a first-order approximation to your hearing response by fitting to a 3 dB cut-off ω_c to your hearing curve, as shown in Figure 18.2.

Discrete-time model: Construct a discrete-time model of the form $H(z) = \frac{1-a}{z-a}$ where $a = 1 - \omega_c * \Delta$ and where Δ = sampling period. (Note that the sampling frequency is 44.1 kHz).

18.4 Nearest Neighbour Quantisation

The effect of the nearest neighbour quantiser will next be examined, see Figure 18.3.

18.4.1 Experiment: Testing the Nearest Neighbour Quantiser

This experiment studies the effects of noise quantisation:

1. By clicking on the Quantiser Block choose the number of bits for the quantisation, say 3 bits, which allows 8 levels, since $2^3 = 8$.
2. Do the experiment by pressing 'Requantise Input'. This will sample the sound clip and quantise each sample into the allowable values.
3. Listen to the original and requantised audio and comment on the qualitative differences.

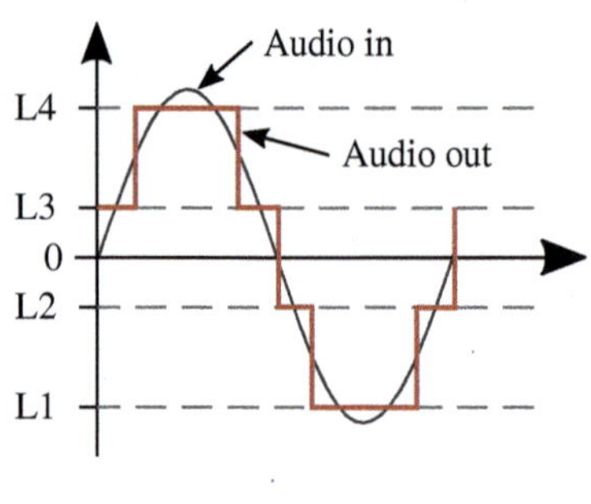

Figure 18.3 Nearest neighbour 2-bit quantisation.

4. Plot the spectrum of the errors. The errors are the differences between the quantised version of the amplitudes and the actual amplitude of the sound clip at each sampling point. Note that you can zoom in within the graph by either right clicking on the graph and using the zoom menu, or by pressing the left mouse button down and dragging the mouse diagonally to the right to draw a box around the feature of interest.
5. Add some 'dither' to the audio and then repeat.

18.5 Optimal Noise Shaping Quantiser

18.5.1 Feedback Quantiser

The reader probably found that with a low number of quantisation levels the nearest neighbour quantiser performs poorly. Of course, with a large number of levels, the nearest neighbour quantiser seems to be satisfactory. However, Hi-Fi enthusiasts expect the best. Thus, in practice, simple nearest neighbour quantisation is rarely used. Instead, a more sophisticated quantisation method is utilised. This method utilises the *noise shaping quantiser* depicted in Figure 18.4. (Here, and in the remainder of this laboratory, q denotes the forward shift operator, $qv_k = v_{k+1}$, where v_k is any sequence and k denotes the sampling index.) Also, the symbol ε_k will be used for the errors introduced by the quantiser, i.e.

$$\varepsilon_k \triangleq u_k - i_k. \tag{18.1}$$

Since ε_k is not available until the quantisation has taken place, the filter $F(q)$ must be strictly proper, i.e. it includes (at least) a unitary sample delay.

When a large number of quantisation levels are used, then the quantisation error can often be well approximated by white noise which is independent of a_k, leading to a linear quantisation model. In Figure 18.4, the output of the system satisfies

$$u_k = a_k + (1 - F(q))\varepsilon_k, \tag{18.2}$$

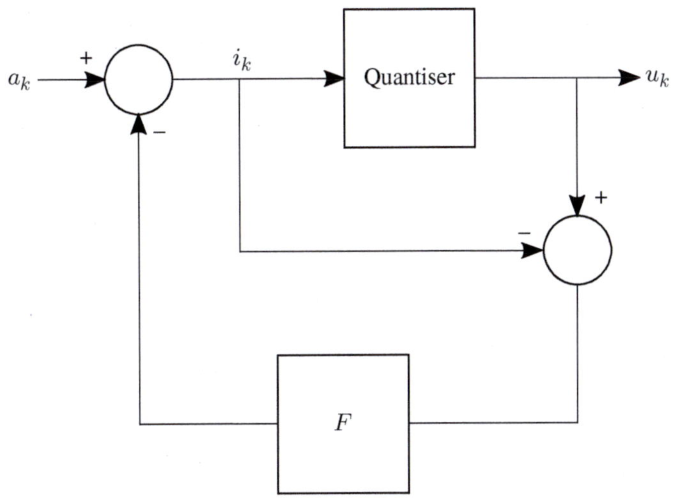

Figure 18.4 Feedback quantiser.

so one may treat $1 - F(q)$ as a sensitivity function which shapes the frequency content of the quantisation noise present in $u(k)$.

One can actually give an interesting interpretation of the circuit in Figure 18.4 via the Youla–Kucera parameterisation of all stabilising feedback circuits. Indeed, if one thinks of the 'plant' as having the (stable) transfer function $G(q) = 1$, then all stabilising controllers are known to be as in Figure 18.7 where F is any stable proper transfer function.

18.5.2 Experiment: Test the Feedback Quantiser

Experiment with the F Filter.

1. Run the laboratory program.
   Set the Quantiser Architecture to the one named 'F Only - Noise Shaping Quantiser'.
2. Enter your F Filter.
   Click on the 'F' block to bring up the filter dialog. In the dialog the 'Select F Filter' tab should be selected. Now click (small circle) next to 'Enter Your custom F Filter' and enter your own F Filter. Remember that the transfer function must be strictly proper.
   Begin with the default Filter as a starting point. Click on the 'Use the Default Filter (such that $1 - F$ is Minimum Phase)'. Now alter the numbers, say, set the coefficient accompanying the 'z' term in the numerator to -0.6, then click OK. You can leave all the other settings as their defaults. Now run the converter by clicking on the 'Requantise Input' button.
3. Listen and describe how it sounds.
   Press the audio output play button on the bottom right of the program window.
4. Plot the spectrum.
   Click on the Quantiser block and select Frequency Domain plots of Quantiser Error and Requantised Audio, click OK and then run the quantiser again. The plots will then appear.
5. Add some dither to the audio and repeat.
   What changes do you notice in the sound quality and in the plots?

18.6 Utilising Your Own Hearing Sensitivity

From Figure 18.4, the input to the quantiser satisfies

$$i_k = a_k - F(q)[u_k - i_k], \tag{18.3}$$

so that solving for i_k leads to the following transfer function:

$$i_k = \frac{1}{1 - F(q)} a_k - \frac{F(q)}{1 - F(q)} u_k.$$

Hence, the system of Figure 18.4 can be redrawn as the two-degree-of-freedom feedback control loop shown in Figure 18.5. Since $1 - F(q)$ is commonly designed to be of minimum phase, both transfer functions in this figure are stable. (The non-minimum phase case will be studied later.)

Then, if the quantiser is thought of as introducing noise, then conceptually the feedback quantiser can view Figure 18.5 as in Figure 18.6.

We see that the transfer function from a to u is

$$\frac{u}{a} = 1. \tag{18.4}$$

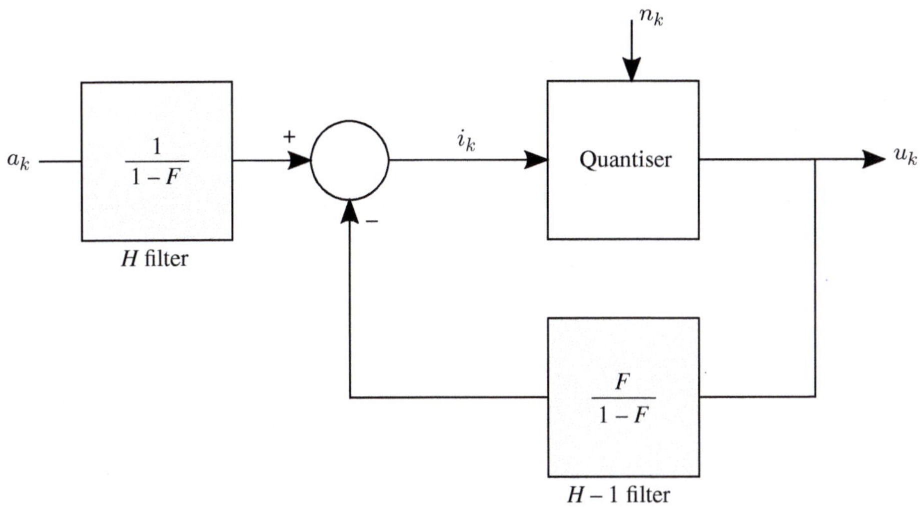

Figure 18.5 Noise shaping quantiser as a two-degree-of-freedom feedback system.

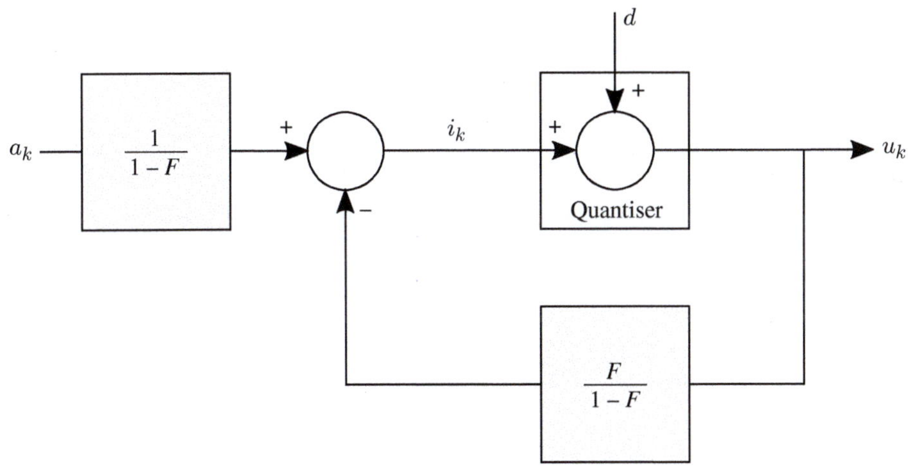

Figure 18.6 The quantiser as a noise source.

whilst the transfer function for d to u is

$$\frac{u}{d} = 1 - F(q). \tag{18.5}$$

The work on psychoacoustics in Section 18.3 suggests that it may be desirable to make the noise-to-output transfer function the inverse of the human perception (i.e. to allow more 'quantisation noise' at frequencies where they have lower perception by your ear).

Thus, a good choice for F would seem to be such that

$$H = \frac{1}{1 - F}. \tag{18.6}$$

Solving for F gives

$$F = \frac{H - 1}{H}. \tag{18.7}$$

18.6.1 Experiment: Test the Feedback Quantiser Using Your Hearing Sensitivity

Experiment with your *H* Filter by using your *H* found in Section 18.3:

1. Run the laboratory program. Set the Quantiser Architecture to the one named 'F Only - Noise Shaping Quantiser'.
2. Enter Your H Filter.
   Run the program as described in Section 18.5.2, this time the 'Select H Filter' tab should be selected. Enter your H Filter, as stated on the dialog the H Filter's Transfer Function should be Biproper.
3. Listen and describe how it sounds.
4. Add dither.
5. Plot the spectra.

18.7 Audio Quantisation from a Bode Sensitivity Integral Perspective

The reader will have noticed when doing the above experiments that, when a filter is used to reduce the impact of quantisation noise in the audible range, then the noise reappeared with greater intensity above the audible range. Of course, noise outside the audible range is not perceptible to humans (although it may annoy your pet dog if you have one since they typically hear higher frequency signals than humans).

In this section, the above results are interpreted in the light of the Bode sensitivity integral trade-offs.

The feedback part of the noise shaping quantiser is as shown in Figure 18.6, where d models quantisation noise.

The feedback system in Figure 18.6 can be interpreted in terms of the Bode integral. In particular, note that the feedback quantiser cannot give an arbitrary response from quantisation noise to output.

Indeed, it can be seen that the loop-transfer function in Figure 18.6 is

$$L(\rho) = \frac{F(\rho)}{1 - F(\rho)}. \tag{18.8}$$

Hence, the sensitivity function is

$$\begin{aligned} S_z(e^{j\omega}) &= \frac{1}{1 + L(e^{j\omega})} \\ &= \frac{1}{1 + \frac{F(e^{j\omega})}{1 - F(e^{j\omega})}} \\ &= \left[1 - F(e^{j\omega})\right]. \end{aligned} \tag{18.9}$$

This can be interpreted in terms of the results given in Chapter 17.

Specifically, from Eq. (17.29),

$$\int_0^{\pi} \ln |S_z(e^{j\omega})| d\omega = \pi \sum_{i=1}^{q} \ln |p_i^z|, \tag{18.10}$$

where $\{p_i^z,\ i = 1, \ldots, q\}$ is the set of open-loop poles of $L(\rho)$ outside the unit circle. Thus, exactly as described in Chapter 17, reducing the disturbance at lower frequencies will cause it to reappear at high frequencies.

18.7.1 Experiment: Spectrum of Errors

Choose a filter $F(\rho)$ so that $1 - F(\rho)$ is minimum phase. Apply the quantiser to an audio signal. Plot the quantisation errors in the frequency domain, showing the spectrum of the errors between the original signal and its quantised version.

You may use your own F filter or the minimum phase one supplied in the program's dialog.

This can be achieved as follows:

1. Run the laboratory program.
2. Set the quantiser architecture to the one named 'F only - noise shaping quantiser' by clicking the 'set quantiser architecture' button and choosing 'noise shaping quantiser - F only', then click the OK button.
3. Enter your F filter by clicking the F block. In the window that appears click the 'select F filter' tab at the top of the dialog. In the body of the tab in that dialog window click on the 'enter your custom F filter' item and enter the filter coefficients and then click OK when done.
   Note you must choose a filter $F(\rho)$ so that $1 - F(\rho)$ is minimum phase.
4. Run the system to quantise the input audio and produce a plot of the quantisation errors in the frequency domain.
   (a) Click the quantiser block.
   (b) In the dialog window that appears go to the 'plot quantisation error' section and click 'in the frequency domain'.
   (c) Click OK.
   (d) To run the system using your F filter click the run button.

The Bode integral can be used to better understand these results. Indeed, when $1 - F(\rho)$ has all its zeros inside the unit disc, i.e. $L(\rho)$ is stable, it follows that

$$\int_0^{\pi} \ln |S_o(e^{j\omega})| d\omega = 0. \tag{18.11}$$

This is seen to be approximately true for the quantiser error.

18.7.2 Experiment: Testing Bode Sensitivity Integral

Adjust $F(\rho)$ to a different value, making sure that $1 - F(\rho)$ remains minimum phase. Check that the "Water Bed Effect" holds.

Note, the F filter supplied by the program is such that $1 - F$ is minimum phase.

18.8 Audio Quantisation for More Complex Cases

So far it has been assumed that $L(p)$ is open-loop stable. However it is possible for $L(\rho) = \frac{F(\rho)}{1-F(\rho)}$ to be open-loop unstable. This will be explored below.

The reader is invited to explore cases where $1 - F(\rho)$ has zeros outside the stability region, i.e. when the loop-transfer function is open-loop unstable. A word of caution is that the architecture shown in Figure 18.6 cannot be used when $1 - F(\rho)$ is not minimum phase. The reason is that the block outside the loop will be unstable in this case. However, it is to be noted that $(1 - F)$ appears in the denominator in both the feedback path and the feedforward path. Hence, the same transfer functions are obtained if $\frac{1}{(1-F)}$ is moved inside the loop as shown in Figure 18.4.

18.8.1 Experiment: More Complex Case

The reader is asked to choose a filter such that $1 - F(\rho)$ has zeros outside the stability region and to compare the allocation of closed-loop noise to the case when $1 - F(\rho)$ has all its zeros inside the unit disc. The reader should find the trade-off is more severe. However, this may not be easy to hear due to the limitations of human hearing.

18.9 Revision Questions

1. For the feedback loop of Figure 18.7 what is the sensitivity transfer function between d and y? What is the result when $G = 1, Q = F$?
   Aside: The purpose of feedback quantisers, such as the one shown in Figure 18.4, is to shape the noise spectrum so that the main 'noise power' appears in less audible regions of the frequency domain. This is aimed at improving the resulting sound quality, as perceived by the human ear. Quantisation is especially audible at low frequencies, hence, it seems reasonable to tune $1 - F(z)$ such that its frequency response moves the errors to higher frequencies.
   (a) For the feedback loop shown in Figure 18.7, what is the relationship linking y to y^* and d?
   (b) What is the result in Question 1 if $G = 1, Q = 1$?
2. Derive Eq. (18.2).
3. Say that one wishes to achieve a quantisation noise sensitivity function of $S(z) = (1 - z^{-1})^2$. What filter $F(z)$ is required?
4. Repeat Question 3 for $S(z) = \frac{(z-1)^4}{(z-\alpha)^4}$.
5. What values of α make $F(z)$ found in Question 4 stable?
6. Consider again $S(z)$ in Question 4. For small frequencies, how does the magnitude of $S(z)$ decay as the frequency decreases?
7. For the system in Question 3, compute the maximum of the sensitivity.
8. Repeat Question 7 for S given in Question 5. What is max $|S|$ when $\alpha = 0.9$?
9. Use your answers to Questions 6 and 8 to give an explanation of why having $\alpha \neq 0$ may be desirable.

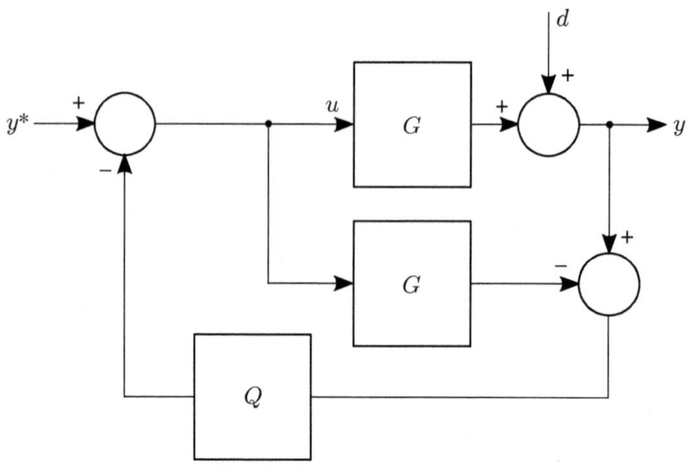

Figure 18.7 Feedback loop.

Further Reading

The following references give more information on optimal noise-shaping questions:

1 M. Akune, R.M. Heddle and K. Adagiri, "Super Bit Mapping: Psychoacoustically Optimized Digital Recording", *presented at the 93rd Convention of the Audio Engineering Society*, October 1992, preprint 3371.

2 R.J. Beaton, J.G. Beerends, M. Keyhl and W.C. Treurniet, "Objective Perceptual Measurement of Audio Quality", in *Collected papers on digital audio bit-rate reduction*, N. Gilchrist and C. Grewin, Eds., pp. 126–152, Audio Engineering Society, Inc., New York, 1996.

3 C. Dunn and M. Sandler, "Psychoacoustically optimal sigma-delta modulation", *Journal of the Audio Engineering Society*, 45:212–223, 1997.

4 M. Gerzon and P.G. Craven, "Optimal Noise Shaping and Dither of Digital Signals", *presented at the 87th Convention of the Audio Engineering Society*, October 1989, preprint 2822.

5 G.C. Goodwin and D.E. Quevedo, "A moving horizon optimal quantiser for audio signals", *Journal of the Audio Engineering Society*, 51(3):138–149, 2003.

6 R.M. Gray and T.G. Stockham, "Dithered quantisers", *IEEE Transactions on Information Theory*, 39:805–812, 1993.

7 S.P. Lipshitz, J. Vanderkooy and R.A. Wannamaker, "Minimally audible noise shaping", *Journal of the Audio Engineering Society*, 39:836–852, 1991.

8 S.P. Lipshitz, R.A. Wannamaker and J. Vanderkooy, "Quantisation and dither: a theoretical survey", *Journal of the Audio Engineering Society*, 40:355–375, 1992.

9 B.C.J. Moore, "Masking in the Human Auditory System", in *Collected papers on digital audio bit-rate reduction*, N. Gilchrist and C. Grewin, Eds., pp. 126–152, Audio Engineering Society, Inc., New York, 1996.

10 K.C. Pohlmann, *Principles of digital audio*, 4th Edition, McGraw-Hill, New York, 2000.

11 L. Schuchman, "Dither signals and their effect on quantisation noise", *IEEE Transactions on Communications*, 12:162–165, 1964.

12 A.B. Sripad and D.L. Snyder, "A necessary and sufficient condition for quantisation errors to be uniform and white", *IEEE Transactions on Acoustics, Speech and Signal Processing*, 25:442–448, 1977.

13 R.A. Wannamaker, "Psychoacoustically optimal noise shaping", *Journal of the Audio Engineering Society*, 40:611–620, 1992.

14 R.A. Wannamaker, S.P. Lipshitz, J. Vanderkooy and J.N. Wright, "A theory of non-subtractive dither", *IEEE Transactions on Information Theory*, 48:499–516, 2000.

15 G. Mirzaeva and G.C. Goodwin, "The use of Feedback Quantizer PWM for shaping inverter noise spectrum", in *15th International Power Electronics and Motion Control Conference (EPE/PEMC)*, 2012.

16 G. Mirzaeva and G.C. Goodwin, "Feedback Quantizer vs Sigma-Delta Modulator for Voltage Source Inverters", in *38th Annual Conference on IEEE Industrial Electronics Society (IECON)*, 2012.

17 G. Mirzaeva and G.C. Goodwin, "Introducing Feedback Quantizer – A Novel pulse-width modulation strategy for voltage source inverters", 22nd Austrialian Universities Power", *Engineering Conference*, 2012.

18 G. Mirzaeva and G.C. Goodwin, "Advanced noise shaping and filter design with Feedback Quantizer PWM", *IEEE International Conference on Industrial Technology (ICIT)*, 469–474, 2013.

Part V

Simple Multivariable Control Problems

The first three parts of the book have focused on single-input single-output systems. This part of the book represents a transition from these simple single-input single-output problems to multivariable control problems. In this part, relatively simple techniques will be described. Chapter 19 describes how simple cascaded loops can be used. Chapter 20 returns to the wind energy generation problem discussed earlier in Chapter 11. Chapter 21 studies cross-directional control in paper machines. More general multivariable design tools will be described in Part VI.

Practical Control System Design: Real World Designs implemented on Emulated Industrial Systems,
First Edition. Adrian Medioli and Graham Goodwin.

Companion website: www.wiley.com/go/medioli/practicalcontrolsystemdesign

19

Tools Used for Simple Multivariable Control Problems

19.1 Introduction

This chapter takes a first look at multivariable control. Two simple approaches to multivariable control will be examined in this chapter, namely,

(i) cascade control, and
(ii) imposed SISO architecture.

Several real-world case studies of these ideas will be pursued in Chapters 20 and 21. The results are often perfectly acceptable. However, in some cases, it is not possible to take a SISO view of multivariable control problems. To deal with more complex scenarios, different design methods are needed. These methods will be examined in Part VI.

19.2 Cascade Control

In many control problems, it makes sense to think of a control system in terms of a sequence of loops that surround inner loops. This is commonly termed 'cascade control'.

A key example where this idea is universally applied is in flow control problems.

Thus, consider the heating control problem in Figure 19.1. In this figure it is implicitly assumed that the valve's closing and opening is directly proportional to the delivery of steam flow.

In practice, valves typically exhibit highly non-linear behaviour with relatively simple dynamics. In this case, it makes sense, if it can be easily done, to measure the steam flow, $u_2(t)$, and then to change the control law to one requesting steam flow.

This leads to the introduction of an inner control loop giving the system a cascaded control architecture. The original model linking $u_1(t)$ to $y(t)$ is shown in Figure 19.2.

An inner loop is added around the valve dynamics block $G_v(s)$ consisting of a feedback loop and controller $C_2(s)$ as shown in Figure 19.3.

Hence, the addition of the inner control loop results in the cascaded architecture of the heating problem (see Figure 19.1) shown in Figure 19.4. The advantage of using the cascaded architecture is that variations in supply pressure, valve nonlinearities etc. are compensated by the fast inner control loop. Hence, the design of the outer loop can assume that the valve is close to ideal.

Cascade architectures of the type shown in Figure 19.4 are particularly effective when the time constants in the inner loop are much faster than in the outer loop.

Practical Control System Design: Real World Designs implemented on Emulated Industrial Systems,
First Edition. Adrian Medioli and Graham Goodwin.

Companion website: www.wiley.com/go/medioli/practicalcontrolsystemdesign

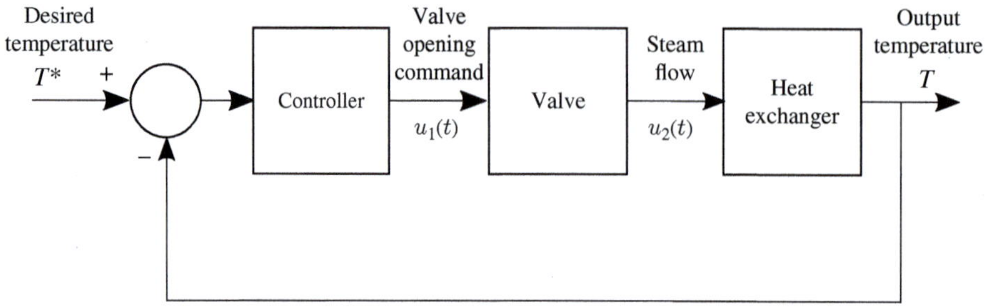

Figure 19.1 Heating problem.

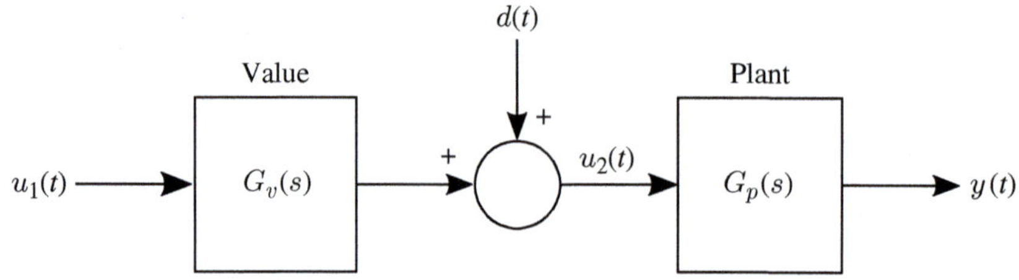

Figure 19.2 Original control actuator and plant.

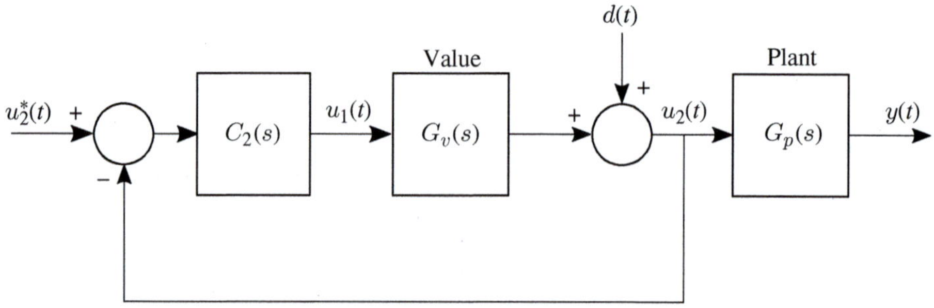

Figure 19.3 Added inner control loop.

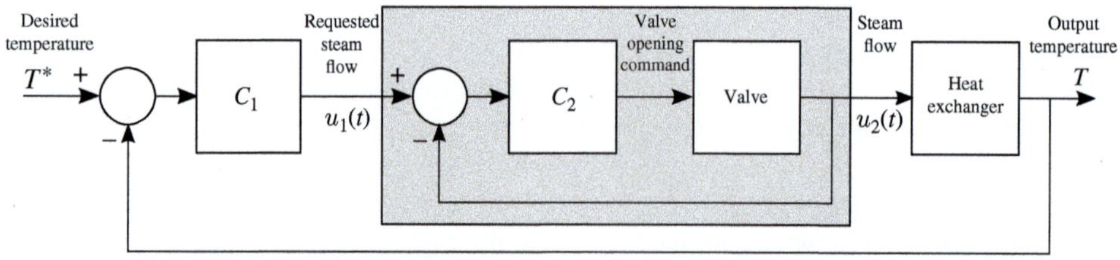

Figure 19.4 Cascade architecture.

19.2.1 Example of Cascade Control

Consider the control problem as shown in Figure 19.2.

Assume that the plant has nominal model given by

$$G_p(s) = \frac{e^{-s}}{2s+1}. \tag{19.1}$$

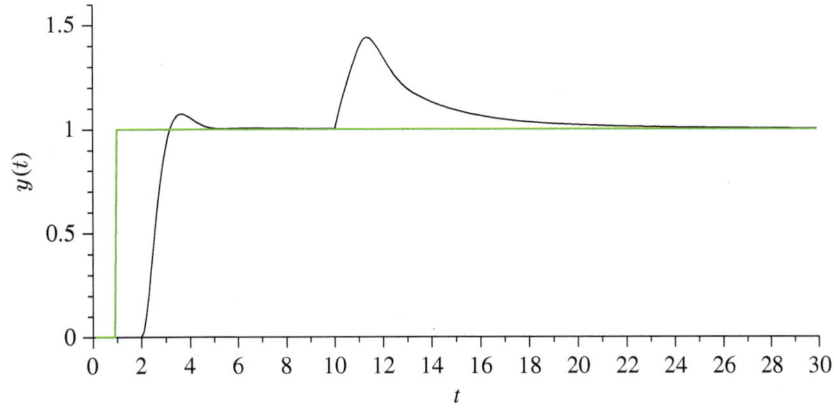

Figure 19.5 Response to disturbance.

Also, the valve is assumed to have nominal dynamics:

$$G_v(s) = \frac{1}{s+1}. \tag{19.2}$$

Hence, the transfer function from $u_1(t)$ to $y(t)$ is

$$G(s) = \frac{e^{-s}}{(s+1)(2s+1)}. \tag{19.3}$$

Say one designs a controller for this system utilising a Smith predictor as outlined in Chapter 12. Unfortunately the design will be compromised by the relatively slow dynamics of the valve to the disturbance $d(t)$.

Figure 19.5 shows the response to a unit step reference change at $t = 1$ and a unit step disturbance at $t = 9$.

It is to be noted that the valve (inner dynamics) is relatively slow yet has simple dynamics. Hence, it is prudent to design an inner loop which 'speeds-up' the valve response and compensates the disturbance $d(t)$ near it's source.

To illustrate, say that one designs an inner PID controller as shown in Figure 19.3. A PI controller is chosen of the form:

$$C_2(s) = \frac{8(s+1)}{s}. \tag{19.4}$$

This yields a closed-loop transfer function from $u_2^*(t)$ to $u_2(t)$ of

$$G_a^{CL}(s) = \frac{1}{0.125s+1}. \tag{19.5}$$

Also, the transfer function from $d(t)$ to the output $u_2(t)$ now becomes

$$S_a^{CL}(s) = \frac{0.125s}{0.125s+1}. \tag{19.6}$$

The overall transfer function from $u_2^*(t)$ to $y(t)$ now becomes

$$G_a^{CL}(s)G_p(s) = \frac{8e^{-s}}{2s^2+17s+8}. \tag{19.7}$$

Again a Smith predictor is used for the controller $C_1(s)$ as shown in Figure 19.4.

Figure 19.6 gives the resulting response to a unit step in the reference at $t = 1$ and a unit step disturbance at $t = 9$.

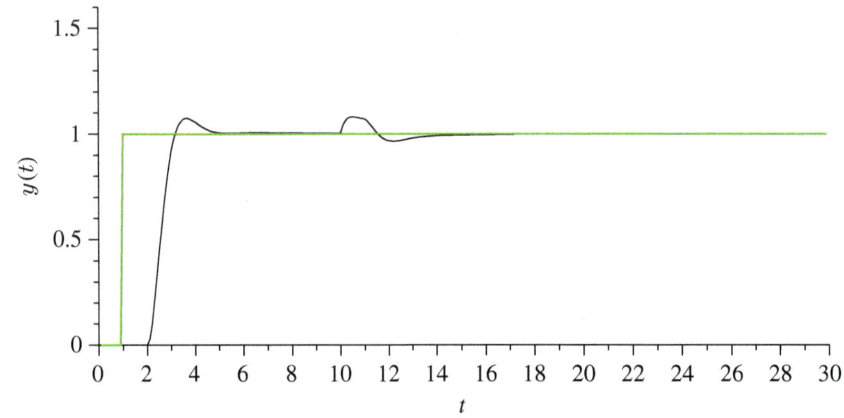

Figure 19.6 Response to disturbance with disturbance feedback.

Comparing the response in Figure 19.6 with that in Figure 19.5 shows that cascade control has led to a greatly improved response.

Cascade control is used very frequently in practice. For example, in the boiler control problem studied in Chapter 30, the regulation of the drum water level at low frequencies is achieved by adjusting the feedwater flow rate. A cascade controller then converts the request for feedwater flow into the opening of a valve in the feedwater line.

A real-world application of cascade control will be presented in Chapter 20 in the context of Wind Power.

19.3 Imposed SISO Architectures

Whilst many real-world control problems are inherently multivariable, there often exists strong motivation to utilise standard SISO controllers since these are easier to understand and implement.

As a real-world example consider the Boiler Control for an Electricity Generation Plant. At a simple level this system comprises two inputs (throttle valve and fuel) and two outputs (drum pressure and generated power) – see Figure 19.7. Note that the drum water level control problem has been ignored since this operates separately to slowly regulate the water level in the drum.

If one adopts a SISO view of this control problem, then one is immediately faced with the question whether y_1 should be controlled by u_1 and y_2 by u_2 or y_1 by u_2 and y_2 by u_1. Actually, in practice both architectures are used.

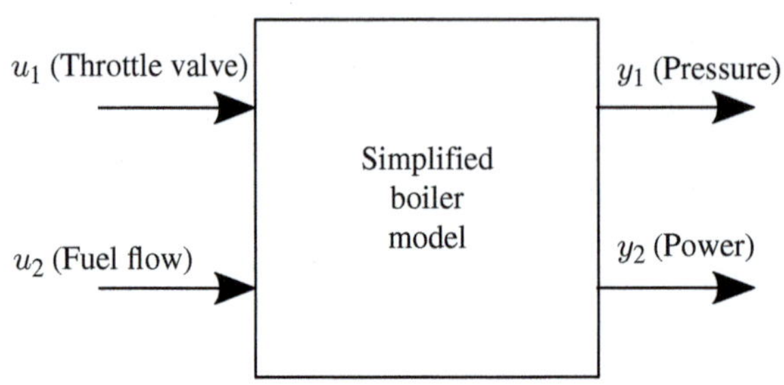

Figure 19.7 Simplified boiler architecture.

Indeed, they have been given specific names as follows:

Boiler following (turbine lead):

$$y_2(\text{Power}) \leftrightarrow u_1(\text{Throttle}), \tag{19.8}$$

$$y_1(\text{Pressure}) \leftrightarrow u_2(\text{Fuel}). \tag{19.9}$$

Turbine following (boiler lead):

$$y_1(\text{Pressure}) \leftrightarrow u_1(\text{Throttle}), \tag{19.10}$$

$$y_2(\text{Power}) \leftrightarrow u_2(\text{Fuel}). \tag{19.11}$$

For the system studied in Chapters 29 and 30, the turbine following architecture will be used.

19.4 Relative Gain Array

A simple tool that is sometimes used in practice to determine MV/PV pairings is the Relative Gain Array (RGA). Assuming a square transfer function $G(s)$, then the RGA is defined by

$$\text{RGA} = G(o)\phi\left[G(o)^{-1}\right]^T, \tag{19.12}$$

where 'ϕ' denotes the Hadamard (or Schur) product (i.e. element by element product. The RGA has several properties, e.g. each row and column of RGA sums to 1.)

A useful rule of thumb is that it is best to arrange the MV in an order such that the diagonal elements of the RGA are maximal, then pair y_i with u_i, $i = 1, m$.

19.5 An Industrial Example

An industrial example of the above problem arises in the under grate airflow control in a boiler system associated with sugar milling.

Preliminary step response testing led to the following model

$$G(s) = \begin{bmatrix} \dfrac{0.2}{8s+1} & \dfrac{-0.175}{10s+1} \\ \dfrac{0.7}{5s+1} & \dfrac{0.5}{6s+1} \end{bmatrix}. \tag{19.13}$$

19.5.1 The Relative Gain Array

It is easily shown that the RGA for this system is

$$\text{RGA} = \begin{bmatrix} 0.45 & 0.55 \\ 0.55 & 0.45 \end{bmatrix}. \tag{19.14}$$

This RGA suggests that a SISO pairing of y_1 to u_2 and y_2 to u_1 is preferable. However, since RGA is certainly not diagonally dominant, then it seems that SISO control of any type will be sub-optimal. Actually, it follows from (19.13) that it may be desirable to define new inputs as $\overline{u}_1 = u_1 - u_2$ and $\overline{u}_2 = u_1 + u_2$. This choice is further explored in the problems given at the end of the chapter.

19.5.2 A Simple MV Transformation

Examination of the physics of the above system suggests that y_1 might be better controlled by $u_1 - u_2$ and y_2 by $u_1 + u_2$. Indeed, letting $\overline{u}_1 = u_1 - u_2$ and $\overline{u}_2 = u_1 + u_2$, leads to the following approximate transfer function linking $\overline{u}_1, \overline{u}_2$ to y_1, y_2.

$$\overline{G}(s) \approx \begin{bmatrix} \frac{0.2}{8s+1} & 0 \\ 0 & \frac{0.7}{5s+1} \end{bmatrix}. \tag{19.15}$$

Then the two loops are approximately decoupled and simple single-input single-output design procedures can be used in the two loops $\overline{u}_1 \to y_1, \overline{u}_2 \to y_2$.

The laboratory in Chapter 21 explores a more complex version of the above design in the context of Cross Directional Control in a Paper Machine.

19.6 Revision Questions

1. Consider the system described in Section 19.2.1. Say that the valve also has a nonlinear gain such that the flow out of the valve is given by

$$f(t) = G_v(s)[u_1(t)]^{\frac{1}{2}}. \tag{19.16}$$

Repeat the simulations given in Figures 19.5 and 19.6 including this nonlinear behaviour.

2. A situation that occurs with cruise controllers in cars is that maintaining constant velocity can cause a car to ram a car in front travelling at a lower speed. Can you suggest a cascade control architecture to address this problem if there is a radar measuring the distance to the vehicle in front?
3. Assuming that the vehicle speed controller has high gain, would you suggest that a pure integral controller be used for the outer loop?

Further Reading

1 G.C. Goodwin, S.F. Graebe and M.E. Salgado, *Control system design*, Prentice Hall, 2000.
2 L. Wang, *PID and predictive control of electrical drives and power converters using Matlab/Simulink*, Wiley IEEE Press, Chapter 2, 2014.

20

Laboratory 9: Wind Power (Electrical Components)

20.1 Introduction

The mechanical aspects of wind power were studied earlier in Chapter 11. In that chapter, it was assumed that the electrical torque, T_e^*, required by the rotational velocity control system could be instantly delivered by the electrical system. This assumption is justified by noting that the time constant of the mechanical system is typically much larger than that of the electrical system. Hence, it makes sense to study the mechanical system in isolation. Note that this is a form of cascade control as described in Section 19.2. The current chapter describes the inner electrical control loops tasked with delivering the required torque.[1]

There are two sides to this problem, namely, the generator side and grid side. The problem is inherently multivariable in nature. However, the two sides are typically decoupled (at least approximately) by the inclusion of a relatively large DC link capacitor between the two sides. Thus, it makes sense to study the two problems separately. By doing this, a complex multivariable problem is converted into two simple single-input single-output systems.

In the material presented below, the model aspects require background knowledge related to 3-phase electrical systems. However, for the design of the control laws, it suffices to assume that the model is given. Indeed, the control laws are relatively simple and use PID controllers. A screen shot of the virtual laboratory interface is shown in Figure 20.1.

20.2 Generator Choices

There are many choices that could be used for the generator. The decision is based on many factors, especially rated power. For a microsystem, it would be typical to use a single-phase squirrel cage induction generator. For very large systems (Megawatts), a typical generator would be a doubly fed induction machine or synchronous generator.

20.3 Physical Parameters for the Laboratory Wind Turbine

The laboratory is based on a medium power range where a 3-phase squirrel cage induction generator is used.

1 Galina Mirzaeva and Maria Seron (University of Newcastle) and Chris Townsend (University of Western Australia) contributed significantly to this chapter.

Practical Control System Design: Real World Designs implemented on Emulated Industrial Systems, First Edition. Adrian Medioli and Graham Goodwin.

Companion website: www.wiley.com/go/medioli/practicalcontrolsystemdesign

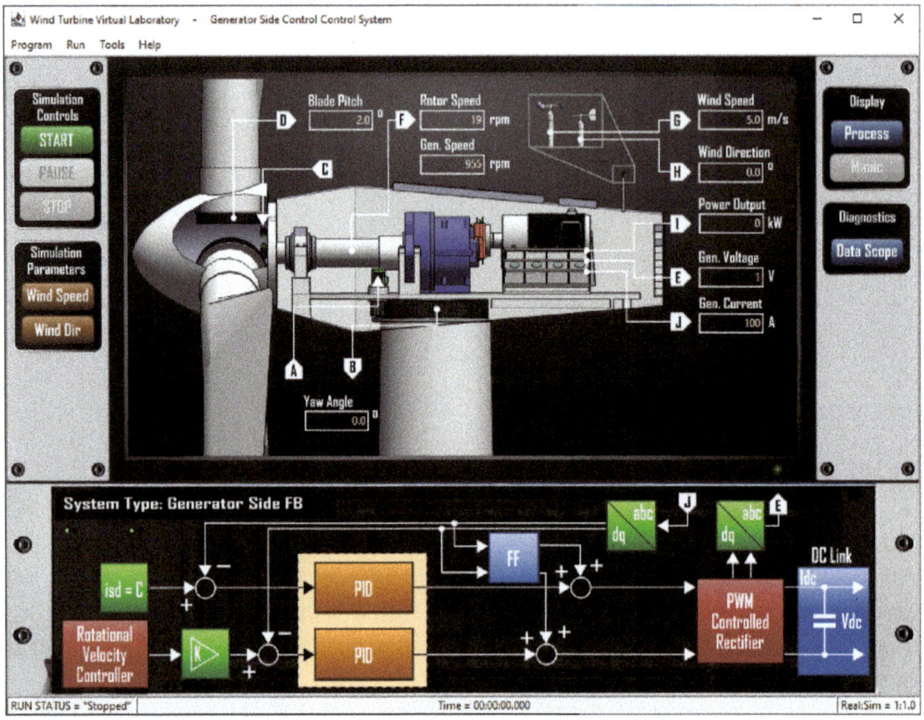

Figure 20.1 Wind turbine lab screenshot showing the generator side controller.

Physical parameters of the installation are summarised below:

Assumed air density: 1.2 kg/m^3	Blade tip radius: 20 m
Turbine cut in speed: 4 m/s	Turbine rated speed: 12 m/s
Turbine cut out speed: 25 m/s	TSR: 8
Operating point wind speed: 5 m/s	Yaw motor current to torque ratio: 2 : 1
Yaw motor maximum current: 6 A	Rated maximum power: 500 kW
Turbine inertia: 3000 kg m^2	Damping: 1000 kg m^2/s
Motor back emf constant: 2 : 1	Motor armature resistance: 0.5 Ω
Blade pitch min: 2°	Blade pitch max: 59°
Blade pitch slew rate limit: 10 °/s	Rotor to generator gear ratio: 1 : 50
Induction generator pole pairs: 3	Magnetising inductance: $L_n = 3.96 \times 10^{-3}$ H
Stator winding resistance: $R_s = 7.9 \times 10^{-3}\ \Omega$	Stator winding leakage inductance: $L_{s\ell} = 9.1 \times 10^{-5}$ H
Rotor winding refined resistance: $6.8 \times 10^{-3}\ \Omega$	Rotor winding leakage inductance: $L_{r\ell} = 1.15 \times 10^{-4}$ H
Nominal grid voltage: 400 V	DC link voltage reference: 650 V
Grid side inductance: 1.546×10^{-4} H	Grid side resistance: 0.01571 Ω

Note that for example with a steady-state wind of 5 m/s and a TSR of 8 we have that the blade tip speed should be 40 m/s giving a rotor speed of 0.318 rev/s ($\approx$2 rad/s or $\approx$19 rpm). The steady-state generator speed under these conditions is translated through the gearbox which multiplies it by 50 resulting in 954 rpm ($\approx$100 rad/s).

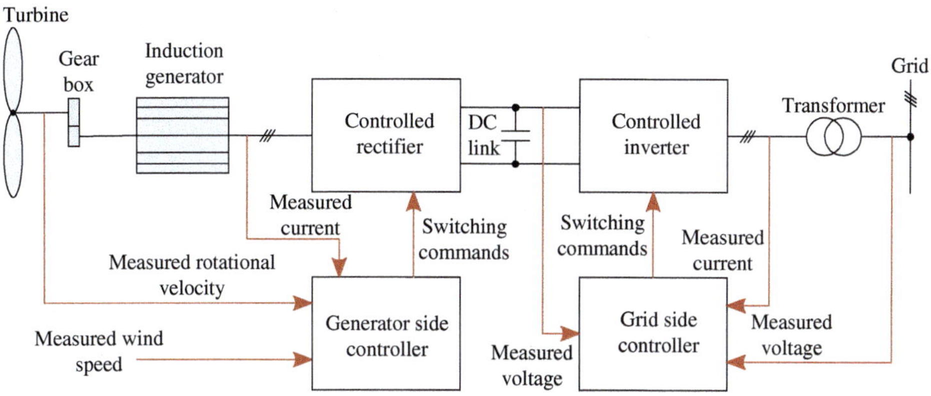

Figure 20.2 Block diagram of electrical system.

20.4 The Generator and Grid Side Architectures

The basic control architecture is illustrated in Figure 20.2 (red lines denote signal paths). As shown in Figure 20.2, the induction generator output is connected to a controlled rectifier. The output of the controlled rectifier feeds current into the DC link capacitor. On the grid side, the DC link capacitor is connected to a controlled inverter via a grid side controller.

Mechanical energy from wind is converted into 3-phase AC electrical energy by the induction generator. This is then converted into DC electrical energy which is temporarily stored on the DC link capacitor. Finally, the energy is converted back to 3-phase AC energy that is supplied into the grid via a transformer.

The reason for double conversion (AC–DC–AC) is that the frequency of the AC voltage and current from the induction generator is variable. (It depends on the changing turbine speed which, in turn, is (approximately) matched to the wind velocity as discussed earlier in Chapter 11.) On the other hand, the frequency on the grid side is typically constant (50 Hz in Australia and the UK or 60 Hz in North America). Also, the voltage should match the grid voltage (say 415 V).

The arrangement shown in Figure 20.2 is called a 'symmetrical back-to-back converter' structure. The advantages of using this symmetrical structure include:

- Power can flow in both directions, i.e. to and from the grid. Under these conditions, the generator can be started as a motor (in a controlled manner) using power from the grid and then moved into generation mode. This simplifies generator starting issues.
- The symmetrical back-to-back structure also has desirable features under fault conditions. For example, there can be a sudden decrease in power required by the grid. In response, the system can simply reduce power coming from the turbine and continue operating. This is said to provide a 'fault ride-through' capability. Systems that do not have this capability need to be shut down, making it more difficult to recover from the fault.

20.5 Background Theory

This section reviews basic ideas from the theory of 3-phase circuits.

In a 3-phase system, the voltages and currents are three signals shifted by $\frac{2\pi}{3}$ radians from each other. As an example, a set of balanced 3-phase voltages could be

$$V_A(t) = M\cos(\omega_0 t), \tag{20.1}$$

$$V_B(t) = M\cos\left(\omega_0 t - \frac{2\pi}{3}\right), \tag{20.2}$$

$$V_C(t) = M\cos\left(\omega_0 t - \frac{4\pi}{3}\right). \tag{20.3}$$

20.5.1 Alpha, Beta Coordinates

For a symmetrical system, it is convenient to use the, so-called, Clarke transformation, to convert to α, β, γ coordinates. The transformation is

$$\begin{bmatrix} V_\alpha \\ V_\beta \\ V_\gamma \end{bmatrix} = \frac{2}{3}\begin{bmatrix} 1 & -\frac{1}{2} & -\frac{1}{2} \\ 0 & \frac{\sqrt{3}}{2} & -\frac{\sqrt{3}}{2} \\ \frac{1}{2} & \frac{1}{2} & \frac{1}{2} \end{bmatrix}\begin{bmatrix} V_A \\ V_B \\ V_C \end{bmatrix}. \tag{20.4}$$

The above transformation can be expressed in vector form as

$$\mathbf{V}^{\alpha\beta\gamma} = K_1 \mathbf{V}^{ABC}. \tag{20.5}$$

For V_A, V_B, V_C as in (20.1)–(20.3), it follows that

$$V_\alpha = M\cos\omega_0 t, \tag{20.6}$$

$$V_\beta = M\sin\omega_0 t, \tag{20.7}$$

$$V_\gamma = 0. \tag{20.8}$$

20.5.2 *dq* Frame

A further transformation to a dq frame generates DC voltages which 'rotate' with the $\alpha\beta$ voltages. The required transformation is

$$\begin{bmatrix} V_d \\ V_q \end{bmatrix} = \begin{bmatrix} \cos\omega_0 t & \sin\omega_0 t \\ -\sin\omega_0 t & \cos\omega_0 t \end{bmatrix}\begin{bmatrix} V_\alpha \\ V_\beta \end{bmatrix}. \tag{20.9}$$

When the transformation (20.9) is applied to the voltages given in (20.6) and (20.7), the result is

$$V_d = M, \tag{20.10}$$

$$V_q = 0. \tag{20.11}$$

When the transformation is applied to

$$V_\alpha = M\cos(\omega_0 t - \phi), \tag{20.12}$$

$$V_\beta = M\sin(\omega_0 t - \phi), \tag{20.13}$$

then the result is

$$V_d = M\cos\phi, \tag{20.14}$$

$$V_q = -M\sin\phi. \tag{20.15}$$

A third component, V_z, is defined to be V_γ. Note that V_z is zero for a symmetrical system. The transformation can be expressed in vector form as

$$\boldsymbol{V}^{dqz} = K_2 \boldsymbol{V}^{\alpha\beta\gamma}. \tag{20.16}$$

To illustrate the above ideas, say that $\boldsymbol{V}^{ABC}$ is a three-phase voltage with magnitude $\overline{V}$ as in (20.1)–(20.3). Let $\boldsymbol{I}^{ABC}$ be the current vector which lags $\boldsymbol{V}^{ABC}$ by ϕ with magnitude $\overline{I}$.

Then

$$\boldsymbol{V}^{dq} = \begin{bmatrix} \overline{V} \\ 0 \end{bmatrix}, \tag{20.17}$$

$$\boldsymbol{I}^{dq} = \begin{bmatrix} \overline{I}\cos\phi \\ -\overline{I}\sin\phi \end{bmatrix}. \tag{20.18}$$

The advantage of using the dq frame is that AC signals are converted to DC signals. However, when building models, it is necessary to account for the rotation of the dq frame (see Section 20.5.4).

The two vectors in (20.17) and (20.18) can be represented in the dq frame as shown in Figure 20.3.

20.5.3 The Inverse Transformation

One can go from the dq frame back to the original ABC frame by applying the inverse transformation.

$$\boldsymbol{V}^{dqz} = K_2 K_1 \boldsymbol{V}^{ABC}, \tag{20.19}$$

$$\boldsymbol{V}^{ABC} = K_1^{-1} K_2^{-1} \boldsymbol{V}^{dqz}. \tag{20.20}$$

20.5.4 First-Order Dynamics in *dq* Frame

Consider two decoupled first-order dynamical systems expressed in $\alpha\beta$ coordinates as

$$L\frac{di_\alpha}{dt} = -Ri_\alpha + V_\alpha, \tag{20.21}$$

$$L\frac{di_\beta}{dt} = -Ri_\beta + V_\beta. \tag{20.22}$$

The dq frame is non-stationary. Thus, one needs to account for the moving frame when evaluating derivatives in the dq frame. Specifically, using (20.9),

$$\begin{bmatrix} i_d \\ i_q \end{bmatrix} = \begin{bmatrix} \cos\omega_0 t & \sin\omega_0 t \\ -\sin\omega_0 t & \cos\omega_0 t \end{bmatrix} \begin{bmatrix} i_\alpha \\ i_\beta \end{bmatrix}. \tag{20.23}$$

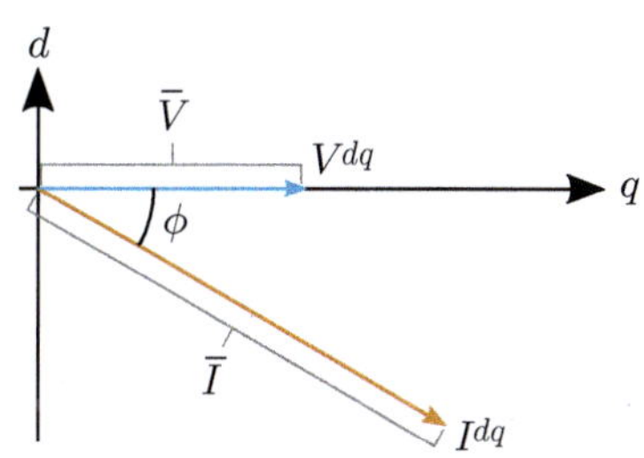

Figure 20.3 Voltage and current vectors in dq frame.

Then applying the chain rule for differentiation:

$$
\begin{aligned}
\frac{d}{dt}\begin{bmatrix} i_d \\ i_q \end{bmatrix} &= \frac{d}{dt}\begin{bmatrix} \cos\omega_0 t & \sin\omega_0 t \\ -\sin\omega_0 t & \cos\omega_0 t \end{bmatrix}\begin{bmatrix} i_\alpha \\ i_\beta \end{bmatrix} + \begin{bmatrix} \cos\omega_0 t & \sin\omega_0 t \\ -\sin\omega_0 t & \cos\omega_0 t \end{bmatrix}\frac{d}{dt}\begin{bmatrix} i_\alpha \\ i_\beta \end{bmatrix}, \\
&= \begin{bmatrix} -\omega_0\sin\omega_0 t & \omega_0\cos\omega_0 t \\ -\omega_0\cos\omega_0 t & -\omega_0\sin\omega_0 t \end{bmatrix}\begin{bmatrix} i_\alpha \\ i_\beta \end{bmatrix} + \begin{bmatrix} \cos\omega_0 t & \sin\omega_0 t \\ -\sin\omega_0 t & \cos\omega_0 t \end{bmatrix}\frac{d}{dt}\begin{bmatrix} i_\alpha \\ i_\beta \end{bmatrix}, \\
&= \begin{bmatrix} \omega_0 i_q \\ -\omega_0 i_d \end{bmatrix} + \begin{bmatrix} \cos\omega_0 t & \sin\omega_0 t \\ -\sin\omega_0 t & \cos\omega_0 t \end{bmatrix}\begin{bmatrix} -\frac{R}{L}i_\alpha + \frac{1}{L}V_\alpha \\ -\frac{R}{L}i_\beta + \frac{1}{L}V_\beta \end{bmatrix}, \\
&= \begin{bmatrix} \omega_0 i_q \\ -\omega_0 i_d \end{bmatrix} + \begin{bmatrix} -\frac{R}{L}i_d + \frac{1}{L}V_d \\ -\frac{R}{L}i_q + \frac{1}{L}V_q \end{bmatrix}.
\end{aligned} \tag{20.24}
$$

In summary, the corresponding dynamics in the dq frame are

$$L\frac{di_d}{dt} = -Ri_d + V_d + \omega_0 L i_q, \tag{20.25}$$

$$L\frac{di_q}{dt} = -Ri_q + V_q + \omega_0 L i_d. \tag{20.26}$$

Note that the dynamics are coupled in the dq frame. Feedforward is sometimes used to cancel the coupling terms when designing control laws. This leads to two decoupled first-order systems.

20.6 Generator Side Model

It is assumed that the 3-phase currents and voltages are balanced, i.e. the magnitudes are the same and the phases differ by 120°. This means that, when transformed to α, β, γ coordinates, the γ component will be zero (see Section 20.5).

Two frequencies are of interest, namely, the mechanical frequency, ω_m, and the electrical frequency, ω_e.

The electrical frequency ω_e will differ from ω_m (scaled by the number of pole pairs) due to slip. In particular,

$$\omega_e = P_p\omega_m + \omega_{slip}, \tag{20.27}$$

where

$$
\begin{aligned}
P_p &= \text{the number of pole pairs in the induction generator} \\
\omega_m &= \text{the angular velocity of the induction motor shaft} \\
\omega_{slip} &= \text{the slip frequency.}
\end{aligned}
$$

A transformation to a dq frame aligned with the rotor flux simplifies the model and controller design. Assume the rotor flux satisfies

$$\begin{bmatrix} \lambda_{r\alpha} \\ \lambda_{r\beta} \end{bmatrix} = \begin{bmatrix} \cos\theta_r \\ \sin\theta_r \end{bmatrix}|\lambda_r|\,;\theta_r = \omega_e t + \delta_r \tag{20.28}$$

Then the required tansformation to dq coordinates is

$$\begin{bmatrix} \lambda_{rd} \\ \lambda_{rq} \end{bmatrix} = \begin{bmatrix} \cos\theta_r & \sin\theta_r \\ -\sin\theta_r & \cos\theta_r \end{bmatrix}\begin{bmatrix} \lambda_{r\alpha} \\ \lambda_{r\beta} \end{bmatrix} = \begin{bmatrix} |\lambda_r| \\ 0 \end{bmatrix} \tag{20.29}$$

In the same coordinate system, the stator current is

$$\begin{bmatrix} i_{sd} \\ i_{sq} \end{bmatrix} = \begin{bmatrix} \cos\theta_r & \sin\theta_r \\ -\sin\theta_r & \cos\theta_r \end{bmatrix} \begin{bmatrix} i_{s\alpha} \\ i_{s\beta} \end{bmatrix} \tag{20.30}$$

where

$$\begin{bmatrix} i_{s\alpha} \\ i_{s\beta} \end{bmatrix} = \begin{bmatrix} \cos\theta \\ \sin\theta \end{bmatrix} |I_s|. \tag{20.31}$$

In this coordinate system, a reasonably accurate model for the stator currents is

$$\sigma L_s \frac{di_{sd}}{dt} = -R_s i_{sd} + v_{sd} + \omega_e \sigma L_s i_{sq}, \tag{20.32}$$

$$\sigma L_s \frac{di_{sq}}{dt} = -R_s i_{sq} + v_{sq} - \omega_e \sigma L_s i_{sd} - \omega_e \frac{L_m^2}{L_r} i_{sd}, \tag{20.33}$$

where $v_{sd}, v_{sq}, i_{sd}, i_{sq}$ denote the stator voltages and currents referred to the dq frame, R_s is the stator resistance and ω_e is the electrical frequency. The other symbols appearing in (20.32) and (20.33) are defined as follows:

$$L_r = L_m + L_{\ell r}, \tag{20.34}$$

$$L_s = L_m + L_{\ell s}. \tag{20.35}$$

Here, $L_m, L_{\ell r}, L_{\ell s}$ denote the magnetising inductance, leakage inductance in the rotor, and leakage inductance in the stator, respectively. Also, σ is the leakage factor defined as

$$\sigma = 1 - \frac{L_m^2}{L_r L_s}. \tag{20.36}$$

In the dq frame it can be shown that the following results hold

$$\omega_{slip} = \frac{R_r i_{sq}}{L_r i_{sd}}, \tag{20.37}$$

$$\frac{i_{sq}}{i_{sd}} = \left(\frac{\tau_r}{\tau_r s + 1} \right) \omega_{slip}, \tag{20.38}$$

$$\tau_r = \frac{L_r}{R_r} \tag{20.39}$$

where s denotes the Laplace transform variable.

For an induction generator i_{sq} and ω_{slip} will be negative.

Also note that for a generator the third term in the right hand side of (20.33) will have greater significance than i_{sq}.

20.7 Generator Side Control Law

There are three components to the generator side control law, namely,

(i) regulation of i_{sd} to a (fixed) set point i_{sd}^*
(ii) regulation of i_{sq} to a (variable) set point i_{sq}^*
(iii) alignment of the dq frame with the rotor flux in the air gap.

These three aspects are discussed below.

20.7.1 Regulation of i_{sd}

The current i_{sd} is used as a magnetising current. The manipulated variable used in this case is v_{sd} which is manipulated by the conduction times in the voltage-controlled rectifier. A simple PI feedback control law is used to regulate i_{sd} to the set point i^*_{sd}. Since the last term in (20.32) is small, then i_{sd} is only marginally impacted by changes in i_{sq}. Hence, it is reasonable to assume that i_{sd} deviates little from its set point. Therefore, in the future development, it will be assumed that $i_{sd} = i^*_{sd}$.

20.7.2 Regulation of i_{sq}

With $i_{sd} = i^*_{sd}$, then the torque produced by the generator can be shown to be

$$T_e = \frac{3}{2}\frac{L_m^2}{L_r} i^*_{sd} i_{sq} P_p, \tag{20.40}$$

where P_p is the number of pole pairs in the generator.

Hence, given a desired torque T^*_e (requested by the rotational velocity control law), the above equation can be inverted to give a set point, i^*_{sq}, for i_{sq}:

$$i^*_{sq} = \frac{2T^*_e L_r}{3L_m^2 i^*_{sd} P_p}. \tag{20.41}$$

A simple PI feedback control law can be used to regulate i_{sq} to i^*_{sq} using v_{sq} as the manipulated variable. The switching of the controlled rectifier will be used to deliver v_{sq}. This inverter operates very fast and hence it is reasonable to assume that v_{sq} is set equal to the value requested by the control law.

20.7.3 Alignment of *dq* Frame

The final step is to align the *dq* frame with the rotor flux vector. This requires a rotor flux angle estimator. Correspondingly, $(\hat{i}_{sd}, \hat{i}_{sq})$ are defined by an equation analogous to (20.30)

$$\begin{bmatrix} \hat{i}_{sd} \\ \hat{i}_{sq} \end{bmatrix} = \begin{bmatrix} \cos\hat{\theta}_r & \sin\hat{\theta}_r \\ -\sin\hat{\theta}_r & \cos\hat{\theta}_r \end{bmatrix} \begin{bmatrix} \hat{i}_{s\alpha} \\ \hat{i}_{s\beta} \end{bmatrix}, \tag{20.42}$$

where $\hat{\theta}_r$ is determined by the following estimator

$$\frac{\mathrm{d}}{\mathrm{d}t}\hat{\theta}_r(t) = \omega_m + \hat{\omega}_{slip}, \tag{20.43}$$

where $\hat{\omega}_{slip}$ is defined using $\hat{i}_{sq}$ in place of i_{sq} in (20.37), i.e.

$$\hat{\omega}_{slip} = \frac{R_r \hat{i}_{sq}}{L_r i^*_{sd}}. \tag{20.44}$$

Some properties of the estimator (20.42), (20.43), (20.44) are explored in the Revision Questions at the end of the chapter.

In particular it is shown that $\hat{\theta}_r$ is an asymptotically unbiased estimate of θ_r.

20.7.4 Conversion of v_{sd}, v_{sq} Back to Time Domain

The control laws defined in Sections 20.7.1 and 20.7.2 lead to voltages v_{sd}, v_{sq}. These are converted back to the time domain via

$$\begin{bmatrix} v_{s\alpha} \\ v_{s\beta} \end{bmatrix} = \begin{bmatrix} \cos\hat{\theta}_r & -\sin\hat{\theta}_r \\ \sin\hat{\theta}_r & \cos\hat{\theta}_r \end{bmatrix} \begin{bmatrix} v_{sd} \\ v_{sq} \end{bmatrix} \tag{20.45}$$

20.8 The Link Capacitor Model

There are two components to this model, namely, (i) the current into the capacitor, denoted I_{DC}^{in} and (ii) the equations describing the evolution of the voltage V_{DC}. Note that the current out of the capacitor I_{DC}^{out} will be used as the manipulated variable when regulating the voltage on the DC-link capacitor.

20.8.1 Current into the Capacitor

The current i_{sq} is rectified and fed into the link capacitor as I_{DC}^{in}. To calculate the value of I_{DC}^{in}, one uses the assumption that the controlled rectifier is lossless. Hence, one can equate the electrical power coming from the generator to the power entering the capacitor.

Let the voltage on the capacitor be V_{DC}. Then the power delivered to the capacitor is

$$P_{in} = V_{DC} I_{DC}^{in}. \tag{20.46}$$

The electrical power, P_e, coming from the generator is given by the product of the torque by the electrical angular velocity ω_e. Hence, electrical power can be estimated using (20.40)

$$\hat{P}_e = \frac{3}{2}\omega_e \frac{L_m^2}{L_r} i_{sd}^* \hat{i}_{sq}. \tag{20.47}$$

By equating P_{in}, as given in (20.46), to, P_e, as given in (20.47) one obtains

$$I_{DC}^{in} = \frac{\frac{3}{2}\omega_e L_m^2 i_{sd}^* \hat{i}_{sq}}{L_r V_{DC}}. \tag{20.48}$$

I_{DC}^{in} will be treated as a disturbance when designing the regulator for the voltage on the link capacitor.

20.8.2 Dynamics of the Capacitor

The evolution of the capacitor voltage satisfies a simple first-order equation:

$$C\frac{dV_{DC}}{dt} = I_{DC}^{in} - I_{DC}^{out}, \tag{20.49}$$

where I_{DC}^{in} is the input DC current given in (20.48) and I_{DC}^{out} is the output current. The latter quantity will be used as a manipulated variable in a cascade loop as described in Section 20.9.

20.9 Regulation of the Capacitor Voltage

A simple PI control law can be used to request a current flow $(I_{DC}^{out})^*$ to regulate V_{DC} to a set point V_{DC}^*. See Figure 20.5b.

20.10 Model for the Grid Side Transformer

The capacitor voltage is switched by a controlled inverter to produce AC voltages. Note that these voltages can again be expressed in another 'rotating' dq frame. Here, the relevant frequency is ω_{grid} (the frequency of the grid). Also, the reference frame is here aligned with the grid voltage, v_g. In the dq frame, the voltages obtained by switching the Link Capacitor Voltage can be represented by two DC voltages, namely, v_{cq} and v_{cd}.

It is assumed that the voltage, v_g, of the grid is given and fixed. In the dq frame the grid voltage is described as follows:

$$v_{gd} = \text{direct axis grid voltage (given)}$$

$$\begin{aligned} v_{gq} &= \text{quadratic axis grid voltage} \\ &= 0 \text{ (By definition of the alignment of the rotating frame used on the grid side.)} \end{aligned}$$

Let L and R denote the inductance and resistance of the cabling (and transformers) connecting the inverter output to the grid. Then the equations for the grid current delivery system are:

$$L\frac{di_{cd}}{dt} = -Ri_{cd} + v_{gd} - v_{cd} + L\omega_g i_{cq}, \tag{20.50}$$

$$L\frac{di_{cq}}{dt} = -Ri_{cq} + v_{gq} - v_{cq} - L\omega_g i_{cd}. \tag{20.51}$$

The terms depending on ω_g in (20.50) and (20.51) arise due to the use of dq frame – see Section 20.5.4.

The electrical power delivered to the grid is given by

$$\begin{aligned} P_e^{grid} &= V_{DC} I_{DC}^{out}, \\ &= \frac{3}{2} v_{cd} i_{cd}, \end{aligned} \tag{20.52}$$

where v_{cd}, i_{cd} are the magnitude of the AC voltage and current out of the inverter in the rotating dq frame.

20.11 The Grid Side Control Law

There are two components of this law, namely, (i) the regulation of i_{cq} to a set point i_{cq}^*, (ii) the regulation of i_{cd} to a set point i_{cd}^*. The basic grid side control architecture is shown in Figure 20.4.

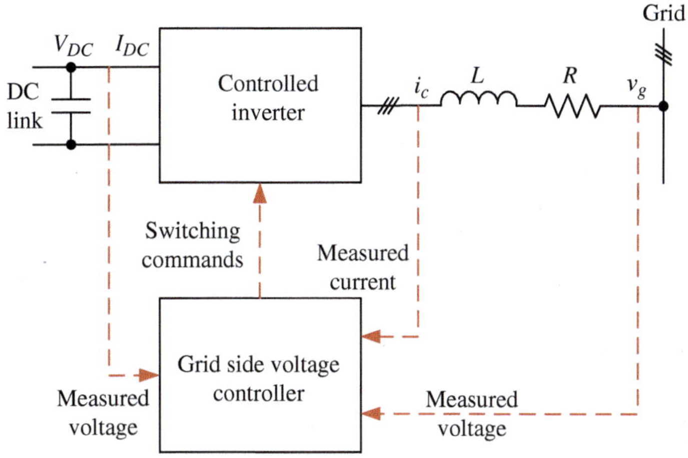

Figure 20.4 Grid side control.

20.11.1 Regulation of i_{cq}

To illustrate the ideas, it is assumed that the wind power system is required to deliver power to the grid with unity power factor. Since the dq coordinates are aligned with the grid voltage, then $v_{gq} = 0$. The unity power factor requirement leads to the following set point for i_{cq}:

$$i_{cq}^{*} = 0. \tag{20.53}$$

The current i_{cq} is regulated to the set point (20.53) by use of v_{cq}.

20.11.2 Regulation of i_{cd}

Finally, the voltage v_{cd} is used to regulate the current i_{cd} to a set point i_{cd}^{*}. The set point, i_{cd}^{*}, is determined by the required DC current flow $(I_{DC}^{out})^{*}$. Again the switches are assumed to be lossless so that i_{cd}^{*} is obtained by using a Power Balance across the inverter. Using (20.52), this leads to

$$i_{cd}^{*} = \frac{V_{DC}(I_{DC}^{out})^{*}}{\frac{3}{2}v_{grid}}, \tag{20.54}$$

where $(I_{DC}^{out})^{*}$ is the DC output current requested by the DC Link Voltage regulator as described in Section 20.9.

Conversely, if i_{cd} is the actual current, then I_{DC}^{out} is given by

$$I_{DC}^{out} = \frac{\frac{3}{2}v_{cd}i_{cd}}{V_{DC}}. \tag{20.55}$$

This current goes into the model (20.49). The required voltage v_{cd} used to regulate i_{cd} to the set point i_{cd}^{*} is obtained by using PWM switches on the DC Link Voltage.

20.12 Complete Electrical System Control Law

A summary of the final control system is given in Figure 20.5.

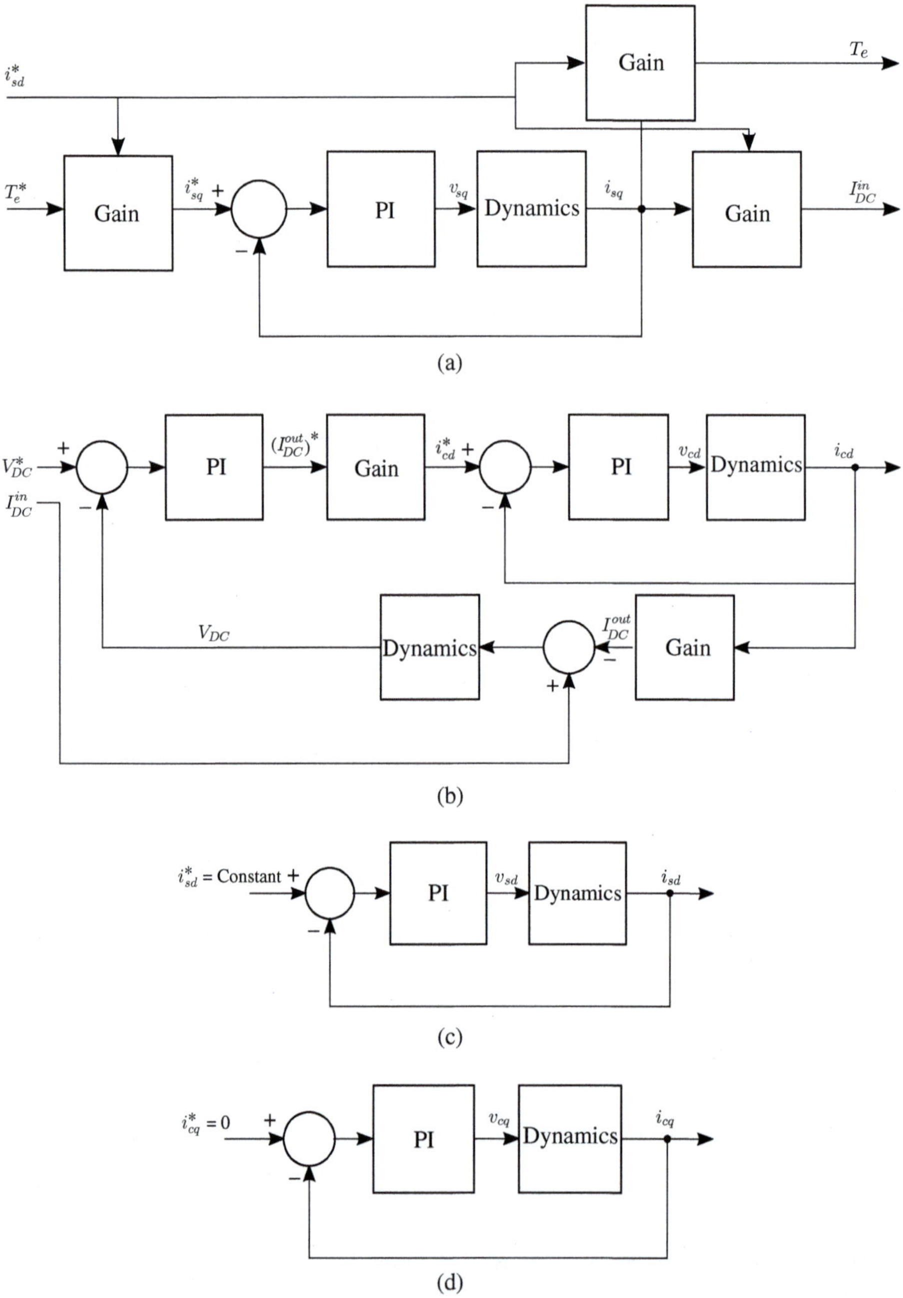

Figure 20.5 The full electrical control system. (a) Electrical torque loop, (b) DC link capacitor loop, (c) generator magnetising loop, and (d) power factor loop.

20.13 Testing the Electrical Control Laws

20.13.1 Generator Side

(a) Go to the Program > Load Controller menu item and select the Generator Side Control. Configure the C_i controller parameters to be

$$K_p = 4.0, \tag{20.56}$$

$$K_i = 2.0. \tag{20.57}$$

N.B. This will change the parameters for both C_i controllers as they are linked.
Click on the Wind Speed button and set a constant wind speed with offset = 5.0 m/s.
Read the steady-state values of rotor speed, generator speed, generator velocity, generator current, generator power, power delivered to DC link capacitor and grid power.

(b) Change the wind speed to 6.0 m/s.
Read the new steady-state values.

(c) By what factor has the power increased?

20.13.2 Grid Side

Load the grid side controller.
The Link Capacitor Voltage set point is 650 V.
In the Link Capacitor Control Law use

$$K_p = 1.33, \tag{20.58}$$

$$K_i = 133. \tag{20.59}$$

In the two grid-side control, laws use

$$K_p = 1.44, \tag{20.60}$$

$$K_i = 160. \tag{20.61}$$

Check Link Capacitor Voltage and other relevant operating conditions.

20.14 Experiments on the Complete System

Up to this point, the mechanical and electrical parts of the wind power turbine system have been considered separately. In reality, all of the systems necessarily operate simultaneously. This fact leads to further issues regarding control of the system.

Previously, when considering each part of the turbine separately, the assumption was made that the wind would always blow directly into the turbine. Unfortunately, this is not always true, which is why a yaw controller is used. Another issue is that, if one demands an angular velocity proportional to wind speed then, even in conditions when the turbine is facing perpendicular to the wind direction - that is, when no wind is actually providing mechanical torque - the electrical torque will still be attempting to keep the same angular velocity as if it were facing directly towards the wind direction.

Most of the time the above issues will cause no problem, as wind direction will usually change very slowly over short periods of time (see Wind Rose given in Figure 20.6). However, sudden gusts

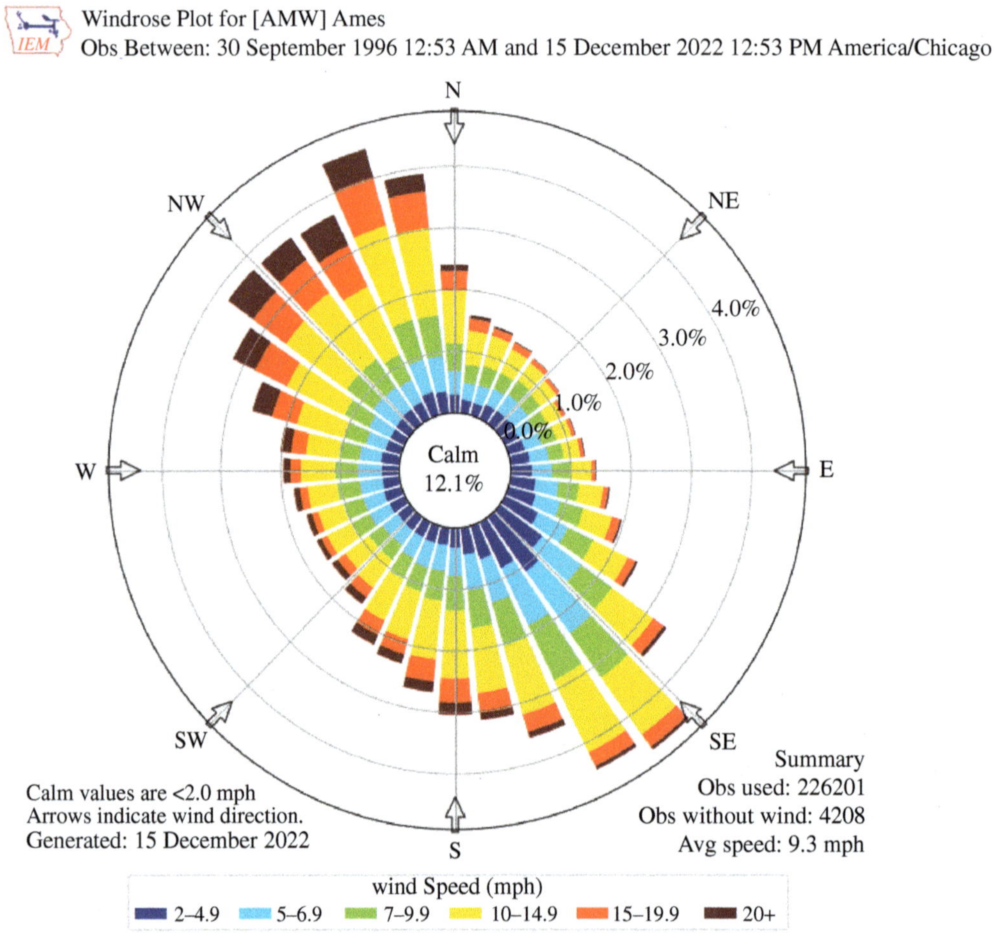

Figure 20.6 Distribution graph of wind direction – wind rose.

of wind may cause problems to occur. If the speed of the yaw control is slow, then electrical torque may increase to very large values, putting an unnecessary drain on the grid.

Hence in the design of the Turbine Speed Controller, rather than only considering absolute wind speed, one should instead consider the wind speed moving perpendicular to the turbine. This can be easily achieved by calculating:

$$V_{perp} = V \cos(\Theta), \tag{20.62}$$

where Θ is the angle between turbine orientation and wind direction. In (20.62), V_{perp} is the projection of the wind velocity in the direction of the turbine. Earlier it was shown how the desired angular velocity for maximum power can be found by using the wind speed. However, to prevent T^* from increasing, it is often desirable to use V_{perp} instead.

20.14.1 Experiment: Testing the Impact of Wind Direction

Set the yaw controller parameter to $K_p = 1, K_d = 3$.

Set the wind direction to 60° from the turbine direction with constant wind speed.

Set the set-point for yaw to 0°.
Use wind speed of 6 m/s.
After testing the system, change the yaw set point to the measured wind direction.
Set the wind direction to a sine wave □□□

20.14.2 Experiment: Testing the Impact of Wind Speed

You may also notice that with a less responsive rotational velocity controller and lower wind speeds, the rotational velocity may sometimes become negative.

The reason this occurs is that when the angular velocity is small, then squaring angular velocity will result in a very small mechanical torque. If the angular velocity controller is a less responsive controller, then T^* will not change very much over a short period of time and this will result in a reduction of angular velocity. This extra angular velocity drop will cause an even larger drop in mechanical torque, until mechanical torque becomes close to 0. Because the electrical torque will be greater than 0, this will push the turbine in the opposite direction until the controller takes effect.

This can be counteracted by using a more responsive controller, since then T_e^* will change more quickly and therefore angular velocity will not reduce. However, this may result in relatively severe fluctuations in the supplied grid power, as the desired electrical torque can change very quickly over a short period of time. It is important to find a compromise between these two extremes.

Unfortunately, wind speed and direction are not constant in practice. Typical wind direction changes are graphically represented in the form of a 'wind rose', see Figure 20.6 (from https://mesonet.agron.iastate.edu/onsite/windrose/IA_ASOS/AMW/AMW_yearly.png).

From the diagram one can see that wind changes in both speed and direction. However, over the course of a few minutes, it is unlikely that the wind direction will change much. In fact, the wind direction can remain relatively unchanged over the course of several days. The only times when wind direction will change rapidly are during strong storms. During these periods it is desirable that the turbine be shut down. It is therefore easy to model wind direction as a small Gaussian noise centred around a particular angle.

(a) Test the complete system by fixing the wind direction and use a sinusoidal wind speed with off-set 6.0 m/s and frequency 0.1 Hz. Choose amplitude 0.1, 0.5.
Record rotation velocity and grid power.

(b) Reduce the gain of the rotational velocity controller to say

$$K_p = 15{,}000, \tag{20.63}$$

$$K_i = 2400, \tag{20.64}$$

and repeat the above test with sinusoidal wind speed amplitude set to 0.5.

(c) Comment on the importance of tuning the rotational velocity control law.

Readers should find that, when less proportional control is used in the rotational velocity controller, then a smoother response is obtained in the supplied power to the grid.

20.15 Revision Questions

1. A device commonly used in Power Electronics and Telecommunications is a phase-locked loop (PLL). This and subsequent questions explore the characteristics of PLLs. Additional information is contained in reference [1] given in Further Reading.

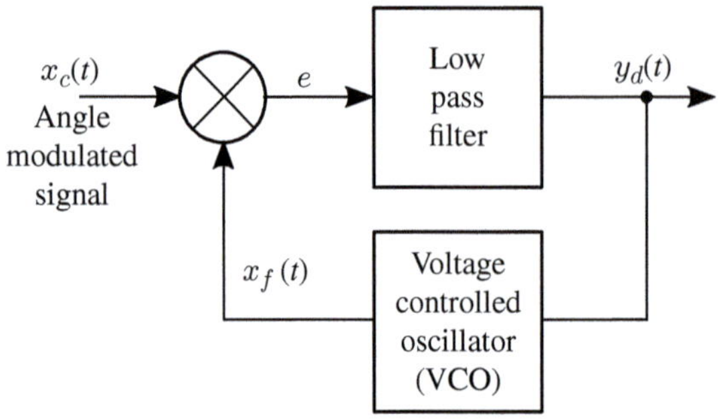

Figure 20.7 Phase locked loop.

A basic PLL is shown in Figure 20.7.
Say

$$x_c(t) = A_c \sin(\omega_c t + \theta_c(t)), \tag{20.65}$$

$$\theta_c(t) = k_f \int_{-\infty}^{t} x(\tau)d\tau. \tag{20.66}$$

When $y_d = 0$; $x_f(t) = A_f \cos(\omega_c t)$.
When $y_d \neq 0$;

$$x_f(t) = A_f \cos(\omega_c t + \theta_f), \tag{20.67}$$

$$\theta_f = g_u \int_{-\infty}^{t} y_d(\tau)d\tau. \tag{20.68}$$

(a) Obtain an expression for e.
(b) What is approximate value of the output of the low pass filter?
(c) Obtain an expression for $y_d(t)$ if $(\theta_t - \theta_f)$ is small.
(d) Using your answer to part (c) redraw the PLL in terms of phases.

2. (a) Use your answer to Question 1 part (d) to obtain the transfer function linking θ_i to y_d.
   (b) Say that the phase θ_i from of constant off-set $\tilde{\omega}$ in the input frequency. Calculate the steady-state value of y_d and the phase error between input and output, Ψ.
3. Return to the function $\sin\left(\theta_i - \theta_f\right)$. Use this function to describe the capture range of the PLL.
4. A commonly used embellishment of a PLL is to include a filter $H(s)$ prior to y_d.
   (a) Why might it be useful to include an integrator in $H(s)$?
   (b) Say $H(s)$ is chosen as $\left[\frac{1+s\tau_2}{s\tau_1}\right]$.
   Find an expression for the transfer function linking $\theta_i(s)$ to Y_d.
   (c) What is the damping and natural frequency of the new closed loop?
5. This question shows that the algorithm for aligning the phase with the rotor flux vector described in Section 20.7.3 can be viewed as a PLL.
   Recall from (20.4) that $\hat{\omega}_{slip} = K\hat{i}_{sq}$ where $K = \frac{R_r}{L_r i^*_{sd}}$.
   Also, (20.43) gives

$$\hat{\theta}_r(t) = \int_{-\infty}^{t} \left(\omega_m + \hat{\omega}_{slip}\right) d\tau. \tag{20.69}$$

   (a) What is the relationship of ω_m to ω_c in Question 1?
   (b) Interpret the equation for $\hat{i}_{sq}$ in (20.42) as a phase comparator.

(c) Hence, validate the estimate $\hat{\theta}_r$ by showing that

$$\lim_{t\to\infty} (\theta - \hat{\theta}_r) = \lim_{t\to\infty} (\theta - \theta_r) \tag{20.70}$$

where

$$\begin{bmatrix} i_{sd} \\ i_{sq} \end{bmatrix} = |I_s| \begin{bmatrix} \cos\theta \\ \sin\theta \end{bmatrix} \tag{20.71}$$

Further Reading

1 K.S. Shanmugam, *Digital communication systems*, John Wiley & Sons, Inc. 1979.

2 A. Grauers, "Efficiency of three wind energy generator systems", *IEEE Transactions on Energy Conversion*, 11(3):650–657, 1996.

3 M.M. Hand, K.E. Johnson, L.J. Fingersh, and A. D. Wright, "Advanced Control Design and Field Testing for Wind Turbines at the National Renewable Energy Laboratory", *World Renewable Energy Congress VIII*, Denver, Colorado, USA, 2004, available www.nrel.gov/docs/fy04osti/36118.pdf.

4 L.H. Hansen, P.H. Madsen, F. Blaabjerg, H.C. Christensen, U. Lindhard and K. Eskildsen, "Generators and Power Electronics Technology for Wind Turbines", *Annual Conference of the IEEE Industrial Electronics Society*, Denver, Colorado, USA, 2001.

5 S.S. Murthy, B. Singh, P.K. Goel and S.K. Tiwari, "A Comparative Study of Fixed Speed and Variable Speed Wind Energy Conversion Systems Feeding the Grid", *International Conference on Power Electronics and Drive Systems*, Bangkok, Thailand, 2007.

6 L.Y. Pao and K.E. Johnson, "A Tutorial on the Dynamics and Control of Wind Turbines and Wind Farms", *Proceedings of the American Control Conference*, St. Louis, Missouri, USA, 2009, pp. 2076–2089.

7 H. Stiesdal, Bonus Info Autumn 1999, "The Wind Turbine Components and Operation", Bonus Energy, 1999, available http://www.windmission.dk/workshop/BonusTurbine.pdf.

8 E. Valsera-Naranjo, A. Sumper, O. Gomis-Bellmunt, A. Junyent-Ferre, and M. Martinez-Rojas, "Design of a Pitch Control of a Wind Turbine to Improve System Frequency Response," *2009 13th European Conference on Power Electronics and Applications*, Barcelona, Spain, 2009.

9 J. Zhang, M. Cheng, Z. Chen and X. Fu, "Pitch Angle Control for Variable Speed Wind Turbines", *International Conference on Electric Utility Deregulation and Restructuring and Power Technologies*, 2008.

21

Laboratory 10: Cross-Directional Control in Paper Machines: PID Control

21.1 Introduction

In this chapter, a realistic multivariable industrial control problem (namely, cross-directional control in a paper machine) will be studied. A screen shot of the virtual laboratory is given in Figure 21.1. The problem will be transformed so that simple PID control can be applied. This will seem to give excellent results. However, the reader is warned that multivariable interactions have a profound, and unexpected, impact on the control problem. Chapter 27 will deal with these more complicated issues.

21.2 Web-Forming Process

A paper machine is a quintessential example of a web-forming process. Web-forming processes represent a wide class of industrial processes with relevance in many different production areas such as paper making, plastic film extrusion, steel rolling, coating and laminating. These are large-scale processes, involving complex machinery, which exhibit challenging difficulties for the design of efficient control systems. As such, they have been the object of much investigation (see references given in Further Reading).

The processes typically give rise to two types of control problems. One is concerned with regulation of a single variable representing the average properties across the strip. This control problem is usually called the 'Machine Direction' (MD) control problem. This is a single-input single-output problem with significant delay in the measurements – see Chapter 12. The other control problem relates to the maintenance of the profile across the strip. This is referred to as the 'Cross Directional' (CD) control problem. The latter problem is more challenging since it is highly multivariable in nature. The CD problem will be the focus of this laboratory.

Figure 21.2 shows the main components of a general web-forming process. In order to control the cross-directional profile of the web, an array of actuators is typically deployed evenly distributed across the sheet. The number of actuators can vary from as few as 30 up to as many as 300. The exit profiles, are typically measured via an array of sensors placed in a downstream position. Alternatively, a set of measurements can be obtained via a scanning sensor that moves back and forth in the cross direction as the strip passes under a fixed location.[1] The number of measurements taken by a single scan of the sensor can be up to 1000. For the purpose of the laboratory, it will be

1 Kalman filtering ideas can be used to convert the scanned data into estimates of the CD profile – see Chapter 23.

Practical Control System Design: Real World Designs implemented on Emulated Industrial Systems,
First Edition. Adrian Medioli and Graham Goodwin.

Companion website: www.wiley.com/go/medioli/practicalcontrolsystemdesign

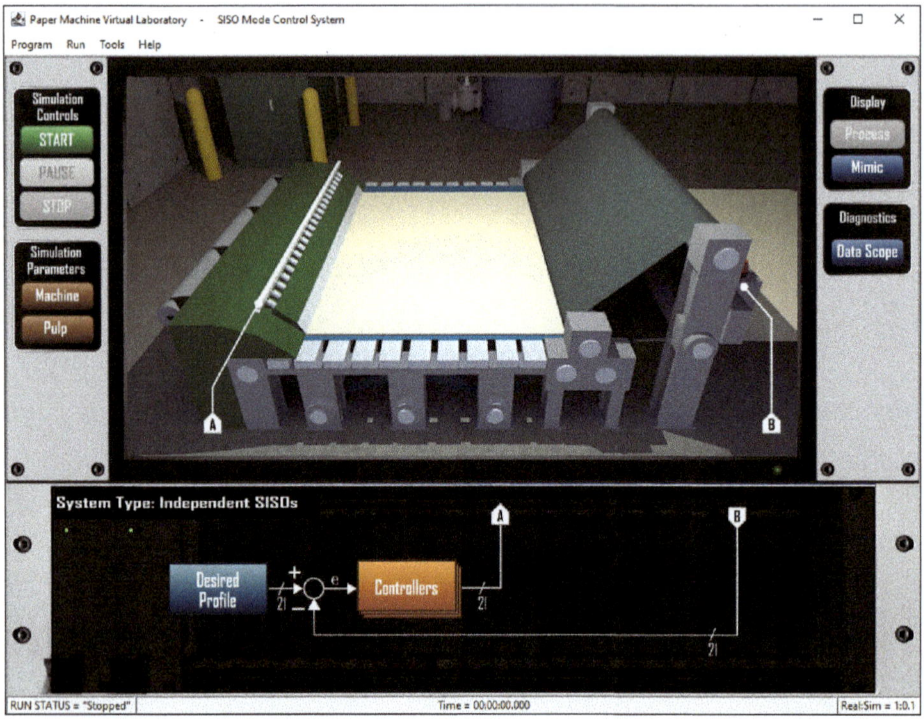

Figure 21.1 Screenshot of program.

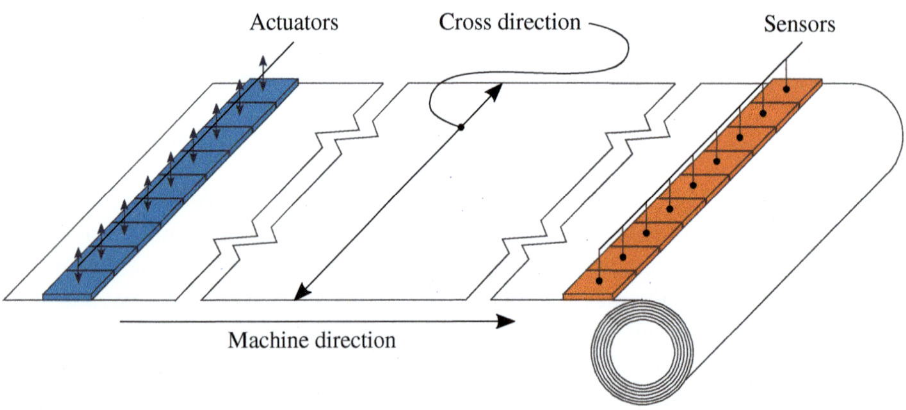

Figure 21.2 Generic web forming process.

assumed that the number of measurements is equal to the number of actuators and that actuators are aligned with the sensing points.

If the number of sensors differs from the number of actuators, then it is possible to project the sensor data onto the actuator space by preliminary signal processing techniques, e.g. by using the Kalman filter – see Chapter 23.

Some of the difficulties associated with the cross-directional control problem include:

- the high dimensionality of the cross-directional system. It is not uncommon to have several hundred actuators in the cross direction;

- the high spatial interaction between actuators;
- the uncertainty in the model;
- the limited control authority of the actuators.

21.3 Basis Weight Control in a Paper Machine

So as to be specific, the laboratory will focus on CD control problems as experienced in paper machines. Note, however, that essentially identical problems arise in other web-forming processes.

A paper machine typically has five separate sections, namely, head box, wire section, press section, drying section and coating section.

Before the pulp enters the paper machine, the surface of the fibres is roughened by rotating blades. Chalk filler, starch and other chemicals are then mixed with the pulp before it reaches the head box. The head box comprises a large nozzle. Here the mixture is spread onto a rapidly moving wire mesh. The moisture content at this stage is high (approximately 99%). The wire section reduces the water content to approximately 70%. The jet flow at the head box and wire section are important in the paper-making process. In the press section, the web (formed in the wire section) is passed between high-pressure rolls to expel further moisture. The water content is now about 50%. In the drying section, heat is applied to further expel moisture down to about 5%. Finally, the coating section applies additives aimed at improving smoothness and other print related qualities.

Basis weight (i.e. mass of fibres per unit area) is a key property of paper. Basis weight is typically measured by a scanning beta gauge. The CD profiles of basis weight are scanned every few minutes. Basis weight distribution is primarily affected by the slice lip profile on the head box.

The actuators used for the basis weight control are a series of motorised screws attached to the slice lip at points across the machine. By adjusting the position of each individual screw, it is possible to locally modify the gap in the slice lip so as to vary the amount of pulp deposited at that position on the mesh of wire. If the gap in the lip is reduced, then less pulp is deposited on the wire screen and a reduced paper density profile is obtained. If the gap in the lip is increased, then more pulp is deposited and an increased paper density profile is obtained. An important issue is that moving one actuator affects the slice over neighbouring regions. This interaction is a major cause of difficulty and means that one needs to take a multivariable view of the associated control problem if one wants to achieve high performance.

21.4 Process Model

As usual, it is desirable to begin with a model which describes those elements of the process which are relevant to controller design. Fortunately, for most web-forming processes, it is possible to build a fairly accurate model by considering the following simplifying assumptions (see references given in Further Reading):

- The effect of one single actuator movement is almost instantaneous in the cross direction.
- The dynamic response in the machine direction is essentially the same for every actuator–sensor pair (save for effects at the edges).

This information will be used to build a simple model for the process. It will be assumed that the system has m input–output pairs. First, suppose that the system is resting in a steady-state condition having a flat paper density profile and all the actuators at their default positions. Next, assume that

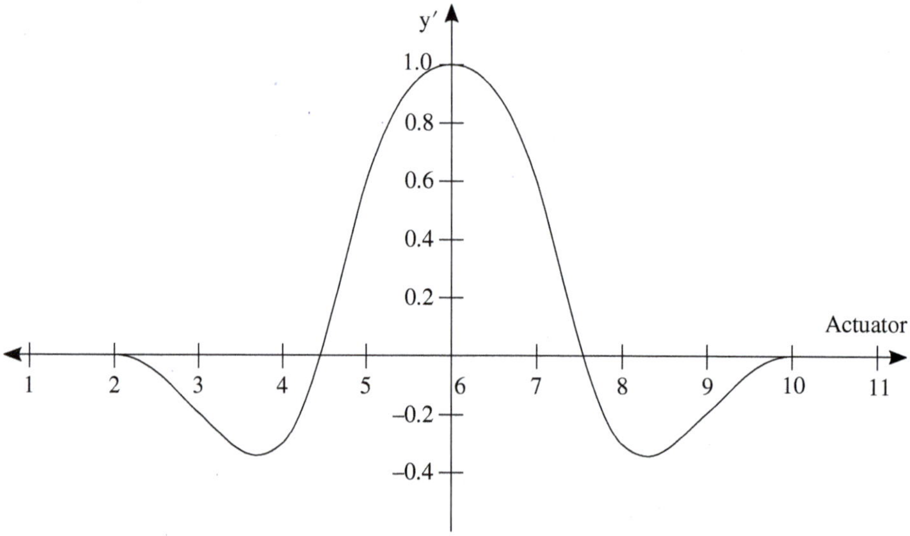

Figure 21.3 CD profile for a unit step movement in actuator No. 6: generic web-forming process with 11 input–output pairs.

a unit step is applied to one actuator (screw). Based on the first simplifying assumption, one will observe a change of the paper density profile *across the entire cross direction* of the web. This change can be measured and arranged in one single-column vector as follows:

$$\mathbf{y}' = \begin{bmatrix} y_1 \\ y_2 \\ \vdots \\ y_m \end{bmatrix}. \tag{21.1}$$

Figure 21.3 shows a typical example of the CD profile obtained when applying a unit step in actuator number 6 when there are 11 sensor actuator pairs, $m = 11$. Note that the cross-directional effect of one single screw displacement affects the CD profile at neighbouring measurement points.

For the example shown in Figure 21.3, the corresponding vector $\mathbf{y}'$ is given by

$$\mathbf{y}' = \begin{bmatrix} 0 & 0 & -0.1 & -0.3 & 0.6 & 1 & 0.6 & -0.3 & -0.1 & 0 & 0 \end{bmatrix}^T. \tag{21.2}$$

It seems reasonable to hypothesise that if a unit step experiment was to be repeated for, say, actuator number 3, then the resulting CD profile should be a shifted version (centred at the position of screw No. 3) of the profile shown in Figure 21.3.[2] The measured CD profile would be, in this case:

$$\mathbf{y}' = \begin{bmatrix} -0.3 & 0.6 & 1 & 0.6 & -0.3 & -0.1 & 0 & 0 & 0 & 0 & 0 \end{bmatrix}^T. \tag{21.3}$$

In a similar fashion, one could repeat the unit-step experiment for each of the 11 actuators and record the corresponding values of the vector $\mathbf{y}'$. One can then interpret the values of $\mathbf{y}'$ as the (vector) gain between a certain actuator and the resulting CD profile. To quantify the total effect of all the actuators, movements on the resulting CD profile, the information collected from all the unit-step experiments needs to be combined. To do this use will be made of matrix notation.

2 Note that there are important practical issues with what happens at the edges, but we will ignore these for the time being.

First define $\mathbf{u}(t)$ to be a column vector containing the value of the position of each screw on the slice lip:

$$\mathbf{u}(t) = \begin{bmatrix} u_1(t) \\ u_2(t) \\ \vdots \\ u_m(t) \end{bmatrix}. \tag{21.4}$$

The net effect on the CD profile at each position is the sum of the partial effects generated by each actuator (i.e. it is assumed that superposition holds). Thus, let $\mathbf{y}_T(t)$ to be a vector containing the net effect, then

$$\mathbf{y}_T(t) = \overline{B}' \, \mathbf{u}(t), \tag{21.5}$$

where

$$\overline{B}' = \begin{bmatrix} 1 & 0.6 & -0.3 & 0.1 & 0 & 0 & 0 & 0 & 0 & 0 & 0 \\ 0.6 & 1 & 0.6 & -0.3 & 0.1 & 0 & 0 & 0 & 0 & 0 & 0 \\ -0.3 & 0.6 & 1 & 0.6 & -0.3 & 0.1 & 0 & 0 & 0 & 0 & 0 \\ 0.1 & -0.3 & 0.6 & 1 & 0.6 & -0.3 & 0.1 & 0 & 0 & 0 & 0 \\ 0 & 0.1 & -0.3 & 0.6 & 1 & 0.6 & -0.3 & 0.1 & 0 & 0 & 0 \\ 0 & 0 & 0.1 & -0.3 & 0.6 & 1 & 0.6 & -0.3 & 0.1 & 0 & 0 \\ 0 & 0 & 0 & 0.1 & -0.3 & 0.6 & 1 & 0.6 & -0.3 & 0.1 & 0 \\ 0 & 0 & 0 & 0 & 0.1 & -0.3 & 0.6 & 1 & 0.6 & -0.3 & 0.1 \\ 0 & 0 & 0 & 0 & 0 & 0.1 & -0.3 & 0.6 & 1 & 0.6 & -0.3 \\ 0 & 0 & 0 & 0 & 0 & 0 & 0.1 & -0.3 & 0.6 & 1 & 0.6 \\ 0 & 0 & 0 & 0 & 0 & 0 & 0 & 0.1 & -0.3 & 0.6 & 1 \end{bmatrix}. \tag{21.6}$$

$\overline{B}'$ is a matrix whose columns are the measurements collected after each unit step experiment. Observe that the 5th and 3rd columns of the matrix $\overline{B}'$ above are precisely the vector $\mathbf{y}'$ in (21.2) and (21.3), respectively.

Next, the issue of the machine direction dynamics will be considered. Let $\mathbf{y}(t)$ be a column vector that contains the measurements of the paper density at a downstream position from the array of actuators as shown in Figure 21.2. When a unit step is applied to one of the actuators, the CD profile changes according to the entries in the corresponding column of $\overline{B}'$ in (21.6). However, the density of the paper (as measured by the set of sensors in the downstream position) will change only *gradually*, until it settles at a certain steady state-value. This is a result of the dynamic response of the system in the *machine direction.*

A simple first-order model with time delay can be used to model the machine direction dynamics. This leads to the following transfer function:

$$h(s) = \frac{g_0 \, e^{-s\beta}}{\tau s + 1}, \tag{21.7}$$

where g_0 is the transfer function's DC gain, τ is a time constant and β is a pure time delay. The time delay $e^{-s\beta}$ accounts for the physical separation that exists between the actuators and the sensors – see Figure 21.2.

Assuming that the dynamic response in the machine direction (given by the transfer function in Eq. (21.7)) is the same for every actuator–sensor pair, then

$$\mathbf{y}(s) = \begin{bmatrix} \dfrac{g_0\, e^{-s\beta}}{\tau s + 1} & & 0 \\ & \ddots & \\ 0 & & \dfrac{g_0\, e^{-s\beta}}{\tau s + 1} \end{bmatrix} \mathbf{y}'_T(s). \tag{21.8}$$

The last step is to combine the machine direction dynamics modelled in Eq. (21.8) with the cross-directional interactions modelled by Eq. (21.6). The final mathematical model for the system is then:

$$\mathbf{y}(s) = \begin{bmatrix} \dfrac{g_0\, e^{-s\beta}}{\tau s + 1} & & 0 \\ & \ddots & \\ 0 & & \dfrac{g_0\, e^{-s\beta}}{\tau s + 1} \end{bmatrix} \overline{B}'\, \mathbf{u}(s). \tag{21.9}$$

For convenience, the model in (21.9) can be normalised so that the machine direction dynamics all have a DC gain of 1. Then, the resulting system model becomes:

$$\mathbf{y}(s) = \begin{bmatrix} \dfrac{e^{-s\beta}}{\tau s + 1} & & 0 \\ & \ddots & \\ 0 & & \dfrac{e^{-s\beta}}{\tau s + 1} \end{bmatrix} \overline{B}\, \mathbf{u}(s). \tag{21.10}$$

The matrix $\overline{B}$ is called the system's *normalised interaction matrix*.

The *normalised interaction matrix* $\overline{B}$ can be expressed in terms of the matrix $\overline{B}'$ in (21.9). The result is:

$$\overline{B} = g_0\, \overline{B}'. \tag{21.11}$$

Let $\overline{b}_k$ be the kth column of the normalised interaction matrix $\overline{B}$. If a unit step is applied to actuator number k, then the steady-state output satisfies

$$\mathbf{y}_{ss} = \mathbf{y}(0) = \begin{bmatrix} 1 & & & \\ & 1 & & \\ & & \ddots & \\ & & & 1 \end{bmatrix} \overline{B}\, \mathbf{u}(0), \tag{21.12}$$

where

$$\mathbf{u}(0) = \begin{bmatrix} 0 & \cdots & 0 & 1 & 0 & \cdots & 0 \end{bmatrix}^T, \tag{21.13}$$

i.e. $\mathbf{u}(0)$ is a zero vector with a 1 in the kth position only. From this, the resulting steady-state output profile is

$$\mathbf{y}_{ss} = \overline{b}_k. \tag{21.14}$$

Hence to estimate the system's interaction matrix it suffices to perform a sequence of unit step experiments selecting each of the actuators and recording the resulting steady-state CD profile. This will provide the entries of the corresponding column of $\overline{B}$.

□□□

The following experimental tests are designed to estimate the parameters corresponding to the system in the virtual lab. The virtual laboratory system has 21 input–output pairs. The associated normalised interaction matrix $\overline{B}$ has a Toeplitz symmetric structure with exponential profile, i.e.

$$b_{i,j} = e^{-0.1|i-j|}, \quad \text{for } i,j = 1,\dots,21, \tag{21.15}$$

where $b_{i,j}$ are the entries of the interaction matrix $\overline{B}$.

21.4.1 Experiment: Measuring the Cross-Directional Profile

- Load the Manual Mode Controller.
- Select one of the controller's outputs (say No. 11) and set its value to 1 (make sure all the other controller's outputs are set to zero).

Observe the system's open-loop response and, in particular, the evolution of the CD cross-directional basis weight profile. Wait for the system to settle to its new steady-state condition. Record the resulting CD output profile (use the Data Scope to plot the time responses). From this data, build the system's interaction matrix $\overline{B}$ following the pattern shown in Eq. (21.11).

After applying a unit step to actuator No. 11, wait until the output measurements $\mathbf{y}(t)$ settle to their steady-state condition. Record the resulting steady-state CD output profile.

21.4.2 Experiment: Measuring the Machine Direction Dynamics

Next estimate the time constant of the machine direction dynamics.

Apply a unit step test but this time record the MD transient at the output $\mathbf{y}(t)$. (Again using the Data Scope to plot the time response.) Use this to estimate the time constant τ of the machine direction dynamics and delay β.

The time constant used for the machine direction dynamics is approximately

$$\tau = 1. \tag{21.16}$$

Also, the system currently does not have any transport delay. Therefore

$$\beta = 0. \tag{21.17}$$

The reason for using $\beta = 0$ is that the question of transport delay has been extensively treated earlier (Chapters 17 and 18). Hence from this point onwards it will be assumed that the transport delay in the machine direction is negligible, i.e.

$$\beta = 0. \tag{21.18}$$

In practice, a large delay can exist between actuators and sensors due to there being a large physical displacement between them. This delay is normally dealt with by using a Smith predictor type structure (see Section 12.6).

21.5 Simple SISO Design Ignoring Coupling

The first approach to control system design is to simply use a set of SISO PI controllers.

The system's model is assumed to be given by

$$\mathbf{y}(s) = \begin{bmatrix} \frac{1}{s+1} & & 0 \\ & \ddots & \\ 0 & & \frac{1}{s+1} \end{bmatrix} \mathbf{u}(s). \tag{21.19}$$

PI controllers have the following structure

$$C(s) = K_p + \frac{K_i}{s} \tag{21.20}$$

will be used to give a nominal SISO closed-loop transfer function

$$T(s) = \frac{1}{0.1s+1}, \tag{21.21}$$

for each actuator–sensor pair. *(Note this design ignores interactions!)*

The desired closed-loop transfer function is

$$T(s) = \frac{10}{s+10}. \tag{21.22}$$

This implies that $C(s)$ should satisfy

$$C(s)\frac{1}{s+1} = \frac{10}{s} \tag{21.23}$$

$$\Longrightarrow \frac{K_p\left(s+\frac{K_i}{K_p}\right)}{s}\frac{1}{s+1} = \frac{10}{s}.$$

Therefore, it suffices to select

$$K_p = 10$$

and

$$K_i = K_p, \tag{21.24}$$

in order to cancel the plant open-loop pole at $s + 1$.

21.5.1 Experiment: Testing Simple PID Controllers

Click the controller block and enter K_p and K_i then click OK. Click the Disturbance Profile block and change the disturbance from constant to any other form. Observe the closed-loop system response.

You probably found that the settling time is of the order of 10 seconds. However, the nominal closed-loop time constant was designed to be 0.1 seconds with a settling time of around 0.4 seconds. It can thus be seen that the achieved closed-loop response is significantly worse than that chosen in the design.

21.6 Simple SISO Design Accounting for Coupling

It was found in Section 21.5 that simple PID design gave rather poor results.

For this reason, a more sophisticated design strategy is next considered. The core idea used here is to diagonalise the system. The interested reader may like to reread Section 2.2 where this idea of diagonalisation was first introduced (albeit in a very simple setting). Also, Section 19.5 describes a simple two by two industrial example of the idea described below.

Note that the model takes the form shown in Eq. (21.10), i.e.

$$\mathbf{y}(s) = \begin{bmatrix} \frac{e^{-s\beta}}{\tau s+1} & & 0 \\ & \ddots & \\ 0 & & \frac{e^{-s\beta}}{\tau s+1} \end{bmatrix} \overline{B}\,\mathbf{u}(s).$$

It is suggested that the reader reviews Section 2.3 before proceeding. The interaction is captured in the non-dynamic matrix $\overline{B}$. The matrix $\overline{B}$ can be diagonalised, e.g.

$$\overline{B} = U\,S\,V^T, \tag{21.25}$$

where S is diagonal and U and V^T are orthonormal matrices, i.e. $U^{-1} = V^T$. Then Eq. (21.10) can be written as

$$\mathbf{y}(s) = \begin{bmatrix} \dfrac{e^{-s\beta}}{\tau s+1} & & 0 \\ & \ddots & \\ 0 & & \dfrac{e^{-s\beta}}{\tau s+1} \end{bmatrix} U\,S\,V^T\,\mathbf{u}(s) \tag{21.26}$$

or

$$\tilde{\mathbf{y}}(s) = \begin{bmatrix} \dfrac{e^{-s\beta}}{\tau s+1} & & 0 \\ & \ddots & \\ 0 & & \dfrac{e^{-s\beta}}{\tau s+1} \end{bmatrix} \tilde{\mathbf{u}}(s), \tag{21.27}$$

where

$$\tilde{\mathbf{y}}(s) = \mathbf{U}^{\mathrm{T}}\mathbf{y}(s). \tag{21.28}$$

$$\tilde{\mathbf{u}}(s) = \mathbf{V}^{\mathrm{T}}\mathbf{u}(s). \tag{21.29}$$

It is then clear that the 'system' connecting $\tilde{\mathbf{u}}(s)$ to $\tilde{\mathbf{y}}(s)$ is diagonal (i.e. non-interacting) Thus, $\tilde{\mathbf{u}}_j(s)$ only affects $\tilde{\mathbf{y}}_j(s)\quad j = 1, \dots, N$.

A decoupled PI controller can then be designed based on the error signal.

$$\tilde{\mathbf{e}}(s) = \tilde{\mathbf{y}}^*(s) - \tilde{\mathbf{y}}(s) = \mathbf{U}^{\mathrm{T}}\left[\mathbf{y}^*(s) - \mathbf{y}(s)\right]. \tag{21.30}$$

Finally, the error $\tilde{\mathbf{e}}(s)$ is passed onto a set of PID controllers to generate $\tilde{\mathbf{u}}(s)$. The input is obtained by inverting Eq. (21.29), i.e.

$$\mathbf{u}(\mathbf{s}) = (\mathbf{V}^{\mathrm{T}})^{-1}\tilde{\mathbf{u}}(s) = \mathbf{U}\tilde{\mathbf{u}}(s). \tag{21.31}$$

21.6.1 Experiment: Testing a Decoupled PID Structure

Click the controller block and go to the MIMO mode controller. Set the Anti-Windup type to 'No Anti-Windup'. Choose the random disturbance profile and the flat reference profile. Run the system and comment.

You should find that the decoupled controller performs much better than the simple SISO PID controller studied in Section 21.5. Indeed, the settling time is now of the order of 1 second as chosen in the original design.

21.7 Summary

This laboratory has shown that CD control problems exhibit severe interaction. The interaction matrix has been measured. A simple centralised PID control has been tested and found to perform poorly. The multivariable interactions can be compensated by converting the inputs and outputs to an alternative space in which each input and output pair can be treated in a decentralised fashion.

A decoupled PID design has been tested. The final decoupled design appears to work extremely well. One might even be tempted to say, 'problem solved!' However, this is by no means the end of the story. Chapter 27 will study the impact of actuator saturation. It will then become clear that there are serious shortcomings with the seemingly simple designs described in the current chapter.

21.8 Revision Questions

1. The Matrix given in (21.6) is symmetrical. What does this imply about the eigenvalues?
2. Consider the matrix $\overline{B}$ given in Eq. (21.6) in the text. Say that the first and last actuators are well inside the edge of the strip. What change would this make to the matrix $\overline{B}'$?
3. Consider the interaction matrix described in Eq. (21.15) in the text and assume that the conditions described in Question 2 apply. Express the steady-state output response as the convolution of the function $e^{0.1k}$ with the input from each actuator if the number of actuator/sensor pairs is large.
4. The spatial two-sided Z-transform of a spatially distributed quantity $h(k)$ is defined as

$$H(z) = \sum_{-\infty}^{\infty} h(k)z^{-k}. \tag{21.32}$$

   What is the spatial Z-transfer of $h(k) = e^{0.1k}$?
5. (a) Use your answer to Question 4 to determine the spatial shape of the corresponding eigenvectors.
   (b) What are the corresponding eigenvalues?
6. Use your answer to Question 5 to show that rapidly changing inputs (in the spatial dimensions) lead to a small output response.
7. Use your answer to Question 6 to show why it will be hard to achieve a rapidly changing shape in the spatial domain.

Further Reading

1 L.G. Bergh and J.F. MacGregor, "Spatial control of film and sheet forming processes", *The Canadian Journal of Chemical Engineering*, 65:148–155, 1987.
2 D. Dochain, G. Dumont, D.M. Gorinevsky and T. Ogunnaike, "Editorial, Special issue on 'control of industrial spatially distributed processes'", *IEEE Transactions on Control Systems Technology*, 11(5):609–611, 2003.
3 S.R. Duncan, "Editorial, special section: cross directional control", *IEE Proceedings Control Theory and Applications*, 149(5):412–413, 2002.
4 S.R. Duncan, J.M. Allwood and S.S. Garimella, "The analysis and design of spatial control systems in strip metal rolling", *IEEE Transactions on Control Systems Technology*, 6(2):220–232, 1998.
5 A.P. Featherstone and R.D. Braatz, "Control-oriented modelling of sheet and film processes", *AIChE Journal*, 43(8):1989–2001, 1997.
6 A.P. Featherstone, J.G. VanAntwerp and R.D. Braatz, *Identification and control of sheet and film processes*, Springer-Verlag, London, 1997.
7 G.C. Goodwin, B. Carney and W.J. Edwards, "Analysis of Thermal Camber Control in Rolling Mills", *Proceedings of the IFAC World Congress*, pp. 160–164, Talinn, USSR, August 1990.

8 W.P. Heath, "Orthogonal functions for cross-directional control of web forming processes", *Automatica*, 32(2):183–198, 1996.

9 M. Hovd, R.D. Braatz and S. Skogestad, "SVD controllers for $H_2-, H_\infty-$ and μ-optimal control", *Automatica*, 33(3):433–439. 1997.

10 D.L. Laughlin, M. Morari and R.D. Braatz, "Robust performance of cross-directional basis-weight control in paper machines", *Automatica*, 29(6):1395–1410. 1993.

11 G.E. Stewart, D.M. Gorinevsky and G.A. Dumont, "Feedback controller design for a spatially distributed system: the paper machine problem", *IEEE Transactions on Control Systems Technology*, 11(5):612–628. 2003.

12 G.E. Stewart, D.M. Gorinevsky and G.A. Dumont, "Two-dimensional loop shaping", *Automatica*, 39:779–792. 2003.

13 P.E. Wellstead and W.P. Heath, "Two-dimensional control systems: application to the CD and MD control problem", *Pulp and Paper Canada*, 95(2):75–78. 1994.

Part VI

Multivariable Control Systems (More General Methods)

In Section V of the book, it has been shown how many multivariable control problems can be converted into simple single-input single-output problems by a preliminary transformation. The resulting control system designs can then be completed by simple methods. This will not always lead to a satisfactory result. Hence, this section of the book will describe methods that are inherently multivariable in nature.

Practical Control System Design: Real World Designs implemented on Emulated Industrial Systems,
First Edition. Adrian Medioli and Graham Goodwin.

Companion website: www.wiley.com/go/medioli/practicalcontrolsystemdesign

22

State Variable Feedback

22.1 Introduction

A particularly insightful way of dealing with multivariable control problems is via the use of state-space ideas. This chapter, and the next, will provide a very brief overview of key state-space design methods. Those readers who seek a more, in-depth, treatment of these topics are referred to the references given at the end of the chapter. The ideas covered in this chapter include state variable feedback and the Linear Quadratic Regulator. The treatment in this chapter will be particularly succinct since the main use of state-space ideas in the Virtual Laboratories is the Kalman filter which will be treated in Chapter 23.

22.2 Sampled-Data Control

For ease of exposition, the sample-data case will be first considered. The corresponding continuous-time results will be briefly described later. For the sequel, it is assumed that the system state is directly measured. (The case when the state is not measured will be the topic of Section 22.9 and Chapter 23.)

22.2.1 Pole Assignment

Design by pole assignment is straightforward. For example, let the system be described in state-space form by the shift-operator model (see Section 17.6 of Chapter 17):

$$qx_k = A'x_k + B'u_k \quad y_k = C'x_k. \tag{22.1}$$

Then a suitable sampled *state-variable feedback* law takes the form:

$$u_k = -K'x_k. \tag{22.2}$$

Subject to regulatory conditions (actually, controllability) one can always find a gain K' to place the poles of the closed-loop system (22.1) and (22.2), i.e. the zeros of $\det(zI - A' + B'K')$ at any desired location. Standard software is available to carry out the required computations.

22.2.2 Linear Quadratic Regulator (LQR)

Instead of designing K' by pole assignment, as outlined in Section 22.2.1, one can instead seek that input which causes x_k to converge to the origin as fast as possible. However, this may use an

Practical Control System Design: Real World Designs implemented on Emulated Industrial Systems, First Edition. Adrian Medioli and Graham Goodwin.

Companion website: www.wiley.com/go/medioli/practicalcontrolsystemdesign

unreasonable amount of input energy. Thus, a trade-off is called for. The trade-off can be captured by a (quadratic) cost function of the form

$$J = \frac{1}{2}\sum_{k=0}^{N-1}\left(x_k^T Q x_k + u_k^T R u_k\right) + \frac{1}{2}x_N^T P_N x_N. \tag{22.3}$$

The above cost function trades off bringing x_k to zero (the first term on the right-hand side) whilst not using too much energy u_k (the second term on the right-hand side). The final term $x_N^T P_N x_N$ describes the arrival cost when the state finally reaches x_N after the inputs, $u_0, \ldots, u_{N-1}$, have all been applied.

22.3 Dynamic Programming

A beautifully insightful property of optimal control problems is that, if one reaches a state x_k at sample k, then the cost to go from that point onwards does not depend on how the state x_k was reached. This property is captured in the 'Dynamic Programming Principle' as outlined below.

Let $V_k(x_k)$ be the optimal cost to go from time k assuming that one is in state x_k at sample k. Note that, at sample $k = N$, it is known from the cost function (22.3) that the arrival cost is

$$V_N\left(x_N\right) = \frac{1}{2}x_N^T P_N x_N. \tag{22.4}$$

Inspection of (22.4) shows that the final cost is a quadratic function of the final state, x_N.

Induction will next be used to show that, for the system (22.1) and with the cost function (22.3), the optimal cost to go from any sample k is also a quadratic function of the state.

To begin the argument by induction, it is assumed that the cost to go from sample time $k+1$ is a quadratic function of the form:

$$V_{k+1}\left(x_{k+1}\right) = \frac{1}{2}x_{k+1}^T P_{k+1} x_{k+1}, \tag{22.5}$$

for some matrix P_{k+1}.

It is then clear that the optimal cost to go from sample k to the end is obtained by minimising the sum of the optimal cost to go from sample $k+1$ plus the incremental cost from sample k to sample $k+1$. Hence using (22.3) it follows that

$$V_k\left(x_k\right) = \min_{u_k}\left[\frac{1}{2}x_k^T Q x_k + \frac{1}{2}u_k^T R u_k + V_{k+1}\left(x_{k+1}\right)\right], \tag{22.6}$$

where x_{k+1} satisfies (22.1).

Substituting (22.1) into (22.6),

$$V_k\left(x_k\right) = \min_{u_k}\left[\frac{1}{2}x_k^T Q x_k + \frac{1}{2}u_k^T R u_k + \frac{1}{2}\left(A'x_k + B'u_k\right)^T Q_{k+1}\left(A'x_k + B'u_k\right)\right]. \tag{22.7}$$

The minimisation in (22.7) can be carried out by setting the derivative, with respect to u_k, to zero. This leads to the following result:

$$Ru_k + B'^T P_{k+1}\left(A'x_k + B'u_k\right) = 0. \tag{22.8}$$

Rearranging (22.8) shows that the optimal input at sample k must be

$$u_k = -\left[R + B'^T P_{k+1} B'\right]^{-1} B'^T P_{k+1} x_k. \tag{22.9}$$

Substituting back into (22.7) yields a quadratic function of x_k. This quadratic function can be expressed as $x_k^T P_k x_k$. Note that P_k will be a function of P_{k+1}. This recursion is called a discrete-time matrix Riccati equation. The equation is solved in reverse time beginning at $k = N$. The boundary condition P_N is part of the problem description in (22.3). Once P_k, for $k = 1, \dots, N-1$ have been calculated then the 'optimal' input, u_k, can be found from (22.9). Standard software tools are available to find $\{P_k\}$ and hence $\{u_k\}$.

This completes the proof by induction, showing that $V_k(x_k)$ is, indeed, a quadratic function of x_k. The Riccati equation gives details of the quadratic function and (22.9) gives the associated optimal input. In practice, one rarely uses the time-varying control law (22.9). Instead, the horizon is usually considered to be infinity. This leads to a time-invariant control law as described in Section 22.4.[1]

22.4 Infinite Horizon Linear Quadratic Optimal Problem

It can be shown that, subject to time-invariant system model and other regulatory conditions, as the horizon, $N \to \infty$, the solution to the Linear Quadratic Regulator problem has the following properties (For proofs see the references given at the end of the chapter):

- For a controllable system, the matrix P_k will converge to a steady-state value P_{ss} as $N \to \infty$.
- The limiting value, P_{ss}, is positive definite.
- The limiting value satisfies the steady-state Matrix Riccati equation of control. This is a nonlinear algebraic equation.
- The steady-state matrix Riccati equation has many solutions. However, there is a unique positive definite solution. It is this particular solution to which the solution of the time-varying Riccati equation converges as $N \to \infty$.
- Given the limiting value, P_{ss}, then infinite horizon control policy is a simple linear time-invariant state-feedback law of the form:

  $$u_k = -\left[R + B'^T P_{ss} B'\right]^{-1} B'^T P_{ss} x_k. \tag{22.10}$$

  This follows by substituting $P_{k+1} = P_{ss}$ in Eq. (22.9). The result in (22.10) can be expressed as

  $$u_k = -K_{ss} x_k, \tag{22.11}$$

  where $K_{ss} = [R + B_{ss}'^{TP} B']^{-1} B'^T P_{ss}$. The reader should note the connection between (22.11) and (22.2). In particular, the Infinite Horizon Linear Quadratic Optimal Control problem has simply led to a particular state feedback gain.
- Subject to the system being controllable, the feedback control design (22.11) is guaranteed to stabilise the system.

22.5 Delta-Domain Result

As explained in Chapter 17, a useful alternative to the shift domain model (22.1) is the delta-domain model:

$$\delta x_k = A'' x_k + B'' u_k. \tag{22.12}$$

The 'optimal' control law still takes the familiar form:

$$u_k = -K' x_k, \tag{22.13}$$

---

1 A case where the time-varying case is appropriate is when the system is, itself, time-varying.

where K' can be found as in Section 22.4, save that divided differences replace the shift operator. If the sampling period is small, then it will not surprise the reader that one can simply use the continuous time result. The continuous case is outlined in Section 22.6.

22.6 Continuous-Time Linear Quadratic Regulator

The above ideas are readily extendable to continuous-state-space models. In this case, the appropriate representation is the following continuous-time state-space model

$$\dot{x} = Ax + Bu, \tag{22.14}$$

$$y = Cx. \tag{22.15}$$

22.6.1 Pole Assignment

Since the state, $x(t)$, summarises all that has happened to the system up to time t, then as in the sampled data case, the best control action at time t is a function of $x(t)$ only, i.e.

$$u(t) = -K(x(t)), \tag{22.16}$$

where $K(x(t))$ is some (possible nonlinear) function.

For linear systems, it suffices to use a linear feedback control law, in which case (22.16) becomes

$$u(t) = -Kx(t), \tag{22.17}$$

where K is a matrix of gains. Equation (22.17) is the continuous-time version of equation (22.2).

The closed-loop system resulting from the use of (22.17) is as follows:

$$\dot{x} = [A - BK]x. \tag{22.18}$$

The closed-loop poles of (22.18) are the zeros of the following polynomial:

$$A_{cl}(s) = det(sI - A + BK). \tag{22.19}$$

As for the sampled-data case, subject to certain technical conditions (controllability), it is possible to choose K such that the zeros of $A_{cl}(s)$ lie at any specified location in the left-half plane.

Standard software (e.g. MATLAB®) can be used to find K such that the zeros of $A_{cl}(s)$ are assigned to desired locations.

22.6.2 Continuous-Time Linear Quadratic Regulator

It is suggested that the reader reread Section 22.2.2 before proceeding.

In the continuous case, the trade-off between bringing the state x at time t to the origin as quickly as possible versus the 'size of' the control effort needed can be captured by an integral quadratic cost function of the form.

$$J = \frac{1}{2}\int_t^{\infty} x(\tau)^T Qx(\tau) + u(\tau)^T Ru(\tau)d\tau, \tag{22.20}$$

where Q is positive semi-definite and R is positive definite. Note that this is similar to (22.3) save that the summation in (22.3) has been replaced by an integral in (22.20) and, for simplicity, the horizon has been assumed to be ∞.

As for the sample-data case, the control that minimises (22.20) is a feedback law of the form (22.17) where the gain, K, now takes a special value. In the continuous-time case, it can be shown that the 'optimal' value of K is given by

$$K = R^{-1}B^TP, \tag{22.21}$$

where P satisfies a continuous-time matrix algebraic Riccati equation:

$$Q - PBR^{-1}B^TP + PA + A^TP = 0. \tag{22.22}$$

Software tools are available (e.g. MATLAB®) to solve (22.22) and hence to find the continuous-time optimal state-feedback gain K as in (22.21).

22.7 Regulation to a Fixed Set-Point

The feedback control law given in (22.17) (or (22.21)) is based on the hypothesis that one wants to 'steer' the state to the origin. However, in most control problems, it is desired to bring the output to some trajectory. A common situation is where one has a constant set-point, say y^*. The latter case can be converted into the problem of convergence to the origin by simply redefining the origin. To show how this can be achieved, let the output be expressed as

$$y(t) = Cx(t). \tag{22.23}$$

Then the objective is to bring $y(t)$ to y^*. It is assumed that there are as many input signals as outputs, i.e. the system has a 'square' transfer function, $T(s)$, where:

$$T(s) = C(sI - A)^{-1}B. \tag{22.24}$$

At steady-state, the input, u^*, required to bring $y(t)$ to y^* satisfies $y^* = T(0)u^*$. Hence, provided $T(s)$ is nonsingular for $s = 0$, then the required input signal, u^*, is obtained by setting $T(0)u^*$ to y^*. This leads to

$$u^* = [T(0)]^{-1}y^* = \left[C(-A)^{-1}B\right]^{-1}y^*. \tag{22.25}$$

The corresponding state, x^* satisfies $x(s) = (sI - A)^{-1}Bu(s)$ with $s = 0$. Hence,

$$x^* = -A^{-1}Bu^*. \tag{22.26}$$

Finally, the control problem can be thought of as bringing the error, $x - x^*$, in the state to the origin using acceptable control effort, expressed as a departure from u^*. To describe this, trade-off the cost function (22.20) should be changed to the following one expressed in terms of increments of x and u about x^* and u^*, respectively. This leads to the following cost function:

$$J = \int_t^\infty \left[x(\tau) - x^*\right]^T Q \left[x(\tau) - x^*\right] + \left[u(\tau) - u^*\right]^T R \left[u(\tau) - u^*\right] d\tau. \tag{22.27}$$

Note that (22.17) is similar to (22.32) but now expressed in increments. Indeed the problem can now be described entirely in the increment space as

$$\tilde{J} = \frac{1}{2}\int_t^\infty \tilde{x}(\tau)^TQ\tilde{x}(\tau) + \tilde{u}(\tau)^TR\tilde{u}(\tau)d\tau, \tag{22.28}$$

where $\dot{\tilde{x}} = A\tilde{x} + B\tilde{u}$.

The control law (22.17) also changes to the following incremental form:

$$\tilde{u}(t) = -K\tilde{x}(t), \tag{22.29}$$

where K is found as in Section 22.6.2. Equation (22.29) can be expressed in terms of $x(t)$ and $u(t)$ as follows:

$$u(t) - u^* = -K(x(t) - x^*), \tag{22.30}$$

where, as usual, K is found by solving a (continuous-time) matrix Riccati equation.

22.8 Frequency Domain Insights into the Linear Quadratic Regulator

Insights into the cost function (22.27) can be obtained by converting to the frequency domain. To illustrate the idea, say that y and u are scalars and that the cost function (22.27) is taken to be

$$\begin{aligned} J &= \int_t^\infty (x(\tau) - x^*)^T C^T C(x(\tau) - x^*) + \lambda(u(\tau) - u^*)^2 d\tau, \qquad (22.31) \\ &= \int_t^\infty (y(\tau) - y^*)^2 + \lambda(u(\tau) - u^*)^2 d\tau. \end{aligned}$$

Using Parseval's Theorem,[2] this can be converted into the frequency domain, yielding

$$J = \int_0^\infty |Y(j\omega) - Y^*(j\omega)|^2 + \lambda\, |U(j\omega) - U^*(j\omega)|\, d\omega. \tag{22.32}$$

Also observe that $Y(j\omega)$, for zero initial conditions, is given by

$$Y(j\omega) = G(j\omega)U(j\omega), \tag{22.33}$$

where $G(j\omega)$ is the system frequency response. The desired steady-state frequency response is denoted $Y^*(j\omega)$. Substituting (22.33) into (22.32) then yields

$$J = \int |G(j\omega)|^2 \left|U(j\omega) - G(j\omega)^{-1} Y^*(j\omega)\right|^2 + \lambda |U(j\omega) - U^*(j\omega)|^2. \tag{22.34}$$

Thus, the Linear Quadratic regulator has the heuristic interpretation that it trades off bringing $U(j\omega)$ to $G(j\omega)^{-1}Y^*(j\omega)$, i.e. bringing the first term on the right-hand side of (22.34) to zero whilst ensuring that $U(j\omega)$ does not depart too much from $U^*(j\omega)$ as captured by the second term on the right-hand side of (22.34). Note that bringing the first term on the right-hand side (22.34) to zero is equivalent to inverting the plant (see Chapter 2).

Further insights can be obtained by considering the "closed-loop bandwidth'. It is convenient to define the 'closed-loop bandwidth', ω_{clb}, as that frequency where one transition from tracking (i.e. inverting the plant, achieved by setting the first term on the right-hand side of (22.34) to zero) and having a constant control (i.e. $u = u^*$ which brings the second term on the right-hand side of (22.34) to zero). From (22.34) it can be seen that ω_{clb} is approximately that frequency, ω_{clb}, where

$$\left|G(j\omega_{clb})\right|^2 \simeq \lambda. \tag{22.35}$$

Thus, if one plots the magnitude of the Bode diagram for the plant, then the closed-loop bandwidth is approximately given by the frequency where $|G(j\omega_{clb})| = \lambda^{\frac{1}{2}}$.

Analogous ideas hold for the sampled-data case.

---

2 Quadratic costs in the time domain are quadratic costs in the frequency domain.

22.9 Output Feedback

Up to this point in the chapter it has been assumed that the state, x, can be measured. Under this condition, control laws of the type (22.22) or (22.17) can be readily implemented. However, it is commonly the case that one does not measure the full state vector, x, but only measures an output y which will typically be a subset of the full state.

22.9.1 A State Estimator (or Observer)

The output can typically be expressed as $y = Cx$ for some matrix, C of dimension $m \times n$ where usually $m < n$. In this case, it is possible to estimate the current state by using an 'observer'. An observer uses feedback to incrementally improve a given state estimate, $\hat{x}(t)$, as extra measurements of $y(t)$ become available. A typical observer (in continuous time) takes the form:

$$\dot{\hat{x}}(t) = A\hat{x}(t) + Bu(t) + L\left(y(t) - C\hat{x}(t)\right), \tag{22.36}$$

where L is an $m \times n$ matrix of 'observer gains'. Notice that (22.36) is a linear dynamical system driven by the observations $y(t)$.

Specifically (22.36) uses the error between the observation, $y(t)$ and the predicted output based on the current state estimate, i.e. $C\hat{x}(t)$, to 'shift' $\hat{x}(t)$ towards a better estimate.

It is insightful to form the equation governed by the error $\tilde{x}(t) = x(t) - \hat{x}(t)$. In particular, subtracting (22.36) from (22.14) and using (22.14) gives

$$\dot{\tilde{x}}(t) = [A - LC]\tilde{x}(t). \tag{22.37}$$

Hence, the dynamics of the state error depend upon the matrix $A_{obs}(s)$, where

$$A_{obs}(s) = \det(sI - A + LC). \tag{22.38}$$

As for the state-feedback problem, subject to certain technical conditions (observability of the pair (C, A)), it is always possible to find a matrix L that assigns the zeros of (22.38) to any desired location in the complex plane. Software is readily available (e.g. in MATLAB®) to carry out the design.

The same ideas apply, mutatis-mutandis, to the sampled-data case.

In the sampled data care, the observer takes the form:

$$\hat{x}_{k+1} = A'\hat{x}_k + B'u_k + L'\left[y_k - c\hat{x}_k\right]. \tag{22.39}$$

22.9.2 Certainty Equivalence

If the state $x(t)$ is not measured, then it is natural to use the state estimate $\hat{x}(t)$ to define the required state-variable feedback in place of $x(t)$. Then, the appropriate control law becomes:

(i) in continuous time

$$u(t) = -K\hat{x}(t), \tag{22.40}$$

(ii) in sampled data

$$u_k = -K'\hat{x}_k. \tag{22.41}$$

Equations (22.40), (22.41) are said to be based on 'Certainty Equivalence' since the unmeasured state, $x(t)$, has been replaced by the estimate, $\hat{x}(t)$.

22.10 Separation

When an observer is used to define the feedback control law as in (22.40) or (22.41), then an interesting separation occurs between the poles of the system governed by state-variable feedback and those governed by state-estimate feedback. To show this, consider the continuous-time case where the control law (22.41) is used. The full system is then governed by the following set of equations which jointly describe the plant plus the observer:

$$\text{Plant: } \dot{x} = Ax + Bu, \tag{22.42}$$

$$\text{Control Law: } u = -K\hat{x}, \tag{22.43}$$

$$\text{Observer: } \dot{\hat{x}} = A\hat{x} + Bu + L\left(y - C\hat{x}\right), \tag{22.44}$$

$$\text{Output: } y = Cx. \tag{22.45}$$

Substituting (22.37) and (22.42) in (22.45) shows that the closed-loop system satisfies the following combined set of equations:

$$\frac{d}{dt}\begin{bmatrix} x \\ \tilde{x} \end{bmatrix} = \begin{bmatrix} A - BK & BK \\ 0 & A - LC \end{bmatrix}\begin{bmatrix} x \\ \tilde{x}. \end{bmatrix} \tag{22.46}$$

where $\tilde{x} = \hat{x} - x$.

It can be seen from (22.46) that the closed-loop poles of the combined system are simply the union of the separate poles achieved by state-variable feedback (i.e. the zeros of $A_{cl}(s)$ as in (22.19) and the observer poles) (i.e. the zeros of $A_{obs}(s)$ as in (22.38)). This is an amazing result since it shows that state-estimate feedback leads to a closed-loop system having poles that are the union of the poles that would result from state-variable feedback and the poles of the observer. Note that this result depends upon the model of the system being exact. Section 22.11 explains how to treat cases where the model is known to be an inexact representation of the true system.

22.11 Achieving Integral Action

22.11.1 The Problem

Consider the practical case when the model (22.42) is only approximately true. Under these conditions, the result in Section 22.10 will no longer hold. Moreover, it is now no longer true that the regulator (22.30) will cause $y(t)$ to converge to y^* even if $x(t)$ is measured. The reason is that x^* and u^* as defined in (22.25) and (22.26) depend upon the validity of the model, i.e. model errors will cause $y(t)$ to not be equal to y^* in steady state. Fortunately, there is a simple remedy to this problem as described below.

For simplicity, it is assumed that the system has a single input and single output but extensions to the multivariable case are straightforward.

22.11.2 The Remedy

The remedy to the problem described above is to hypothesise that the output actually satisfies a more general model, i.e. instead of (22.45), it is hypothesised that the output, $y(t)$ satisfies:

$$y(t) = Cx(t) + d, \tag{22.47}$$

where d is an *unknown* constant offset. The offset, d, can be used to express the fact that the model (22.45) does not strictly hold (even in steady state). When d is a constant, then it can be described (in continuous time) by a simple extra model of the form:

$$\dot{d} = 0. \tag{22.48}$$

Subject to the total system being observable, one can then design a composite observer to estimate both x and d given measurements of y. The appropriate observer is a simple embellishment of (22.36) but now having two gain matrices L_1 and L_2. The required observer equations are

$$\dot{\hat{x}} = A\hat{x} + Bu + L_1\left(y - C\hat{x} - \hat{d}\right), \tag{22.49}$$

$$\dot{\hat{d}} = 0 + L_2\left(y - C\hat{x} - \hat{d}\right). \tag{22.50}$$

The observer matrix now takes the form:

$$A_{obs} = \begin{bmatrix} A - L_1C & -L_1 \\ -L_2C & -L_2. \end{bmatrix} \tag{22.51}$$

As usual, subject to observability of the combined system (22.42), (22.48), and (22.47) one can choose L_1, L_2 to assign the observer poles to any desired location.

The control system design is next modified to account for $\hat{d}$. In particular, it seems sensible to bring $C\hat{x}$, not to y^*, but to $y^* - \hat{d}_{ss}$ where $\hat{d}_{ss}$ is the steady-state estimate of d provided by (22.50). Under these conditions, $\hat{x}_{ss}$ satisfies (22.25), (22.26) and hence u^* should be replaced by solving

$$C\hat{x}_{ss} = y^* - \hat{d}_{ss}. \tag{22.52}$$

22.11.3 Properties

Independent of the true system it follows, provided only that a steady state is reached, that y will converge to y^*. All that is required is that the system settles to *something*. This is easily established as follows: If a steady state is reached, then it follows from (22.50) that this equation cannot be driven by a signal having a constant component since, otherwise, this would cause $\hat{d}$ to diverge, (i.e. a steady state would not have been reached). Hence, the steady state values of $\hat{y} = C\hat{x}$ and $\hat{d}$ must be such that the term $(y - C\hat{x} - \hat{d})$ in (22.49), (22.50) has no DC component, i.e. the steady-state value, y_{ss}, of y must satisfy the following equation:

$$y_{ss} - C\hat{x}_{ss} - \hat{d}_{ss} = 0, \tag{22.53}$$

where, since $\hat{x}_{ss}$ exists (by assumption) then from (22.42), it must be true that, in steady state, the last term on the right-hand side of (22.49) is zero. Then the steady-state value, $\hat{x}_{ss}$, of $\hat{x}$ must satisfy:

$$A\hat{x}_{ss} + Bu^*_{ss} = 0. \tag{22.54}$$

That is

$$\hat{x}_{ss} = -A^{-1}Bu^*_{ss}. \tag{22.55}$$

Recall that u^*_{ss} was defined via (22.52). Hence,

$$y_{ss} = -CA^{-1}Bu^*_{ss} + \hat{d}_{ss}. \tag{22.56}$$

Then using (22.52) for $\hat{d}_{ss}$, it follows from (22.56) that the steady-state value, y_{ss}, of y must satisfy:

$$\begin{aligned} y_{ss} &= -CA^{-1}Bu^*_{ss} + \left(y^* + CA^{-1}Bu^*_{ss}\right), \\ &= y^*. \end{aligned} \tag{22.57}$$

Hence, based only on the assumption that a steady state, of some sort or another, has been reached, then

$$y_{ss} = y^*. \tag{22.58}$$

In summary, the inclusion of the extra observer (22.50) gives *integral action* and ensures zero tracking error even if the system model (22.14), (22.15) does not exactly describe the true system response. Of course, there may be robustness issues with actually achieving a steady state. These are similar to the robustness issues studied earlier in Section 6.10 of Chapter 6. In particular, one must be careful not to choose the closed-loop bandwidth too high if there exit model errors.

Again the extension to the sampled-data case is immediate. Instead of (22.48) the disturbance is assumed to satisfy:

Shift domain

$$d_{k+1} = d_k. \tag{22.59}$$

Delta domain

$$\delta d_k = 0. \tag{22.60}$$

22.12 All Stabilising Control Laws Revisited

The idea of the parameterisation of all stabilising control laws was described earlier in Section 7.4 of Chapter 7 (for open-loop stable systems) and in Section 14.3 of Chapter 14 (for open-loop unstable systems). This circle of ideas has an alternative interpretation when state-space ideas are used. This idea was first described in reference [1] given in Further Reading. See also reference [2].

Consider a linear plant described in state-space form by

$$\rho x = Ax + Bu \tag{22.61}$$

$$y = Cx, \tag{22.62}$$

where ρ is any suitable operator (e.g. $\frac{d}{dt}$ in continuous time or z in discrete time).

Also, consider an observer:

$$\rho\hat{x} = A\hat{x} + Bu + L[y - C\hat{x}], \tag{22.63}$$

and an associate state-estimate feedback law:

$$u = -K\hat{x}. \tag{22.64}$$

It is assumed that the control law (22.63), (22.64) leads to a closed-loop system having 'satisfactory' closed-loop poles. The error process, $\nu = y - C\hat{x}$ in (22.63), is given a special name. It is called, 'The Innovative Process'.

It then turns out that, all stabilising control laws can be described by the following augmented feedback law

$$u = -K\hat{x} + Q(\rho)\nu, \tag{22.65}$$

where $Q(\rho)$ is any proper-stable transfer function. Note that (22.65) is as in (22.40) with the addition of a stable-proper transfer function, $Q(\rho)$, operating on the innovative process, ν.

Further insights into the choice of $Q(\rho)$ are given below.

22.12.1 Stable Open-Loop Plants

Say that the open-loop plant is stable. (Or more generally has 'satisfactory' open-loop poles.) In this case, a suitable choice for L in (22.63) is $L = 0$, leading to the open-loop observer:

$$\rho\hat{x} = A\hat{x} + Bu. \tag{22.66}$$

Also, K in (22.64) can be chosen as zero. In this case, the expression given in (22.65) for all stabilising control laws takes the form:

$$\begin{aligned} u &= Q(\rho)\nu \\ \nu &= y - C\hat{x}. \end{aligned} \tag{22.67}$$

The reader should verify that (22.66), (22.64) are equivalent to (a multivariable version) of the parametrisation given in Figure 7.5 of Chapter 7.

22.12.2 Adding Stable Uncontrollable Disturbance States

Another interpretation of all stabilising control laws can be obtained by adding a disturbance, d, to the model.

Say that the disturbance satisfies a model of the form:

$$\begin{aligned} \rho x_d &= A_d x_d \\ d &= C_d x_d, \end{aligned} \tag{22.68}$$

where A_d is used to describe the dynamics of the disturbance.

The disturbance can be included in the model description, e.g. as an input disturbance as follows:

$$\rho x = Ax + B(u + d) \tag{22.69}$$

$$y = Cx, \tag{22.70}$$

In the event that A_d is a stability matrix (e.g. if it is used to describe a stable noise process), then a stable observer can be obtained by using:

$$\rho \begin{bmatrix} \hat{x} \\ \hat{x}_d \end{bmatrix} = \begin{bmatrix} A & BC_d \\ 0 & A_d \end{bmatrix} \begin{bmatrix} \hat{x} \\ \hat{x}_d \end{bmatrix} + \begin{bmatrix} B \\ 0 \end{bmatrix} u + \begin{bmatrix} L_1 \\ L_2 \end{bmatrix} \left[y - C\hat{x}\right]. \tag{22.71}$$

In this case one can use the control law

$$u = -K\hat{x} - C_d\hat{x}_d. \tag{22.72}$$

Using (22.71), the term $C_d\hat{x}_d$ can be expressed in transfer function form as

$$C_d\hat{x}_d = C_d\left[\rho I - A_d\right]^{-1} L_2 \left[y - C\hat{x}\right]. \tag{22.73}$$

Note that (22.73) again has the form of (22.65) where $Q(\rho)$ now is the stable transfer function:

$$Q(\rho) = C_d\left[\rho I - A_d\right]^{-1} L_2. \tag{22.74}$$

22.12.3 Adding Non-stabilisable Disturbance States

In many (arguably most) real-world applications, one will want to include non-stabilisable disturbance states in the model. (A specific case is where one wishes to use the disturbance observer to achieve integral action as described earlier in Section 22.11.) The observer (22.71) can still be used together with the control law (22.72). However, $Q(\rho)$ as defined in (22.74) will now not be a stable transfer function. Indeed, the instability of (22.68) was specifically exploited in Section 22.11 to argue that provided $\hat{x}_d$ does not diverge then $y - \hat{y}$ can have no steady-state modes at the eigenvalues of A_d. In turn this was used to argue that the system output, y, had no steady-state components corresponding to the eigenvalues of A_d. The reader will have probably noticed that this result does not fir into the same framework as "All Stabilising Controllers". Specifically, $Q(\rho)$ is now not stable. However, this is easily understood since the states used to describe constant disturbances are not stabilisable. Thus one needs to go beyond "All Stabilising Control Laws".

22.13 Model Predictive Control

An extension of the ideas presented in this chapter is to add the constraint that u_k and x_k must belong to convex sets Ω_u and Ω_x, respectively. This leads to a Quadratic Program which is relatively easy to solve. There is a substantial literature on this topic which goes under the name Model Predictive Control (MPC) see for example references [3] and [4] given in Further Reading.

The key idea in MPC is to choose a prediction horizon of N samples. Assuming that x_0 is known (or estimated), one uses constrained optimization to evaluate the control sequence over the prediction horizon $\{u_0, \dots, u_{N-1}\}$. One then applies u_0 to the plant. Next x_1 is measured (or estimated). One then uses constrained optimization to compute $u_1, \dots, u_N$. One then applies u_1 and so on indefinitely. Hence MPC involves Rolling Horizon constrained optimization.

When the state is not measured then an observer should be used and the problem recast in terms of the observed state. To achieve robustness to model errors, then one should add a disturbance to the model as outlined in Section 22.11.

Many of the designs described in this book can be reinterpreted in the MPC framework. This interpretation lies outside the scope of the current book. Thus, the reader is encouraged to pursue these ideas on their own initiative.

22.14 Revision Questions

1. Consider the following simple first-order discrete-time system:

$$x_{k+1} = x_k + u_k; \quad y_k = x_k, \tag{22.75}$$

   with a quadratic cost function as in (22.3) where $Q = 1, R = 1, P_N = 0$.
   (a) What is the Matrix Riccati equation for the cost to go from time step k?
   (b) What is the time-varying control law?
2. (a) For the system of Question 1, what is the equation for the steady-state solution of the Riccati equation steady state?
   (b) Show that the steady-state Riccati equation has two possible solutions. What are they?
   (c) Show that the positive steady-state solution to the Riccati equation leads to a stable closed-loop system.

3. Consider the following simple first-order continuous-time system:

$$\dot{x} = x + u; \; y = x. \tag{22.76}$$

   Repeat Question 1 and 2 using the same values for P_N, Q, R.
4. Consider the system

$$x_{k+1} = 2x_k + u_k. \tag{22.77}$$

   Say that one wishes to control the system to achieve $y = 2$ in steady state.
   What is the required steady-state regulation control law assuming a quadratic cost as in Question 1?
5. Consider the system of Question 2. Say that one wishes to achieve $y = 1$ in steady state. What is the required steady-state control law?

Further Reading

1 T.T. Tay and J.B. Moore, "Enhancement of fixed controllers via adaptive disturbance estimate feedback", *Proceedings of the IFAC Symposium on Identification*, 1:149–154, 1988.
2 B.D.O. Anderson,"From Youla-Kucera to identification, adaptive and nonlinear control", *Automatica*, 34(12):1485–1506, 1998.
3 J.B. Rawlings, D.Q. Mayne and M.M. Diehl, *Model predictive control theory, computation and design*, 2nd Edition, No. 6, Hill Publishing, 2017.
4 G.C. Goodwin, M.M. Seron and J. de Dona, *Constrained control and estimation*, Springer-Verlag, 2004.
5 B.D.O. Anderson and J.B. Moore, *Optimal control: linear quadratic methods*, Dover, 1990.

23

The Kalman Filter

23.1 Introduction

The concept of using an observer to estimate the states of a system was introduced in Section 22.9 of Chapter 22. However, the design of the observer gains described in Chapter 22 was somewhat ad-hoc. This leads to an interesting question, namely, 'Is there a best choice for the observer gains?'

This question has inspired a great deal of research over the years.

The first researchers to tackle the problem in the modern era were Wiener (in the USA) and Kolmogorov and Stratonovich (in the Soviet Union).

The time-varying case was studied by Kalman, Bucy, and Stratonovich in the 1960s using State Space descriptions. The associated observer is usually called the 'Kalman Filter'. The Kalman filter is widely used in a vast number of areas of modern systems science. It is thus appropriate that the reader become familiar with the key ideas. Hence the current chapter will give an outline of the basic principles involved in the Kalman Filter. Those readers wishing to have a more in-depth treatment should consult the references given at the end of the chapter. The ideas described here will be used in a practical problem in the laboratory presented in Chapter 24.

23.2 Periodic Disturbances

The Kalman filter poses, and solves, the question asked in the introduction, namely, what is the 'best' choice for the observer gain matrix L. Of course, 'best' is a subjective measure. Indeed, additional assumptions are necessary so that the term, 'best', can be properly defined.

23.2.1 Continuous-Time Model

Most implementations of the Kalman filter use sampled data. Thus, the sampled-data case will be used to develop the ideas. However, to obtain a better understanding of the model, it is useful to first consider the sampling process applied to an underlying continuous-time model. This model should, in principle, be expressed in incremental form as a Stochastic Differential Equation driven by a Wiener Process. However, for ease of exposition, a formal model based on 'continuous-time-white-noise' will be used. The appropriate continuous-time model then takes the form:

$$\dot{x} = Ax + Bu + \omega, \tag{23.1}$$

$$y = Cx + v. \tag{23.2}$$

Practical Control System Design: Real World Designs implemented on Emulated Industrial Systems,
First Edition. Adrian Medioli and Graham Goodwin.

Companion website: www.wiley.com/go/medioli/practicalcontrolsystemdesign

Here, ω is termed the 'process noise' and v is termed the 'measurement noise'. The processes ω, v are assumed to be 'continuous-white-noise' processes with intensity Q and R, respectively. (Actually, Q and R are the noise Spectral Densities.) Formally, the covariance of $\omega(t)$ and $v(t)$ are given by

$$E\left\{\omega(t)\omega(t-\tau)^T\right\} = Q\delta(\tau), \tag{23.3}$$

$$E\left\{v(t)v(t-\tau)^T\right\} = R\delta(\tau), \tag{23.4}$$

where $E\{\cdot\}$ denotes the expectation operator and $\delta(\tau)$ is the Dirac Delta function. The connection between (23.3), (23.4) and spectral densities is examined in Question 2 of the Revision Questions given at the end of the chapter.

23.2.2 Sampled-Data Process Noise

Using the same ideas as in Section 3.2.2 of Chapter 3, the solution of the model (23.1) over a period of Δ is

$$x((k+1)\Delta) = e^{A\Delta}x(k\Delta) + \int_{k\Delta}^{(k+1)\Delta} e^{A((k+1)\Delta-\tau)}Bu(\tau)d\tau + \int_{k\Delta}^{(k+1)\Delta} e^{A((k+1)\Delta-\tau)}\omega(\tau)d\tau. \tag{23.5}$$

Using the standard sampled-data notation, as in Chapter 17, let k denote the kth sample, so that $t = k\Delta$. Also, it will be assumed that $u(\tau)$ is constant over the sample period (i.e. it is implemented via a zero-order hold). Then $u(t) = u_k$ for $k\Delta \leq t < (k+1)\Delta$. As usual, define

$$x_k = x(k\Delta). \tag{23.6}$$

Then (23.5) can be written in shift operator notation as

$$x_{k+1} = A'x_k + B'u_k + \omega'_k, \tag{23.7}$$

where

$$A' = e^{A\Delta}, \tag{23.8}$$

$$B' = \int_0^{\Delta} e^{A(\Delta-\tau)}Bd\tau, \tag{23.9}$$

$$\omega'_k = \int_{k\Delta}^{(k+1)\Delta} e^{A((k+1)\Delta-\tau)}\omega(\tau)d\tau. \tag{23.10}$$

The discrete noise term, ω'_k, introduced in (23.7) is easily seen to have the following property:

$$E\left\{\omega'_k\left(\omega'_j\right)^T\right\} = E\left\{\left[\int_{k\Delta}^{(k+1)\Delta} e^{A(k\Delta-\tau)}\omega(\tau)d\tau\right]\left[\int_{j\Delta}^{(j+1)\Delta} e^{A(j\Delta-\overline{\tau})}\omega(\overline{\tau})d\overline{\tau}\right]^T\right\}, \tag{23.11}$$

$$= E\left\{\int_{k\Delta}^{(k+1)\Delta}\int_{j\Delta}^{(j+1)\Delta} e^{A(k\Delta-\tau)}\omega(\tau)\omega(\overline{\tau})^T\left(e^{A(j\Delta-\overline{\tau})}\right)^T d\tau d\overline{\tau}\right\}, \tag{23.12}$$

then, using (23.3)

$$E\left\{\omega'_k\left(\omega'_j\right)^T\right\} = \begin{cases} \int_{k\Delta}^{(k+1)\Delta} e^{A(k\Delta-\tau)}Q\left(e^{A(k\Delta-\tau)}\right)^T d\tau & \text{for } k=j, \\ 0 & \text{for } k \neq j. \end{cases} \tag{23.13}$$

Note that, for small Δ, $e^{A\Delta}$ is approximately I. Hence, it can be seen from (23.13) that the sampled-data noise sequence, ω'_k, is a discrete-white noise (i.e. uncorrelated in the time domain) whose variance grows as Δ increases. This makes sense, since more of the 'white noise', $\omega(t)$, gets pulled into the last term on the right-hand side of (23.5).

23.2.3 Sampled-Data Measurement Noise

Next, the sampling of the process, $y(t)$, will be examined. This operation is usually called 'analogue to digital conversion' (A/D). The reader is asked to revise Section 17.2 of Chapter 17. A particular issue of importance in this context is the frequency domain folding that results from time-domain sampling. In particular, one cannot directly sample $y(t)$ since this would give samples with infinite noise variance due to the infinite folding of the continuous noise spectrum. Instead, so as to avoid the folding of high-frequency measurement noise, all A/D converters incorporate some form of anti-aliasing filter so that the sampled value, y_k, is sensibly defined. Actually, all A/D converters use (something like) the average value of $y(t)$ over the interval $((k-1)\,\Delta, k\Delta)$. (The reader is referred to Figure 17.2 of Chapter 17 where the idea of an averaging filter was first described.)

Assuming that the anti-aliasing filter averages the output, then the sampled output satisfies

$$\begin{aligned} y_k &= \frac{1}{\Delta}\int_{(k-1)\Delta}^{k\Delta} y(t)dt, \\ &= \frac{1}{\Delta}\int_{(k-1)\Delta}^{k\Delta} Cx(t)dt + \frac{1}{\Delta}\int_{(k-1)\Delta}^{k\Delta} v(t)dt, \end{aligned} \tag{23.14}$$

where $v(t)$ is the white measurement noise introduced in Eq. (23.2).

To gain insight into (23.14), it is assumed that the $x(t)$ is slowly varying relative to period Δ. Hence, the averaging in the first term of (23.14) can be ignored. Under these conditions, (23.14) can be written more simply as

$$y_k = Cx_k + n'_k, \tag{23.15}$$

where

$$n'_k = \frac{1}{\Delta}\int_{(k-1)\Delta}^{k\Delta} v(t)dt. \tag{23.16}$$

The measurement noise term, n'_k, is also a discrete-white noise sequence as shown below. Its covariance satisfies:

$$\begin{aligned} E\left\{n'_k(n'_j)^T\right\} &= \frac{1}{\Delta^2}E\left\{\int_{(k-1)\Delta}^{k\Delta} v(t)dt \int_{(j-1)\Delta}^{j\Delta} v(\tau)^T d\tau\right\}, \\ &= \frac{1}{\Delta^2}\int_{(k-1)\Delta}^{k\Delta}\int_{(j-1)\Delta}^{j\Delta} R\delta(t-\tau)dtd\tau, \\ &= \begin{cases} \frac{1}{\Delta}R & \text{for } k = j, \\ 0 & \text{for } k \neq j. \end{cases} \end{aligned} \tag{23.17}$$

It can thus be seen that, the 'discrete', measurement noise process, n'_k, is a discrete-white noise process. Also, as can be seen from (23.17), the covariance of n'_k will decrease as Δ increases. (See also Revision Question 2 at end of the chapter.)

23.2.4 The Full Sampled-Data Model

Summarising the results presented in Sections 23.2.2 and 23.2.3, a suitable discrete (sampled data) model incorporating noise processes takes the form:

$$x_{k+1} = A'x_k + B'u_k + \omega'_k, \tag{23.18}$$

$$y_k = Cx_k + n'_k, \tag{23.19}$$

where ω'_k, n'_k are modelled as discrete-time 'white noise' processes having zero mean and covariances given by

$$E\left\{(\omega'_k)(\omega'_k)^T\right\} = Q', \tag{23.20}$$

$$E\left\{(n'_k)(n'_k)^T\right\} = R', \tag{23.21}$$

where Q' and R' are sample period-dependent quantities. Based on the arguments presented in Sections 23.2.2 and 23.2.3, Q' will generally increase as Δ increases whilst R' will generally decrease as Δ increases. A reasonable approximation is $Q' \simeq Q\Delta$ and $R' \simeq \frac{1}{\Delta}R$.

23.3 The Best Observer Gain

With the above formalism in place, the question of the best gain, L, for the sampled-data observer can be examined. As in Chapter 22, an observer of the following form will be assumed:

$$\hat{x}'_{k+1} = A'\hat{x}'_k + B'u_k + L'_k\left\{y_k - C\hat{x}'_k\right\}. \tag{23.22}$$

Note that the usual time-variant observer gain L', used in Chapter 22, has here been generalised in (23.22) so that it depends upon the sample time, i.e. L' has been replaced by L'_k.

Substituting (23.18) into (23.22) shows that the error $\tilde{x} = \hat{x}' - x'$ satisfies:

$$\tilde{x}_{k+1} = \left(A' - L'_kC\right)\tilde{x}_k + L'_kn'_k - \omega'_k. \tag{23.23}$$

The goal is then to choose L'_k to minimise the covariance of $\tilde{x}$, where the covariance of the estimation error is defined to be

$$E\left\{\tilde{x}_k\tilde{x}_k^T\right\} = P'_k. \tag{23.24}$$

As in Chapter 22, induction will be used to obtain an expression for P'_k and from this expression to determine the 'best' observer gain that minimise P'_k.

To begin, say the matrix P'_k is already known, then the design question becomes, 'For given L'_k, what is the value of P'_{k+1} and then what value of the matrix L'_k minimises this covariance?' By squaring both sides of (23.23) and using (23.20), (23.21), it is easily seen that P'_{k+1} satisfies:

$$P'_{k+1} = \left(A' - L'_kC\right)P'_k\left(A' - L'_kC\right)^T + L'_kR'(L'_k)^T + Q'. \tag{23.25}$$

Equation (23.25) is quadratic in L'_k and hence by setting the derivative to zero it is easily shown that P'_{k+1} is minimised by the following 'optimal' choice for L'_k:

$$L'_k = A'P'_kC^T\left[R' + CP'_kC^T\right]^{-1}. \tag{23.26}$$

Substituting (23.26) into (23.23) shows that the best value for P'_{k+1} satisfies:

$$P'_{k+1} = A_kP'_k(A_k)^T - A_kP'_kC^T\left(R' + CP'_kC^T\right)^{-1}CP'_kA_k^T + Q', \tag{23.27}$$

where L'_k is a function of P'_k as in (23.26).

The above argument can be continued for all time.

In summary, Eq. (23.25) gives the optimal observer gain. Also, when this is substituted into (23.25), then the resultant equation for the optimal covariance P'_{k+1} is (23.27).

To complete the specification of the optimisation problem, it is assumed that an initial estimate $\hat{x}_0$ of x_0 is given. Moreover, it is assumed that $E\{(x_0 - \hat{x}_0)(x_0 - \hat{x}_0)^T\} = P_0$ is given as part of the problem set-up. Then Eq. (23.27) can be solved (in forward time) recursively for P'_k; $k > 0$ beginning with the known values P_0.

Hence $P_1, P_2 \ldots$ can be found and the corresponding optimal observer gains can be found by using (23.26).

Equation (23.27) is known as the (Discrete) Matrix Riccati equation of Filtering. The initial condition for this equation is the initial covariance P'_o of $(\hat{x}'_o - x'_o)$.

A great deal is known about Eq. (23.27) dating from the initial work of Riccati in the 17th century CE and continuing well into the late 20th century CE with the work of Kalman and many others. Standard software is available to solve the equation.[1]

In theory, the matrices Q' and R' have precise definitions as the covariance of the process and measurement noise, respectively. However, in practice, these matrices are often considered as 'tuning parameters'. These matrices can be given a heuristic interpretation as follows: Say one fixes Q' and varies R', then as $R' \to \infty$, the filter essentially becomes open-loop (because it has been told that the measurement noise is very large). Alternatively, when $R' \to 0$, the filter essentially becomes a differentiator (because it has been told that the measurement noise is negligible and hence differentiation of the output can be used to estimate the states).

23.4 Steady-State Optimal Estimator

An important case arises when the filter is run for a long time. It can then be shown that, provided the system is time-invariant, then the value of P'_k converges to a constant denoted P_{ss}. Correspondingly, the observer gain will also converge to a steady value, denoted L_{ss}. In this case, the Kalman filter has simply led to a 'best' (time-invariant) observer gain.[2]

The discrete steady-state optimal filter takes the form

$$\hat{x}_{k+1} = A\hat{x}_k + Bu_k + L_{ss}\left(y_k - C\hat{x}_k\right), \tag{23.28}$$

where L_{ss} satisfies an equation similar to (23.26) but with P'_k replaced by P'_{ss}, the steady-state solution of the Riccati equation.

Analogously to the results in Section 22.4 of Chapter 22, the following properties can be shown to hold (For proofs – see the references given at the end of the chapter.):

- For a completely observable system, the matrix P'_k will converge to a well-defined limit as $k \to \infty$.
- The limiting value of P'_k (denoted $\overline{P}'_\infty$) is positive definite.
- The value $\overline{P}'_\infty$ satisfies the algebraic Riccati equation (of filtering):

$$\overline{P}'_\infty = \left(A - L'_\infty C\right)\overline{P}'_\infty\left(A - L'_\infty C\right)^T + L'_\infty R'\left(L'_\infty\right)^T + Q', \tag{23.29}$$

$$L'_\infty = A^T P'_\infty\left(C'\right)^T\left[R' + C'P'_\infty\left(C'\right)^T\right]^{-1}, \tag{23.30}$$

  (These equations follow from (23.25) by setting $P'_{k+1} = P'_k$.)
- Equation (23.30) has many solutions, but there is a unique positive definite solution $\overline{P}'_\infty$.
- When $\overline{P}'_\infty$ is used to define the steady-state observer, then the resultant steady-state observer (as given in (23.28)) is stable.

1 Note that in practice, it is usual to let $k \to \infty$. This leads to a steady-state observer as explained in Section 23.4. A case where the time-varying version is appropriate is when the system is, itself, time varying.

2 In practice this convergence occurs rapidly and hence it is usual to use only the steady-state version of the results.

23.5 Treating Non-White Noise

In some applications, it is important to model the noise more carefully than to simply say 'assume it is white noise'. A useful idea, commonly deployed in practice, is to use another state-space model to turn white noise into correlated noise.

As an illustration, say that, rather than having 'white measurement noise' one wants to describe correlated measurement noise. One can describe correlated measurement noise by passing white noise through a filter (another model).

The above idea leads to an augmented model as follows:

$$y'_k = C'x_k + n_k^C, \tag{23.31}$$

where n_k^C is correlated measurement noise which is assumed to satisfy a second model of the form:

$$x_{k+1}^C = A^C x_k^C + n'_k, \; E\{n'_k\} = 0, \; E\{n'_k n'^T_k\} = Q^C, \tag{23.32}$$

$$n_k^C = C^C x_k^C + n''_k, \; E\{n''_k\} = 0, \; E\{n''_k n''^T_k\} = R^C, \tag{23.33}$$

where n'_k is taken to be white noise and n_k^C, is a correlated noise process. The model (23.32), (23.33) can then be added to the original model to obtain a composite Kalman Filter.

23.6 Dealing with Constant Disturbances

A particular disturbance of great importance in control is that of a (near) constant output disturbance. This idea was introduced earlier in Section 22.11 of Chapter 22. A near-constant disturbance can be modelled as in (23.32) but with $A^C = I$, i.e.

$$x_{k+1}^C = x_k^C + n'_k, \tag{23.34}$$

$$n_k^C = C^C x_k^C + n''_k, \tag{23.35}$$

One can then obtain a Kalman filter for the composite state x_k and x_k^C.

The reader is asked to compare this idea with the ideas in Section 22.11 of Chapter 22.

The noise source $\{n'_k\}$ gives a means of describing how 'constant' one believes the disturbance really is. Making Q^C large implies that $\{x_k^C\}$ can actually 'wander around' quite a bit.

23.7 Periodic Disturbances

Another extension of the idea described in Section 23.6, is to model (near) periodic disturbances. Thus, say that d_k has a (near) period of p samples. This can be modelled by a p state model of the form

$$\begin{bmatrix} x'_{k+1} \\ \\ x^p_{k+1} \end{bmatrix} = \begin{bmatrix} 0 & \cdots & 0 \\ 10 & \cdots & 0 \\ & \ddots & 10 \end{bmatrix} \begin{bmatrix} x'_k \\ \\ x^p_k \end{bmatrix} + n'_k. \tag{23.36}$$

This idea will be explored further in Chapter 24 in connection with a real-world design problem.

23.8 Accounting for Delays

Another common problem met in practice is when the system contains a pure time delay.

One can include a delay in the model (in discrete time) by simply adding states. Hence a delay of three samples (for example) can be described by

$$x_{k+1} = A'x_k + B'u_k + \omega'_k, \tag{23.37}$$

$$x^1_{k+1} = x_k, \tag{23.38}$$

$$x^2_{k+1} = x^1_k, \tag{23.39}$$

$$x^3_{k+1} = x^2_k, \tag{23.40}$$

$$y_k = Cx^3_k + n'_k. \tag{23.41}$$

One can then design an observer for the augmented model.

23.9 Multiple Output Measurements

One can also describe the situation where one has two measurements (with different features) by suitable augmentation of the model, i.e. one might have y_k and $\bar{y}_k$ described as follows:

$$x^a_{k+1} = A^a x_k + B^a u_k + D^a \omega'_k, \tag{23.42}$$

$$y^a_k = \begin{bmatrix} y_k \\ \bar{y}_k \end{bmatrix} = \begin{bmatrix} C \\ \bar{C} \end{bmatrix} x^a_k + n'_k, \tag{23.43}$$

Then, given measurements of both y_k and $\bar{y}_k$ there is a single optimal filter that gives the best estimate of x^a_k. This idea allows one to combine data from multiple sensors. This idea was implicitly used in the question posed at the end of Section 13.10.1. In particular, for the problem in Chapter 13, a useful idea is to use the BISRA gauge for fast dynamic response but to use a traditional thickness measurement for long-term steady-state errors.

The reader may also wonder how the idea of a single best observer can be rationalised with the idea of using the first measurement, i.e., y_k to deal with rapid transients and the second measurement, $\bar{y}_k$, to steady-state corrections.

Actually, a little thought indicates that one needs to be more careful about how one describes the system. If one really believes that y_k has long-term (near constant) disturbances on it, then one should use the ideas of Section 23.6 to augment the model to describe the (near) constant disturbances and the other observe for estimates of the system states for dynamic feedback. The optimal observer then takes the form:

$$\hat{x}_{k+1} = A\hat{x}_k + Bu_k + [L_1, L_2] \begin{bmatrix} y_k - C\hat{x}_k - \hat{d}_k \\ \bar{y}_k - \bar{C}\hat{x}_k \end{bmatrix}, \tag{23.44}$$

$$\hat{d}_{k+1} = A\hat{d}_k + [L^d_1, L^d_2] \begin{bmatrix} y_k - C\hat{x}_k - \hat{d}_k \\ \bar{y}_k - \bar{C}\hat{x}_k \end{bmatrix}. \tag{23.45}$$

In conclusion, it can be seen that, by setting the model up carefully, then $\bar{y}_k$ will be predominantly used for steady-state correction but y_k will be predominately used under transient conditions.

23.10 Continuous-Time Kalman Filter

All of the above ideas can also be extended to continuous-time models. In this case, the appropriate model is as given earlier in (23.1), (23.5), i.e.

$$\dot{x} = Ax + Bu + \omega \tag{23.46}$$

$$y = Cx + n, \tag{23.47}$$

where ω and n are 'white' (i.e. wide band) processes having spectral density Q and R, respectively.

The corresponding continuous-time optimal filter takes the form:

$$\dot{\hat{x}} = A\hat{x} + Bu + L\left(y - C\hat{x}\right), \tag{23.48}$$

where the optimal value of L can now be found by solving a (continuous-time) matrix Riccati equation.

Similarly, the ideas in Section 23.6 can be extended to the continuous-time case by modelling a constant disturbance by

$$\dot{x}^c = \omega^c, \tag{23.49}$$

$$d^c = x^c + n^c. \tag{23.50}$$

The ideas of Sections 23.6 and 23.7 can also be applied in continuous time. For example, a periodic disturbance can (by Fourier analysis) be decomposed into sinusoidal components. Consider one such component of frequency θ say. Then a suitable model for this component is

$$\ddot{y} = -\theta^2 y, \tag{23.51}$$

or in state-space form

$$\begin{bmatrix} \dot{x}_1 \\ \dot{x}_2 \end{bmatrix} = \begin{bmatrix} 0 & 1 \\ -\theta^2 & 0 \end{bmatrix} \begin{bmatrix} x_1 \\ x_2 \end{bmatrix} \tag{23.52}$$

$$y = x_1. \tag{23.53}$$

A real-world example of this will be given in Chapter 24.

23.11 Linking Continuous Kalman Filter and Discrete Kalman Filter

It is also possible to link the discrete- and continuous-time Kalman filters using the divided difference operator.

Specifically, the discrete Kalman filter (23.20) can be expressed in divided difference form as:

$$\left(\frac{\hat{x}'_{k+1} - \hat{x}'_k}{\Delta}\right) = \frac{1}{\Delta}\left[A' - I\right]\hat{x}'_k + \frac{1}{\Delta}B'u_k + \frac{L'}{\Delta}\left[y_k - C\hat{x}'_k\right], \tag{23.54}$$

where

$$A' = e^{A_c\Delta} \tag{23.55}$$

$$B' = \int_0^{\Delta} e^{-A(t-\tau)}Bd\tau. \tag{23.56}$$

Letting $\Delta \to 0$, gives the corresponding continuous filter as

$$\dot{\hat{x}} = A_c\hat{x} + B_c u + L_c[y - C\hat{x}], \tag{23.57}$$

where A_c and B_c appear in the continuous model. Also, from (23.23)

$$L' = A'PC'^T\left[R' + C'PC'^T\right]^{-1}. \tag{23.58}$$

We recall that $R' = \frac{R}{\Delta}$ where R is the continuous-time noise spectral density. Also, $A' \simeq I + A_c\Delta$ and $Q' = Q\Delta$.

Hence, from (23.54), (23.58):

$$L_c = \frac{L'}{\Delta} \simeq \left[I + A_c\Delta\right] P'C'^T\left[\frac{R}{\Delta} + C'PC'^T\right]^{-1}\frac{1}{\Delta}. \tag{23.59}$$

Letting $\Delta \to 0$, gives

$$L_c = PC^TR^{-1}. \tag{23.60}$$

Also, the matrix Riccati equation (23.25) can be expressed (for Δ small) as

$$P'_{k+1} = \left(I + A\Delta - L_c\Delta C\right)P_k\left[(I + A_\Delta) - L_c\Delta C\right]^T + \Delta L_c\frac{R}{\Delta}L_c^T\Delta + Q\Delta. \tag{23.61}$$

Taking the limit $\Delta \to 0$, and setting $\dot{P} = \frac{P'_{k+1} - P'_k}{\Delta}$ gives

$$\dot{P} = \left(A - L_cC\right)P + P\left(A - L_cC\right)^T + L_cRL_c^T + Q, \tag{23.62}$$

or using (23.61) one obtains a differential equation for P.

$$\dot{P} = AP + PA^T - PC^T - PC^TR^{-1}CP + Q. \tag{23.63}$$

23.12 The Linear Quadratic Regulator Revisited

The reader is reminded of the ideas introduced in Section 22.9 of Chapter 22. In particular, when the state is not measured, it makes sense to utilise certainty equivalence, i.e. to replace the state x_k in (22.11) by an estimate of x_k provided by an optimal observer (i.e. the Kalman filter).

Using the steady-state versions, the total system description now becomes

$$x_{k+1} = Ax_k + Bu_k + \omega_k\,;\; y_k = Cx_k + v_k \tag{23.64}$$

$$\hat{x}_{k+1} = A\hat{x}_k + Bu_k + L_{ss}\left[y_k - C\hat{x}_k\right] \tag{23.65}$$

$$u_k = -K_{ss}\hat{x}_k, \tag{23.66}$$

where L_{ss} and K_{ss} could be designed by optimal methods.

In this context, it is suggested that the reader reread Section 22.10 regarding the closed-loop dynamics for the control and estimation problems.

23.13 Quantifying the Performance

Before concluding this chapter it is important to give a word of warning about the word 'optimal' as used in this and Chapter 22. In particular, just because a control law and filter are 'optimal', this does not mean that the overall system has *good* performance. Indeed, the combination of the Kalman filter and optimal-state feedback is simply a fancy way of defining an output feedback control law. The connection is best explored using transfer functions.

In particular, transfer functions can be used to articulate the feedback law obtained by combining the Kalman filter and optimal-state feedback gain. Specifically, from (23.28) the optimal steady-state observer takes the form:

$$\hat{x}_{k+1} = (A - L_{ss}C)\,\hat{x}_k + Bu_k + L_{ss}y_k. \tag{23.67}$$

In transfer function form, (23.67) can be written as

$$\hat{x} = (zI - A + L_{ss}C)^{-1}\,(Bu_k + L_{ss}y_k). \tag{23.68}$$

Also since $u_k = -K_{ss}\hat{x}_k$, then using (23.68),

$$u_k = -K_{ss}\left[zI - A + L_{ss}C\right]^{-1}\,(Bu_k + L_{ss}y_k). \tag{23.69}$$

Bringing the terms in u_k to the left-hand side gives

$$u_k = \left(I + K_{ss}(zI - A + L_{ss}C)^{-1}B\right)^{-1}\left(-K_{ss}(zI - A + L_{ss}C)^{-1}L_{ss}\right)y_k. \tag{23.70}$$

Thus, the combination of 'optimal' control and the 'optimal' Kalman filter has simply led to a particular dynamic feedback control law of the form:

$$u_k = C(z)y_k, \tag{23.71}$$

where $C(z)$ is a transfer function. From (23.70) this transfer function is

$$C(z) = \left(I + K_{ss}(zI - A + L_{ss}C)^{-1}B\right)^{-1}\left(-K_{ss}(zI - A + L_{ss}C)^{-1}L_{ss}\right). \tag{23.72}$$

Thus, all of the comments made in Chapter 16 regarding sensitivity trade-offs remain applicable to the 'optimal' state-space solution.

23.14 Revision Questions

1. Consider a scalar sampled-data system.

$$x_{k+1} = ax_k + bu_k + \omega_k, \tag{23.73}$$

$$y_k = x_k + v_k, \tag{23.74}$$

where ω_k, v_k are white noise sequences with variance Q' and R', respectively.
   (a) Write down the form of the time-varying Kalman filter assuming that an estimate, $\hat{x}_0$, is available satisfying: $E\left\{[\hat{x}_0 - x_0][\hat{x}_0 - x_0]^T\right\} = P_0$.
   (b) What is the corresponding steady-state Riccati equation?
   (c) Show that the steady-state Riccati equation has two solutions.
   (d) Investigate your result in (a) when a is stable and R becomes large and R becomes small.
2. Let $a = 1 + a_c\Delta$, $b = b_c\Delta$, $Q = Q_c\Delta$, $R = R_c/\Delta$.
   (a) Investigate your solution to Question 1 part (d) as $\Delta \to 0$.
   (b) Comment on the relationship to the continuous-time Kalman filter.
3. An extension of the Kalman filtering idea applies to nonlinear systems. Here an approximate linearised model is obtained by linearising the system about the current state estimate. This leads to the, so-called, extended Kalman filter. Consider the following nonlinear system.

$$\dot{x} = f(x, u) + \omega \tag{23.75}$$

$$y = h(x) + u \tag{23.76}$$

where $E\left\{[x(0) - \hat{x}(0)][x(0) - \hat{x}(0)]^T\right\} = P_0$
Derive the form of the extended Kalman filter.

4. An often used application of the extended Kalman filter is for tracking using radar measurements. Typically a radar supplies range and bearing information – see Figure 23.1 where r is the range, θ the bearing.

Figure 23.1 Radar range and bearing.

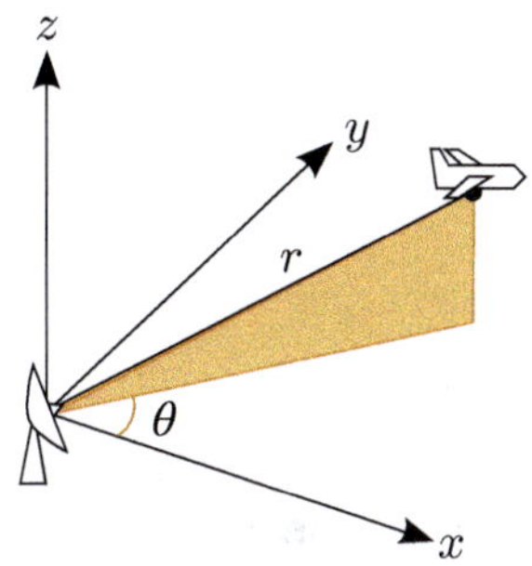

Let the system evolve in x–y space according to

$$\dot{x}_1 = \omega_1 \tag{23.77}$$

$$\dot{x}_2 = \omega_2 \tag{23.78}$$

where $E\begin{bmatrix}\omega_1\\ \omega_2\end{bmatrix}\begin{bmatrix}\omega_1^T & \omega_2^T\end{bmatrix} = \begin{bmatrix}Q_1 & 0\\ 0 & Q_2\end{bmatrix}$.

(a) Derive an expression for x_1, x_2 in terms of r, θ.
(b) Hence write down the appropriate extended Kalman filter for estimating the Cartesian coordinates of the aircraft given range and bearing measurements.

5. Consider the following continuous-state space model:

$$\dot{x} = ax + b(u + d) + \omega_1, \tag{23.79}$$

$$\dot{d} = \omega_2, \tag{23.80}$$

$$y = x + v, \tag{23.81}$$

where

$$E\left\{\begin{bmatrix}\omega_1\\ \omega_2\end{bmatrix}\begin{bmatrix}\omega_1^T & \omega_2^T\end{bmatrix}\right\} = \begin{bmatrix}Q_1^c & 0\\ 0 & Q_2^c\end{bmatrix}, \tag{23.82}$$

$$E\left\{vv^T\right\} = R^c, \tag{23.83}$$

Develop the Kalman filter for estimating x and d.

6. Repeat Question 5 for the sampled-data system

$$x_{k+1} = a'x_k + b'\left(u_k + d_k\right) + \omega_1, \tag{23.84}$$

$$d_{k+1} = d_k, \tag{23.85}$$

$$y_k = x_k + v_k, \tag{23.86}$$

where $E\left\{\begin{bmatrix}\omega_1\\ \omega_2\end{bmatrix}\begin{bmatrix}\omega_1^T & \omega_2^T\end{bmatrix}\right\} = \begin{bmatrix}Q_1 & 0\\ 0 & Q_2\end{bmatrix}; E\left\{v_k v_k^T\right\} = R^c$.

7. Extend Question 5 to the case where d is a sine wave of frequency ω_0.
8. Extend Question 6 to the case where d_k is a sampled sine wave of frequency ω_0.
9. Occasionally in practical problems, prior knowledge indicates that the process noise and measurement noise are correlated. Hence, consider the following sampled data model:

$$x_{k+1} = Ax_k + Bu_k + \omega_k, \tag{23.87}$$

$$y_k = Cx_k + v_k, \tag{23.88}$$

where $E\left\{\begin{bmatrix}\omega_k \\ v_\ell\end{bmatrix}\begin{bmatrix}\omega_k^T & v_\ell^T\end{bmatrix}\right\} = \begin{bmatrix}Q & S \\ S^T & R\end{bmatrix}\delta(k-\ell)$.

Since y_k is measured, then y_k can be added and subtracted to the right-hand side of (23.87) to yield:

$$x_{k+1} = Ax_k + Bu_k + \omega_k - SR^{-1}\left[y_k - y_k\right]. \tag{23.89}$$

Then, using the fact that (23.88) holds:

$$x_{k+1} = Ax_k + Bu_k - SR^{-1}\left(Cx_k + v_k\right) + SR^{-1}y_k + \omega_k, \tag{23.90}$$

(a) Hence show that an equivalent state-space model is

$$x_{k+1} = A'x_k + Bu_k + SR^{-1}y_k + \omega_k' \tag{23.91}$$

$$y_k = Cx_k + v_k \tag{23.92}$$

where $A' = A - SR^{-1}C$; $\omega_k' = \omega_k - SR^{-1}v_k$.

(b) Evaluate $E\left\{\begin{bmatrix}\omega_k' \\ v_k\end{bmatrix}\begin{bmatrix}\omega_\ell^T & v_\ell^T\end{bmatrix}\right\}$

(c) Hence derive the form of the Kalman filter for the model (23.87), (23.88).

10. Define the 'innovations' noise process by

$$e_k = y_k - C\hat{x}_k \tag{23.93}$$

(a) Show that in steady state

$$E\left\{e_k e_k^T\right\} = R + CP_\infty C^T, \tag{23.94}$$

(b) Hence show that the steady-state Kalman filter can be expressed as

$$\hat{x}_{k+1} = A'\hat{x}_k + Bu_k + J_\infty\left[y_k - C\hat{x}_k\right] \tag{23.95}$$

$$y_k = C\hat{x}_k + e_k \tag{23.96}$$

This is commonly called the 'innovations model'.

(c) Express the model for $\hat{y}_k = C\hat{x}_k$ in transfer function form driven by y_k and u_k.

Further Reading

1 B.D.O. Anderson and J.B. Moore, *Optimal filtering*, Prentice Hall, 1979.
2 G.C. Goodwin and K.S. Sin, *Adaptive filtering prediction and control*, Dover, 2009.
3 F.L. Lewis, L. Xie and D. Popa, *Optimal and robust estimation*, CRC Press, 2008.
4 M.E. Salgado, R.H. Middleton and G.C. Goodwin, "Connections between continuous and discrete Riccati equations with applications to Kalman filtering", *IEE Proceedings D (Control Theory and Applications)*, 135(1):28–34, 1988.

24

Laboratory 11: Rolling Mill Revisited (Periodic Disturbances)

24.1 Introduction

Here, the rolling mill problem discussed earlier in Chapter 13 will be studied in greater depth. In particular, the impact of periodic disturbances will be considered. The laboratory interface is shown in Figure 24.1.

It was found in Chapter 13 that the BISRA Gauge solution performed very well. However, in practice the performance is certainly not perfect. Indeed, when the BISRA Gauge is deployed in practice, it is often found that oscillations appear on the thickness of the exit strip. Thus, whilst the BISRA Gauge leads to impressive improvements, these improvements have a ceiling. This situation is common in control system design where successive improvements ultimately reach a limit. These limits can only be remedied by further investigation of the process and formulation of further remedial strategies which again lead to improvements in the system. This cycle is typically repeated many times.

In the case of the BISRA Gauge, the source of the difficulty is not immediately obvious.

Pinpointing the source of the problem was a significant breakthrough. It turns out that the source of the perceived oscillations is due to eccentricity of the rolls. This laboratory will study the impact of eccentricity of the rolls on the performance of the control system based on the BISRA Gauge. Having understood the problem, remedial strategies will be implemented based on observer design which will take the system to an improved performance level.

In keeping with the evolutionary improvement in technology, this virtual laboratory will successively investigate a number of control architectures aimed at dealing with periodic disturbance issues raised in this application.

24.2 Disturbances

The most severe disturbances in terms of off-specification products, are those originating in the strip.[1] Input strip thickness and hardness variations can have a particularly severe waveforms that approximate step functions. These basic disturbances have been addressed in Chapter 13. Another important disturbance, namely, eccentricity, discussed in detail here, originates from deviations in the radius of the rolls due primarily to manufacturing imperfections, heating and wear. It will be shown in this laboratory that these disturbances have a significant effect on the performance

1 Set up errors, hydrodynamic backup roll bearing speed effects and input strip surface roughness are also significant but are beyond the scope of this laboratory.

Practical Control System Design: Real World Designs implemented on Emulated Industrial Systems,
First Edition. Adrian Medioli and Graham Goodwin.

Companion website: www.wiley.com/go/medioli/practicalcontrolsystemdesign

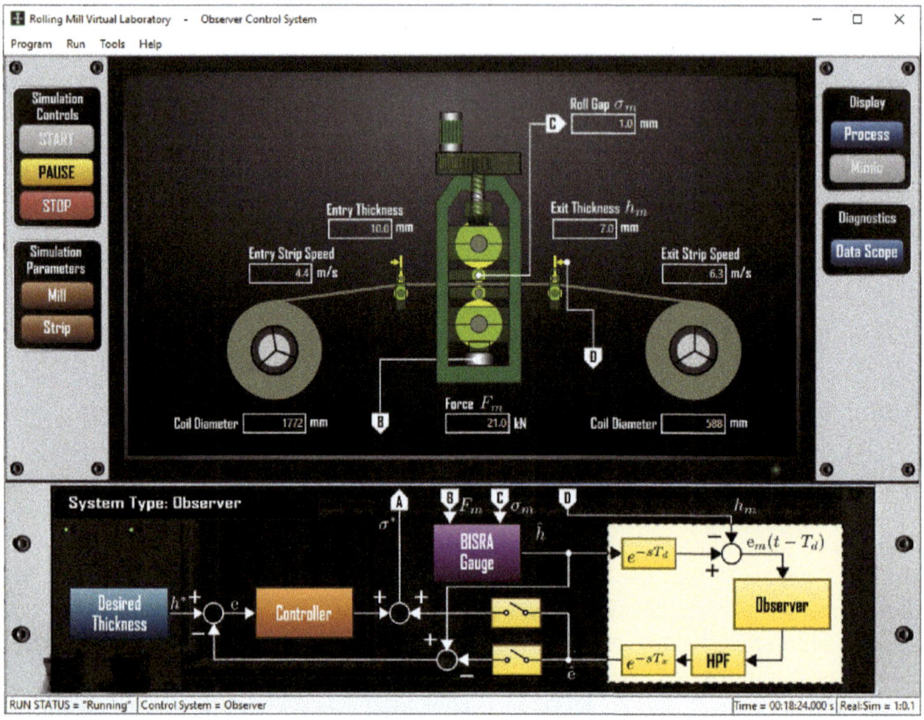

Figure 24.1 Screenshot of program showing the observer control system.

especially when the BISRA soft sensor is deployed. However, a simple remedy is available as will become clear as the chapter evolves.

24.3 Effects of Roll Eccentricity

In practice, the rolls have 'bumps' or imperfections which are known as roll *eccentricity*, see Figure 24.2. Eccentricity may arise due to a number of reasons including imperfection in manufacturing, rolls with imperfectly centred shafts, uneven wear of the roll, axial slipping of the sleeve on the roll arbour, etc.

Eccentricity produces variations in exit thickness. These variations are periodic, repeating with each revolution. This periodic signal may be expressed as a Fourier series (i.e. in terms of sine waves). Alternatively, one can model the periodicity in sampled form – see Section 23.7 of

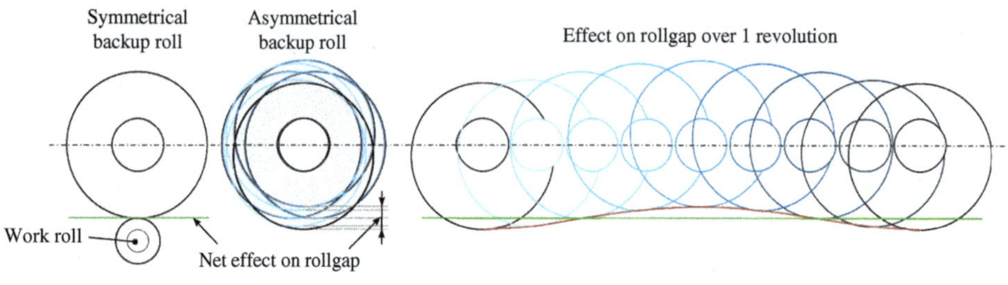

Figure 24.2 Roll eccentricity exaggerated for illustration.

Chapter 23. For the purpose of the current laboratory exercise, eccentricity will be modelled as a single sine wave. In reality, eccentricity resembles a very large collection of sine waves. However, the step from one sine wave to many is quite straightforward. The key design issues are captured by the scalar sine wave case.

The model of the system can be modified to include the effect of eccentricity. The eccentricity signal, e, is defined such that a positive value indicates the roll gap is narrower than would be anticipated by using the nominal unloaded roll gap, σ_s. Then the force on the roll will increase. Equation (13.1) from Chapter 13 becomes

$$F = M(h - \sigma_s + e), \tag{24.1}$$

where σ_s is the nominal (unloaded) roll gap.

Equation (13.2) from Chapter 13 remains unaltered. Solving (24.1) and (13.2) simultaneously shows the exit thickness is now described by:

$$h = \frac{M}{M+W}\sigma_s + \frac{W}{M+W}H - \frac{M}{M+W}e. \tag{24.2}$$

Thus, for the open-loop plant the effect of eccentricity on the true exit thickness is attenuated by the factor $\frac{M}{M+W}$ (which, for nominal values, is usually of the order of $\frac{1}{3}$).

24.3.1 Experiment: Measuring the Impact of Roll Eccentricity

1. Select the BISRA control system in the virtual laboratory.
2. Ensure $M = 3.5$ inside the BISRA gauge.
3. Ensure the feedback loop is open and the controller is $K_p = 1, K_i = 0$.
4. Activate Eccentricity: i) Click on the 'Rolling Mill Parameters' button. ii) Inside the 'Rolling Mill Parameters' dialog click the word Eccentricity.
5. Observe the actual eccentricity, the thickness estimate and the true strip thickness.

24.4 Tight Feedback Control

The feedback loop will next be closed using a 'tight' PI control law applied to the BISRA gauge estimate as in Chapter 13.

When using the BISRA gauge, the unloaded roll gap used in the calculation is the unloaded roll gap, i.e. the gap we believe to be true. This has been previously denoted σ_s. However, the true unloaded roll gap is $\sigma = \sigma_s - e$ due to the fact that positive values of e decrease the gap.

Of course, the BISRA gauge uses σ_s in its calculation since it doesn't 'know about' the eccentricity. Hence, the estimated exit thickness provided by the BISRA gauge is $\widehat{h}$ where

$$\widehat{h} = \frac{F}{M} + \sigma_s. \tag{24.3}$$

However, the measured force is affected by eccentricity, as is (24.1), i.e.

$$F = M[h - \sigma_s + e]. \tag{24.4}$$

Substituting (24.4) into (24.3) yields

$$\widehat{h} = h + e. \tag{24.5}$$

In other words, $\widehat{h}$ differs from the true exit thickness, h, by e.

Next note tight feedback control using the BISRA gauge estimate will have the effect of bringing $\widehat{h}$ to the desired exit thickness, h^*. Thus, the feedback will subtract e from the output. Recall from (24.2) that the open-loop error was only of the order of $-\frac{1}{3}e$. Hence tight feedback control from the

BISRA gauge yields an extra error of $\frac{2}{3}e$ that is impressed on the strip. This represents a significant shortcoming of the BISRA gauge methodology. However, it will be shown below that there is a simple remedy.

24.4.1 Experiment: Testing the Impact of Eccentricity on the BISRA Gauge

1. Implement the closed-loop controller and note the effect of eccentricity on the thickness estimate and the true strip thickness.

Increase the closed-loop bandwidth and examine how eccentricity affects the exit strip for different controller gains.

24.4.2 Analysis of the Effect of Control Law Bandwidth

You probably found in the previous experiment that as the loop bandwidth increases, then so does the error in the exit thickness. The explanation is as follows:

As the closed-loop bandwidth increases, the eccentricity impressed on the output strip also increases. For very low bandwidths, the eccentricity is attenuated by a factor of $\frac{1}{3}$ (as per the open-loop case). As the bandwidth is increased, the eccentricity on the output strip increases until the full eccentricity signal is impressed onto the strip. In particular, since the eccentricity signal e is present in the measurement signal, it appears to the feedback loop as a reference change of $-e$. As the closed-loop bandwidth increases, the ability to track this apparent reference signal improves until it is perfectly tracked and $-e$ is imprinted on the strip.

An alternative explanation is as follows: The loop is trying to make $\hat{h} = h + e$ equal h^*, for high loop bandwidth $\hat{h} \approx h^*$, so $h = h^* - e$.

Thus, it can be seen that one is faced with a design trade-off as follows: One wants a high bandwidth for the loop for fast disturbance rejection and reference tracking. However, one wants smaller loop bandwidth to reduce the effect of eccentricity on the output.

In summary, tight feedforward control significantly increases the size of the eccentricity effect relative to open loop. The reader is asked, at this point, to consider possible remedies.

24.5 Eccentricity Compensation

It was found in Section 24.4.2 that eccentricity produces a design trade-off when using the BISRA gauge. Obviously, one could improve performance if one could remove eccentricity from the thickness estimate.

Unfortunately, no direct physical measurement of eccentricity exists. However, a soft sensor (*an observer*) can be used to estimate the roll eccentricity. This estimate can then be used to improve the estimate of thickness provided by the BISRA gauge. The reader is asked to revise Chapter 23.

In this section, an observer will be used to estimate the eccentricity signal, which, for simplicity, is assumed to be a single sine wave of a fixed frequency and amplitude although the extension to multiple sine waves is straightforward. Trade-offs in observer design will also be examined. Finally, the observer's eccentricity estimate will be used to improve closed-loop performance.

24.5.1 A Simple Eccentricity Predictor

Recall that the BISRA thickness estimate includes the eccentricity signal. Specifically, from (24.3)

$$\hat{h} = \frac{F}{M} + \sigma_s. \tag{24.6}$$

Hence,

$$\hat{h}(t) = h(t) + e(t). \tag{24.7}$$

If this estimate were to be delayed by the same amount as the time taken for the strip to reach the physical thickness estimate, then one would obtain $\hat{h}(t - T_d)$. If the true measured height is subtracted from this signal, then an estimate of the eccentricity T_d seconds ago is obtained, i.e.

$$\hat{e}_m(t - T_d) = \widehat{h}(t - T_d) - h_m(t) = \widehat{h}(t - T_d) - h(t - T_d). \tag{24.8}$$

However, this estimate is T_d seconds too late. What is needed is an estimate of e at the present time. One can exploit periodicity to correct the delay. Specifically, the eccentricity is periodic.[2] Hence, $\hat{e}_m(t - T_d)$ can be used to predict the current eccentricity, $\hat{e}_m(t)$. This can be achieved by further delaying the measurement such that the total delay is an integer multiple of the eccentricity period.

This gives an estimate of eccentricity. This is a good starting point. However, this estimate is likely to be quite noisy. This motivates the design of an eccentricity observer using Kalman filtering ideas.

24.6 Optimal Observer Design

Here, it will be shown how an optimal observer can be used to extract the eccentricity signal from the noisy eccentricity measurement.

Recall the basic idea of a state observer (see Chapters 22 and 23). Consider

$$\dot{x}(t) = Ax(t) + Bu(t), \tag{24.9}$$

$$y(t) = Cx(t). \tag{24.10}$$

Then a linear observer (e.g. the Kalman filter) takes the following form:

$$\dot{\hat{x}} = A\hat{x}(t) + Bu(t) + J(y(t) - C\hat{x}(t)), \tag{24.11}$$

where $\hat{x}(t)$ is the state estimate, J is the observer gain and the term $(y(t) - C\hat{x}(t))$ is called the innovation process, which (for non-zero J) represents the feedback error between the observed and predicted model output. For a large value for J, the estimate will depend more on the measurements. Alternatively, for a small J, the estimates will depend more on the model predictions. This fact can be used to adjust the sensitivity of the observer to measurement noise and modelling errors. Alternatively, Kalman filtering ideas can be used to give a 'best' value of J – see Chapter 23.

To use these ideas to build an observer for the eccentricity signal, it is first necessary to develop a state-space model of the eccentricity signal, which is assumed, here to be a single sine wave of frequency ω_0 rad/s. The response to other disturbances will be treated as "noise". Hence, the output is modelled, for the current purpose, as a single sinewave. Thus consider

$$y(t) = \sin(\omega_0 t). \tag{24.12}$$

It follows that:

$$\dot{y}(t) = \omega_0 \cos(\omega_0 t), \tag{24.13}$$

$$\ddot{y}(t) = -\omega_0\, \dot{y}(t). \tag{24.14}$$

---

2 $sin(\frac{2\pi}{T}t) = sin(\frac{2\pi}{T}(t + kT))$, where $k = 1,2,3 \ldots$.

The states can be chosen as

$$x_1(t) = \sin(\omega_0 t) = y(t), \tag{24.15}$$

$$x_2(t) = \cos(\omega_0 t). \tag{24.16}$$

Then, the following state-space model for the eccentricity signal is obtained

$$\begin{bmatrix} \dot{x}_1(t) \\ \dot{x}_2(t) \end{bmatrix} = \begin{bmatrix} 0 & \omega_0 \\ -\omega_0 & 0 \end{bmatrix} \begin{bmatrix} x_1(t) \\ x_2(t) \end{bmatrix} + \begin{bmatrix} 0 \\ 0 \end{bmatrix} u(t), \tag{24.17}$$

$$y(t) = \begin{bmatrix} 1 & 0 \end{bmatrix} \begin{bmatrix} x_1(t) \\ x_2(t) \end{bmatrix}. \tag{24.18}$$

Rearranging Equation (24.11) and noting B is 0 in the system, gives the following observer

$$\dot{\hat{x}} = (A - JC)\hat{x}(t) + Jy(t). \tag{24.19}$$

Substituting the A and C matrices from (24.17), (24.18) into this general form yields:

$$\begin{bmatrix} \dot{\hat{x}}_1(t) \\ \dot{\hat{x}}_2(t) \end{bmatrix} = \begin{bmatrix} -J_1 & \omega_0 \\ -J_2 - \omega_0 & 0 \end{bmatrix} \begin{bmatrix} \hat{x}_1(t) \\ \hat{x}_2(t) \end{bmatrix} + \begin{bmatrix} J_1 \\ J_2 \end{bmatrix} y(t), \tag{24.20}$$

$$\hat{e}(t) = \begin{bmatrix} 1 & 0 \end{bmatrix} \begin{bmatrix} \hat{x}_1(t) \\ \hat{x}_2(t) \end{bmatrix}. \tag{24.21}$$

Further insights are obtained by converting the observer (24.20), (24.21) into the transfer function form. (Recall that for $\dot{x} = Ax + Bu, y = Cx$, then $G(s) = C(sI - A)^{-1}B$.)

The observer transfer function is then found to be

$$G_{observer}(s) = \frac{J_1 s + J_2 \omega_0}{s^2 + J_1 s + \omega_0 (J_2 + \omega_0)}. \tag{24.22}$$

Setting $s = j\omega_0$ shows that the observer passes the sine wave (with gain 1 and zero phase shift) but attenuates other components.

Increasing J_1 makes the band-pass region wider.

At frequencies above ω_0 there is little effect of changing J. J_2 also changes the low-frequency gain and DC gain. The sign of J_2 also has an effect on the low-frequency phase shift.

24.6.1 Experiment: Testing the Eccentricity Estimator

1. Select the observer control system architecture in the virtual laboratory.
2. Use a high gain controller ($K_p = 3, K_i = 300$). Click on the rolling mill to open the 'Rolling Mill Parameters' dialog. Activate eccentricity. Let the system run for a few simulation seconds while the High Pass Filter 'warms up' (i.e. the states are initialised).
3. Display the estimated eccentricity, e_est, true eccentricity, e and the actual strip thickness at the rolls,[3] h.
4. Note the steady state-peak-to-peak eccentricity amplitude seen on the actual strip thickness.
5. Close the feedback and feedforward compensation switches.

   Choose $J_1 = 126, J_2 = 0$.

   Note that the feedback signal used for control has $\hat{e}$ subtracted from $\hat{h}$. Hence $\hat{h}$ is very close to the real h.

3 Obviously, this measurement is not available in a real mill. The actual thickness can only be verified offline using other instruments on the finished product.

Verify that tight feedback control (using the eccentricity-compensated BISRA gauge) has low sensitivity to the eccentricity signal.

As a final step add noise to the BISRA gauge output. Note the current observer treats this noise in an ad-hoc fashion.

In this relatively simple case, one can think of ad-hoc ways of selecting the observer gain J. For more complex problems it becomes necessary to find a systematic way of determining J which reflects one's relative confidence in the model and expected variance of the noise in the measurements. All of these ideas have been described earlier in Chapter 23. The necessary design steps are outlined in Section 24.7.

24.7 Eccentricity Compensation Using the Kalman Filtering

A Kalman filter selects the optimal[4] J values, given information regarding how noisy the measurements are and how noisy/uncertain/inaccurate the model is. Matrices R and Q are used to capture these specifications. The R matrix is the specification of variance of the measurement noise. The larger R, the noisier the Kalman filter believes that the measurement is. The Q matrix indicates the variance of process noise in our model. A larger Q implies that one is less certain about the fidelity of the model. In the latter case, the observer gain depends less on the model and more on measurements. Sometimes Q and R have known values corresponding to real noise variances (when they are available). In other cases, the Q and R matrices can be thought of as 'tuning parameters' for the Kalman filter. If the R matrix is made larger, then the value of J will be made smaller such that the estimates are less affected by the noisy measurements and instead rely on the model. If the Q matrix is made larger, J will be larger and the estimates will depend more heavily on measurements. As the Kalman filter runs it updates the P, or covariance matrix, which indicates the variance of the state estimate errors.

24.7.1 Experiment: Testing the Kalman Filter for Eccentricity Estimation

In this virtual laboratory, noise will be added to the BISRA gauge output. Then by selecting Q and R in the Kalman filter design you will be able to determine an 'optimal' value for the observer gains.

You should try different values of Q and R and test the resulting performance of the Kalman filter.

Also, try different variances for the noise and deliberately introduce errors in the eccentricity frequency.

Note that the Riccati equation update is included as part of the software. Thus, all the user need to supply is the values of Q and R, the initial value of the state covariance matrix, P_0, initial values for the observer gain, J_0 and initial state estimates.

As you increase the size of R relative to Q, you will see that the Kalman Filter places less emphasis on the measurements. This means smaller values of J and greater frequency selectivity, i.e. narrower observer bandwidth so that only frequencies close to what is expected in the model will influence the estimate.

Setting R to 0 tells the Kalman Filter that the measurements are perfect, thus, in principle, the Kalman Filter does not need to rely on the model. The effect of this is simply to make the observer gains very large, this in turn lets the measurements dominate the estimate. In effect, you are using the raw measurements as your estimate.

---

4 In a least squares sense given assumptions about how the process and measurement noises are described (see Chapter 23).

24.8 Conclusion

This laboratory has studied the use of an observer and the Kalman filter for periodic disturbance compensation. The technique described here has been used in practice. Indeed, it is shown that via the use of an Observer and Kalman Filter then typical rolling mills are able to manufacture a rolled product at a specified thickness to an accuracy in the region of 0.1%.

24.9 Revision Questions

1. Consider a periodic disturbance of period p samples.
   Explain why this disturbance can be modelled as

$$(z^p - 1)d_k = 0. \tag{24.23}$$

2. Based on your answer to Question 1, write the model in terms of distinct frequencies.
3. Show that including $(z^p - 1)$ in the denominator of the control law leads to perfect tracking of any reference signal having period p. This leads to a technique known as "Repetitive Control". For further information see the references given at the end of the chapter.
4. Say that p is even, show that including $z^p - 1$ in the denominator of $C(z)$ always leads to a loop-transfer function having a pole at -1.
5. Say that one has a model error of one sample. Show that this will destabilise any control law that contains $(z^p - 1)$ in the denominator (p even).
6. Consider the feedback system shown in Figure 24.3 where d is a disturbance.
   For $F(z) = z^{-1}$, show that the transfer function, $S(z)$, between d and u has a zero at $z = 1$.
7. Consider again the feedback system of Question 6. Let

$$F(z) = +2z^{-1} - z^2. \tag{24.24}$$

   What is the corresponding value of $S(z)$?
8. What choice for $F(z)$ in Question 6 leads to $(u - u^*)$ having no errors for sinusoids of frequency $\omega_0\Delta$?

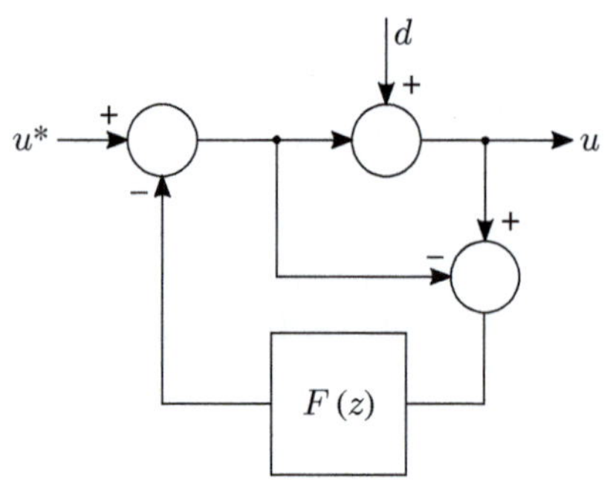

Figure 24.3 Feedback system.

Further Reading

1 The periodic disturbance compensation scheme described in this laboratory is commonly used in rolling mill applications. For example Roll Eccentricity Compensation (REC) schemes can be found in many thickness control systems and these typically use observers and/or Kalman filtering for periodic disturbance estimation. A GOOGLE® search will yield many papers.

2 An early paper on repetitive control is: R.H. Middleton, G.C. Goodwin and W. Longman, "A method for improving the dynamic accuracy of a robot performing a repetitive task", *International Journal of Robotics Research*, 8(5), 1989 https://doi.org/10.1177/027836498900800506.

Part VII

Introduction to the Modelling and Control of Nonlinear Systems

The material presented so far in the book has emphasised linear models and control system design for linear systems. Often nonlinear systems can be treated by using an approximate linear model. Feedback then ensures that the control system, when implemented on the nonlinear system, performs adequately. Indeed, linear methodology is adequate for the majority of control system design problems. Nonetheless, there are some problems in which nonlinear behaviour is central to the problem. Hence, this part of the book will examine nonlinear behaviour in more detail. Chapter 25 will study simple nonlinear behaviour including actuator saturation and actuator minimal movement. Chapter 26 will apply some of these ideas to the continuous casting machine control problem. Chapter 27 will consider the impact of actuator saturation in the cross-directional control problem.

Practical Control System Design: Real World Designs implemented on Emulated Industrial Systems,
First Edition. Adrian Medioli and Graham Goodwin.

Companion website: www.wiley.com/go/medioli/practicalcontrolsystemdesign

25

Modelling and Analysis of Simple Nonlinear Systems

25.1 Introduction

The case studies presented so far in the book have all assumed that the system can be sufficiently well approximated by a linear system and the control laws have been restricted to linear systems. However, all real systems exhibit some form of nonlinear behaviour. For example, even a simple electronic amplifier will be nonlinear and show non-ideal characteristics under some circumstances. This was one of the key stumbling blocks in achieving long-distance telephony. A remarkable contribution in this area was that of Black's feedback amplifier. The key idea was to design an amplifier having a much higher (open-loop) gain than was actually needed. Then feedback was used to reduce sensitivity to nonlinearities and other imperfections. Indeed, the sensitivity to model errors and output disturbances is reduced by a factor equal to approximately the inverse of the loop gain.

Fortunately, in many real-world designs it suffices to treat a nonlinear system as if it were linear. For example, there is no doubt that a real rocket exhibits nonlinear behaviour. Nonetheless, many associated designs are based on approximate linear models. This was the approach adopted in Chapter 15. Thus, a linear approximation is often sufficient to undertake a control system design. However, there are cases where one cannot ignore the nonlinear behaviour. It is beyond the scope of the current introductory book to delve too deeply into the topic of nonlinear control. (Entire books are available devoted to this subject). Thus, the discussion here is limited to simple nonlinear behaviour. This chapter will examine nonlinear behaviour arising from nonlinear actuators such as valves. In particular, the following issues will be examined:

(i) large movement, and
(ii) small movement.

25.2 Errors Arising from Large Actuator Movement

The typical response of a valve (and other actuators) to different demands is shown in Figure 25.1.

It can be seen that the output is a highly nonlinear function of the valve position. Moreover, there comes a point where further input movements produce either little, or no, change in the output. This issue will be addressed below.

Practical Control System Design: Real World Designs implemented on Emulated Industrial Systems, First Edition. Adrian Medioli and Graham Goodwin.

Companion website: www.wiley.com/go/medioli/practicalcontrolsystemdesign

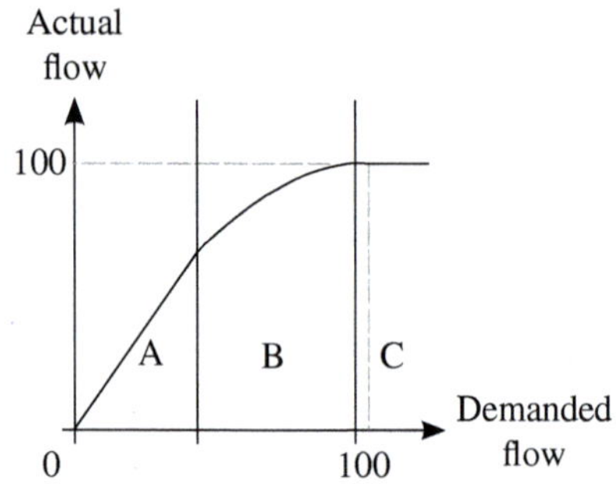

Figure 25.1 Real valve flow dynamics.

25.3 Nonlinear Correction by Gain Change

The simplest way to correct a nonlinear characteristic, when in zone B (Figure 25.1), is to invert the characteristic. Thus, say that

$$u = f(u^*), \tag{25.1}$$

where $f(u^*)$ is a smooth invertible function and where u and u^* are the actual output and desired output, respectively.

Then a simple inverse relationship can be used to linearise the system, i.e. define

$$\overline{u} = f^{-1}(u^*), \tag{25.2}$$

and apply $\overline{u}$ to the valve. The result is

$$\begin{aligned} u &= f(\overline{u}), \\ &= f\left(f^{-1}(u^*)\right), \\ &= u^*. \end{aligned} \tag{25.3}$$

The main problems with this approach are that f needs to be (relatively) accurately known and must be invertible over the required range of u^*.

25.4 Nonlinear Correction by Cascade Control

An alternative to the idea described in Section 25.3 is to use feedback as a mechanism for linearising a valve.

The type of behaviour seen in zones A or B in Figure 25.1 is easily remedied by measuring the output and then placing the valve in a cascaded feedback loop as shown in Figure 25.2.

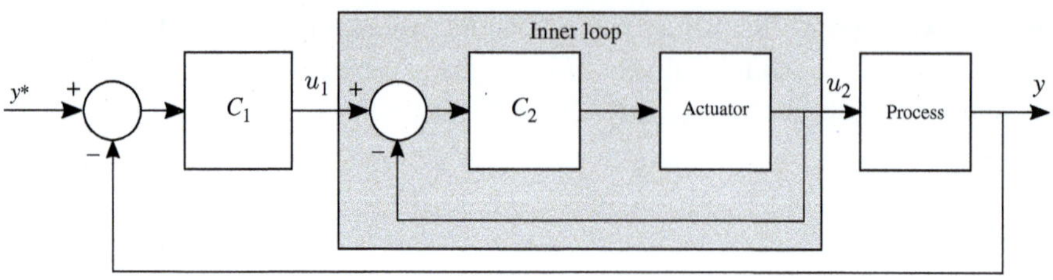

Figure 25.2 Cascaded loops.

Typically, a simple PID controller would be used for C_2. The reader is reminded of the discussion of Black's feedback amplifier design given in Section 25.1. The impact of the nonlinearity is reduced by (approximately) the inverse of the loop gain.

25.5 Saturation

A more difficult problem is encountered in region C (Figure 25.1). Here the demanded value position can be increased indefinitely with no change in the output. This phenomenon is extremely prevalent in industry. A core consequence of saturation of this type is a phenomenon known as 'windup'. Specifically, if a simple integral controller is used around the valve, then it will continually see an error and hence continually increase (via integration) the demand to the valve. However, this increased demand leads to no increase in the output. Hence the values at the output of the integrator can, in principle, increase without bound. This is commonly termed 'integration windup' or, more simple, 'windup'.

Windup was less of a problem with, old style, pneumatic or hydraulic control systems since there was a natural upper bound on the value that the integrator output could reach. Moreover, this is easily captured in modern digital control systems by simply stopping the integrator as soon as the output hits a saturation limit. An alternative strategy for stopping the integrator is described in Section 25.5.1.

25.5.1 Achieving Integral Action via Feedback

The key idea here is to rearrange the control law so that it is evaluated using feedback. To illustrate, let the control law be a PI controller of the form:

$$C(s) = \frac{K(\tau s + 1)}{\tau s}. \tag{25.4}$$

The transfer function $C(s)$ can be implemented by feedback. To see how this can be achieved, let

$$C(s)^{-1} = \overline{C}_o + D(s), \tag{25.5}$$

where $D(s)$ is a strictly proper transfer function.

Then, consider the feedback system shown in Figure 25.3.

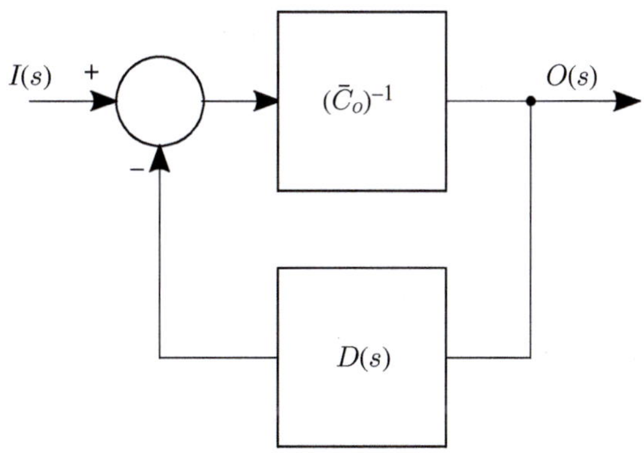

Figure 25.3 Feedback system.

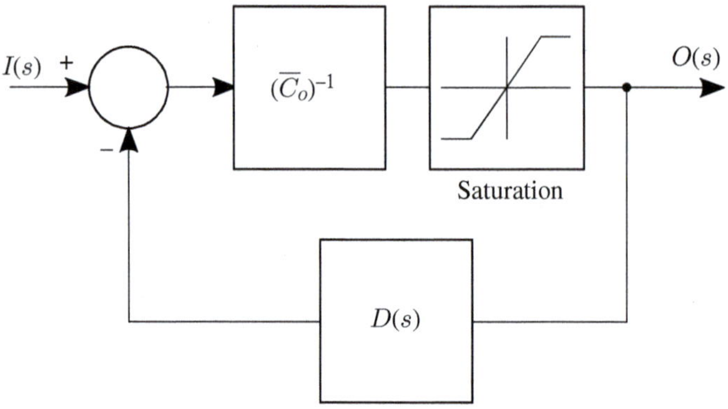

Figure 25.4 Feedback system with saturation.

To see how this works, it is necessary to evaluate the transfer function linking $I(s)$ to $O(s)$ in Figure 25.3. A simple calculation shows that

$$\begin{aligned}\frac{O(s)}{I(s)} &= \frac{\left(\overline{C}_o\right)^{-1}}{1+\left(\overline{C}_o\right)^{-1}D(s)}, \qquad (25.6)\\ &= \frac{1}{\overline{C}_o + D(s)},\\ &= \frac{1}{(C(s))^{-1}},\\ &= C(s).\end{aligned}$$

Thus, the feedback system of Figure 25.3 is an alternative way of implementing the controller transfer function, $C(s)$. In summary, the feedback system in Figure 25.3 implements the transfer function $C(s)$. (See also Questions 1 and 2 given at the end of the chapter.)

With the above as background, the core idea of anti-windup implementation is to include the saturation inside the feedback loop as shown in Figure 25.4.

In other words, the dynamic part of the control law is 'informed' about the saturation.

25.5.2 Introducing Anti-Windup in Control Laws Implemented via the Youla–Kucera Parameterisation

The reader will probably immediately see connections between the ideas discussed above and the ideas described in Section 7.7 of Chapter 7. Indeed, the methodology is identical. The same principle can be extended to the case when a general observer is used as shown in Section 25.5.3.

25.5.3 Anti-Windup When an Observer is Used

The key idea inherent in Figure 25.4 is that all key (dynamic) parts of the control law should be informed about the presence of the saturation nonlinearity. The idea can be applied to the case where an observer is deployed. The key idea is to send the saturated input to the observer as shown in Figure 25.5 so that the observer 'sees' the true input.

Note again that the key dynamics in the control law (here they reside in the observer) are informed about saturation of the input signal when it occurs.

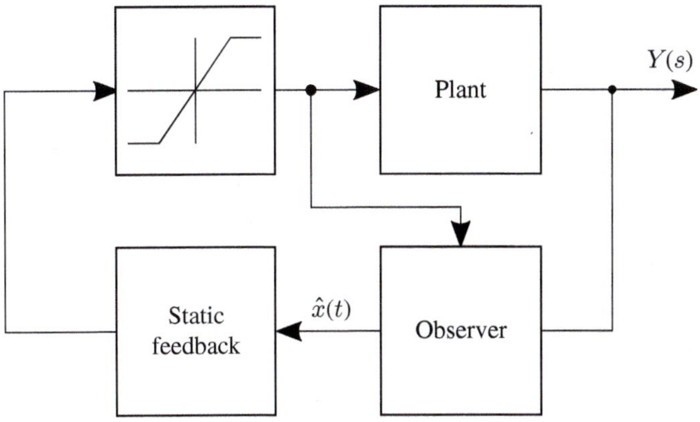

Figure 25.5 Saturation-aware observer.

25.6 Extension to Rate Limitations

The extension of the above ideas to a rate of change limit is explored in Questions 3 and 4 at the end of the chapter.

The ideas described in Sections 25.5 and 25.6 can be interpreted as a simple form of Model Predictive Control (see Section 22.13 of Chapter 22) for the case when the principal constraint of interest is on the set of allowable input signals.

25.7 Minimal Actuator Movement

Another problem that frequently occurs in valves is that of minimum movement. This lies at the other end of the range of values that a valve sees. A common example of this phenomenon is stiction.

These kinds of nonlinearity can lead to self-sustained oscillations. The reason being that the valve is unable to settle at a zero equilibrium and thus oscillates backwards and forwards.

An old, but useful, tool for studying such oscillations is the 'Describing Function Technique'. This idea is briefly outlined in Section 25.8.

25.8 Describing Function Analysis

Nonlinearities will typically lead to harmonics of the driving signal. The core idea associated with Describing Function analysis is to track a single sine wave (the first harmonic) around the feedback loop.

For example, consider a simple relay as shown in Figure 25.6. Say that

$$i(t) = \overline{A}\sin(\omega_1 t); \quad \omega_1 = \frac{2\pi}{T_p}. \tag{25.7}$$

If a sine wave appears at the input to a relay, then a saturated sine wave will appear at the output as shown in Figure 25.7.

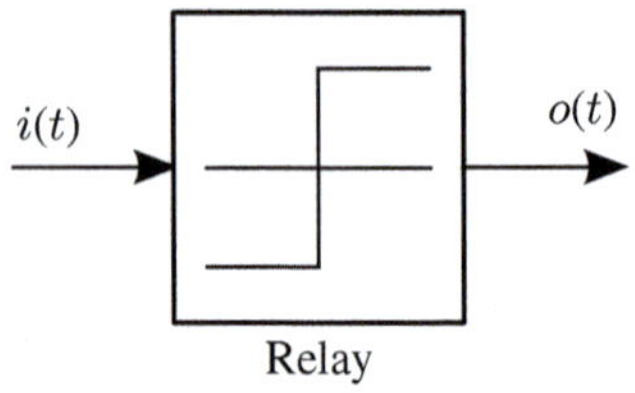

Figure 25.6 Relay function.

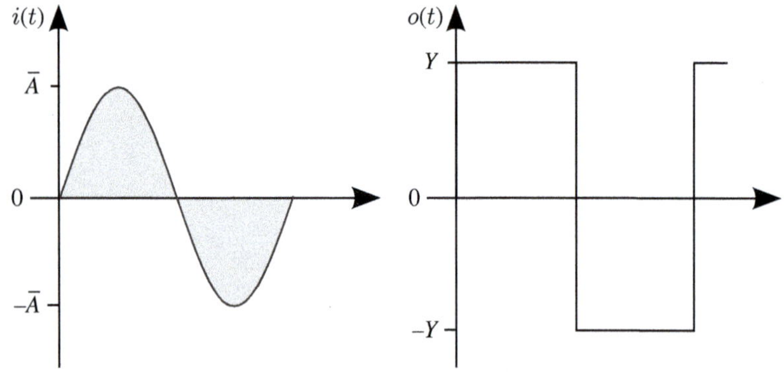

Figure 25.7 Output of relay to a sinusoidal input.

The output waveform shown in Figure 25.7 can be transformed, via Fourier Series, into a harmonically related set of sine waves.

Thus, for $i(t) = \overline{A}\sin(\omega_1 t)$; $\omega_1 = \frac{2\pi}{T_p}$, then the Fourier analysis of $o(t)$ has the form

$$o(t) = \sum_{k=1}^{\infty} A_k \sin\left(\omega_k t\right), \tag{25.8}$$

where $\omega_k = k\omega_1$.

Of particular interest is the fundamental component in the signal $o(t)$. Fourier series analysis shows that the amplitude of the fundamental component is

$$A_1 = \frac{4Y}{\pi}, \tag{25.9}$$

where Y is the amplitude of the square wave appearing at the output of the relay.

If the sinusoids in (25.8) are applied to the input of a 'low-pass' plant, then the output will be dominated by the fundamental component.

Under these conditions, the nonlinear relay block can be thought of as a simple gain between the input (of amplitude $\overline{A}$) and the fundamental component of the output (of amplitude $\frac{4Y}{\pi}$). Hence, the gain of the nonlinear block can be approximated by a gain $\overline{N}$

$$\overline{N} = \frac{4Y}{\pi\overline{A}}. \tag{25.10}$$

$\overline{N}$ is called the 'Describing Function' of the nonlinearity.

The idea of Describing Function analysis can be applied to more general nonlinear elements, e.g. those containing hysteresis. See, for example, references [1, 2] and [3] given in Further Reading. A practical illustration of these ideas will be given in Chapter 26.

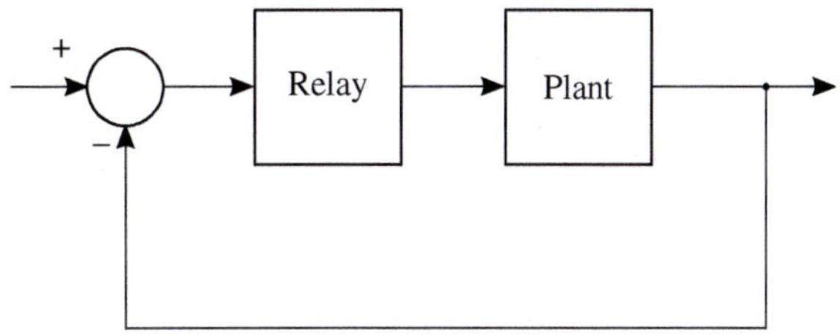

Figure 25.8 Closed-loop with relay.

25.9 Predicting the Period and Amplitude of Oscillations

The describing function idea outlined in Section 25.8 can be used to give an approximate answer to the question of the frequency of oscillation. The idea is illustrated below for a simple case.

Consider the feedback loop shown in Figure 25.8 which includes the relay analysed in Section 25.8.

The loop in Figure 25.8 can be used to study the induced oscillatory behaviour. To illustrate, say that the plant has transfer function $\frac{1}{(s+1)^3}$.

It is known from Nyquist stability theory that a sustained oscillation will occur when the loop phase shift is 180° and the loop gain is 1.

Based on this understanding, the loop in Figure 25.8 will oscillate if there exists a situation where the gain is unity and the phase is 180°. For the plant given above, the frequency where the phase shift is 180° is $\omega = \tan^{-1}(60°) = 1.73 = \sqrt{3}$.

At this frequency, the gain of the plant is

$$\left|G(j\sqrt{3})\right| = \frac{1}{|j\sqrt{3}+1|^3} = \frac{1}{8}. \tag{25.11}$$

From the Describing Function methodology, it is known that for a sustained oscillation, the total loop gain at this frequency should be 1. Then, replacing the nonlinear element by the describing function having gain $\overline{N}$, the total loop gain when the system oscillates will be describing function approximation it must be true that, for an oscillation at $\omega = \sqrt{3}$,

$$\overline{N}\left|G(j\sqrt{3})\right| = \frac{\overline{N}}{8}. \tag{25.12}$$

For a sustained oscillation, this loop gain should equal one.

Using (25.10) for $\overline{N}$ and substituting into (25.12) shows that $\overline{A}$ (the amplitude of the oscillation at the input to the relay) must be

$$\overline{A} = \frac{Y}{2\pi}. \tag{25.13}$$

The above analysis can be generalised to different plants and different nonliearities. Chapter 26 we will explore a real-world application of the ideas described above.

25.10 Revision Questions

1. Consider the implementation of a controller, $C(s)$, as given in Figure 25.2.
   What are the values of $\overline{C}_o$ and $D(s)$ if $C(s) = \frac{K(\tau s+1)}{\tau s}$?

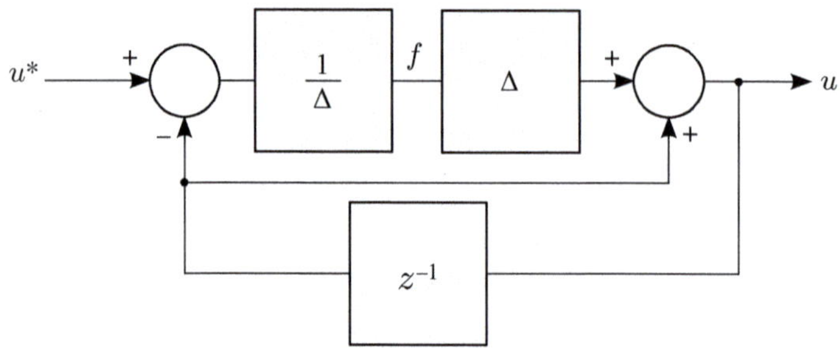

Figure 25.9 Question 3 block diagram.

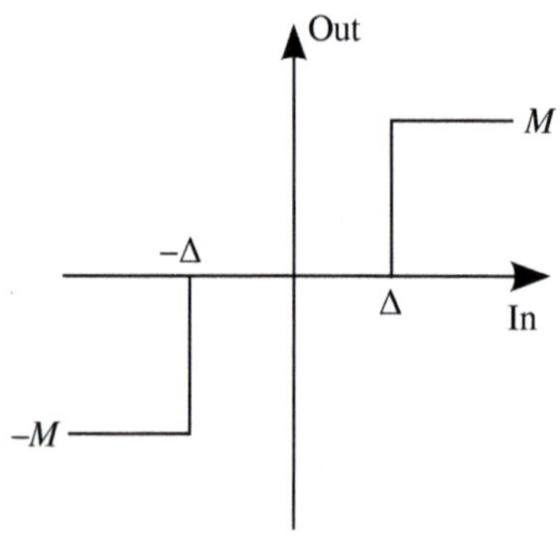

Figure 25.10 Relay function with dead zone.

2. Using your answer to Question 1, confirm by explicit calculation that Figure 25.3 indeed gives an implementation of $C(s)$.
3. Consider the block diagram shown in Figure 25.9, where z^{-1} is a unit delay, Δ is the sample period.
   (a) What is the signal f?
   (b) Hence explain why f is an approximation to the derivative of u.
   (c) Show that placing a saturation on the signal f leads to a circuit that limits the rate of change of the signal u.
4. Modify the system of Question 3 to include both a rate of change and saturation on the signal u. (Hint: Make sure all dynamics know about the saturations.)
5. Extend the analysis given in Section 25.7 to a relay with Dead-zone – as shown in Figure 25.10.

Further Reading

1 J.J. DiStefano, A.R. Stubberud and I.J. Williams, *Feedback and control systems* 2nd, Schaum's Outline Series, McGraw-Hill, 1990.
2 M. Vidyasagar, *Nonlinear system analysis*, Prentice Hall, 1993.
3 A. Gelb and W.V. Velde, *Multiple-input describing functions and nonlinear system design*, McGraw Hill, 1968.

Background on Describing Function Methods can be found in many text books on Control System Design, see for example,

4 C. Bohn and D.P. Atherton, "An analysis package comparing PID anti-windup strategies", *IEEE Control Systems Magazine*, 15(2):34–40, 1995.
5 S. Galeani, S. Tarbouriech, M. Turner and L. Zaccarian, "A tutorial on modern anti-windup design", *European Control Conference (ECC)*, pp. 306–323, 2009.

26

Laboratory 12: Continuous Casting Machine (Nonlinear Considerations)

26.1 Introduction

Control system design for a continuous casting machine was studied earlier in Chapter 9.

In the introduction to the earlier laboratory, it was mentioned that continuous casting machines, throughout the world, exhibit self-sustained oscillations. However, these were not evident in the linear design considerations presented in Chapter 9. In this chapter, attention will be focused on various nonlinear considerations to see if they can explain the oscillations. A screen shot of the virtual laboratory is shown in Figure 26.1.

26.2 The Slide Gate Valve

In Chapter 9, the closed-loop system shown in Figure 26.2 was introduced.

From Chapter 9, the following valve, $K_v = 0.28338$ for the gain of the Slide Gate Valve (SGV) was used.

In developing the simplified model for the system in Chapter 9, it was assumed that the valve area, A_v (mm^2), (and hence the flow into the mould, f_{in}) was directly proportional to the valve position $u(t)$ [%]. It turns out that the relationship between the SGV opening and flow is actually non linear. To explain why this is the case, a more realistic model relating $u(t)$ to the SGV area, A_v will be described below.

The actual SGV in the real system operates by moving one circular cross-sectional area across another (see Figure 26.3).

Let r denote the radius of both circular openings, and x_v the movement of the SGV.

Simple geometry yields,

$$A_v = r^2(a - \sin(a)), \tag{26.1}$$

where

$$a = 2\cos^{-1}\left(1 - \frac{x_v}{2r}\right),$$

and

$$0 \leq x_v \leq 2r.$$

Practical Control System Design: Real World Designs implemented on Emulated Industrial Systems,
First Edition. Adrian Medioli and Graham Goodwin.

Companion website: www.wiley.com/go/medioli/practicalcontrolsystemdesign

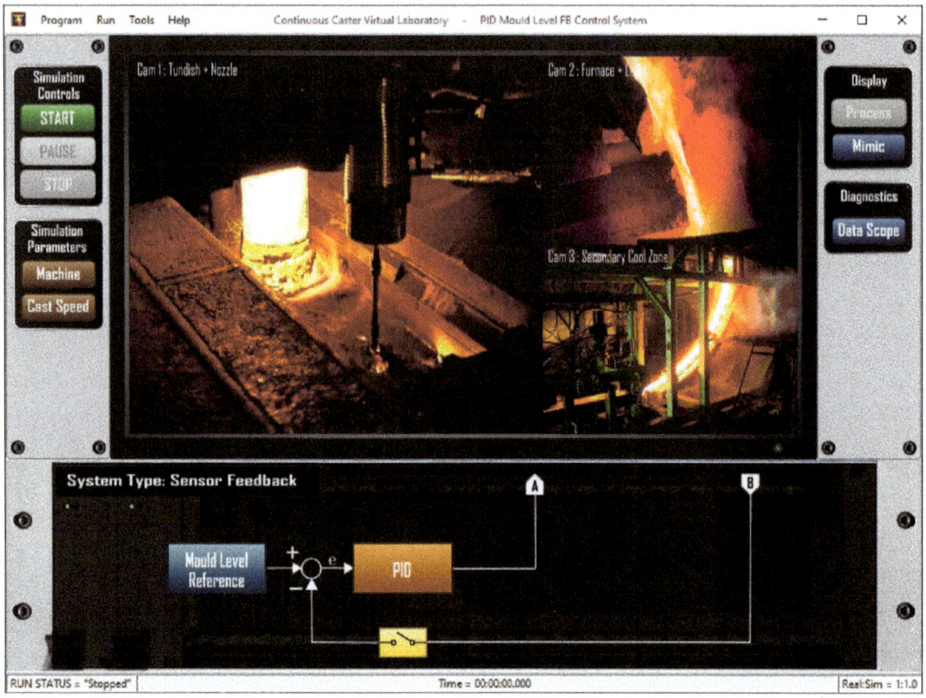

Figure 26.1 Screenshot of continuous Caster laboratory.

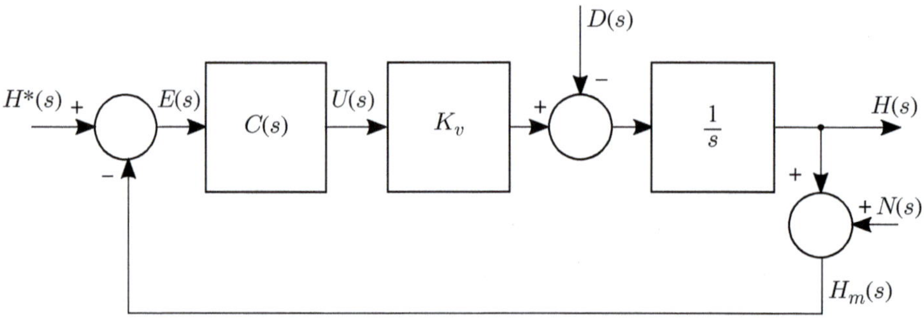

Figure 26.2 Control loop.

Equation (26.1) gives a model for the nonlinear behaviour of the SGV. The impact of this geometry on the systems performance will next be studied.

The nonlinear valve characteristics described above could, in principle, be linearised by the methods described in Section 25.3 of Chapter 25. This approach will not be adopted here. Instead, a check will be made to see if the nonlinearity described above has a major impact on the closed-loop control system performance.

26.3 Investigation of Effect of Nonlinear Valve Geometry

The effect of the nonlinearity described in Section 26.2 will be examined experimentally. Extra features will also be added to the model. Specifically, all real valves are driven by hydraulics and

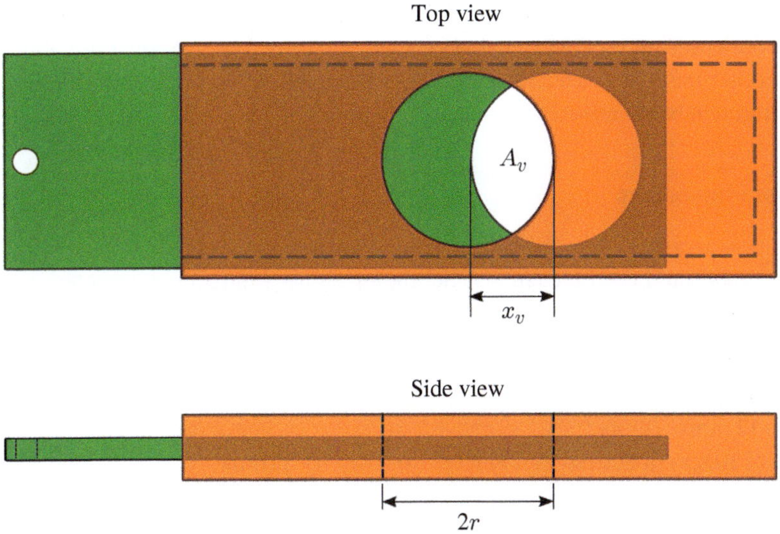

Figure 26.3 Slide gate valve.

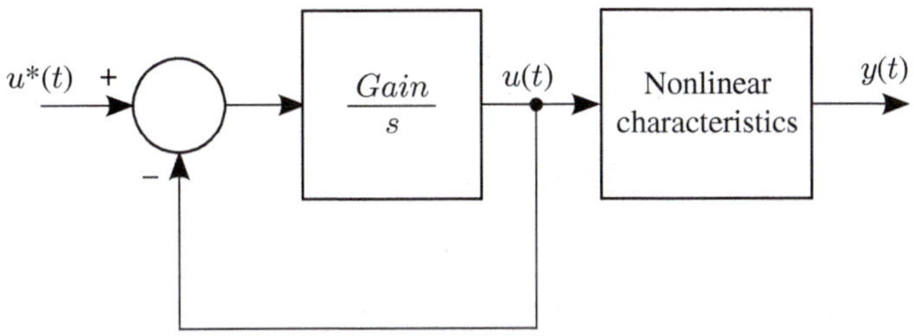

Figure 26.4 Model of the valve with nonlinear characteristics and additional dynamics.

these introduce additional dynamics. These dynamics will be modelled as shown schematically in Figure 26.4 where $u^*(t)$ is the desired valve opening, $u(t)$ is the actual valve opening and $y(t)$ is the resulting molten steel flow. The experiment will check whether, or not, these refinements make any qualitative difference to the observed performance.

26.3.1 Experiment: Testing Impact of the Nonlinear Geometry of the Valve

In the machine configuration parameter box, check the box 'activate valve dynamics', but do not activate 'use realistic valve dynamics'.

(a) Test the response to a set-point change: Observe the system's response by configuring the Mould Level Reference to be a square wave with offset 70 mm, amplitude 10 mm, Frequency 0.022 Hz, casting speed 12 mm/s. Use a PI controller with parameters ($K_p = 3.5288, K_i = 0.8822$). Also, set feedforward gain to 3.5288 and duty cycle 0.5.

(b) Test the response to a casting speed change: To do this, configure the Mould Level Reference to be a constant of 70 mm, then configure the Casting Speed Setpoint to be a square wave with offset 12 mm/s, amplitude 5 mm/s, period 40 s and duty cycle 0.5.

You probably found that only insignificant transient and steady-state effects were introduced by the above considerations. This is a consequence of the use of feedback in our control system design.

(This reinforces 'The Magic of Feedback' to mitigate mild model errors.) Thus, at this point, a satisfactory explanation for the observed oscillatory behaviour has not been found. Many (arguably most) real valves have another nasty characteristics which, so far, has been ignored. Next this, as yet unnamed, nonlinearities will be examined.

26.3.2 Other Nonlinear Phenomena

In the machine configuration parameter box check 'use realistic valve characteristics'.

Set $K_p = 1.488$ and $K_i = 0.298$, and use FF Gain = 0. Set the machine simulation parameters as follows:

- Activate Valve Dynamics
- Use realistic Valve characteristics
- Set the casting speed to 12 mm/s
- Introduce a single step of 10 mm in desired mould level from say 70–80 mm.

Observe that the mould level now exhibits oscillations!

The reader is asked to think about the possible causes of this oscillation. (Do this before proceeding to the next section.)

26.4 An Explanation for the Observed Oscillations

If you have already read Section 25.9 of Chapter 25, you might suspect that the cause of the oscillation may be some form of non-smooth nonlinearity. Actually in the real system that the authors and colleagues worked on, it took a group of engineers significant effort to correctly identify the cause of the problem. However, readers of this book can be assured that valves (and many other practical actuators) often exhibit these kinds of issues.

With the benefit of hindsight, the realistic valve characteristics are modelled by using slip-stick friction of the SGV as shown in Figure 26.5.

Note that this model uses a *dead zone* to describe the fact that the valve cannot sustain small velocities (shown as $v(t) = 1/k(\dot{u}(t)))$) below some threshold, i.e. they 'stick'. Hence, below a cut-in threshold, no movement occurs. This is illustrated by the input and output for a sine wave input into a valve having slip-stick friction are as shown diagrammatically in Figure 26.6.

Figure 26.7 shows the full system describing the mould level controller when slip-stick friction is taken into account.

The ideas presented in Section 25.9 of Chapter 25 will next be applied to verify that the oscillations can, indeed, be modelled in this fashion.

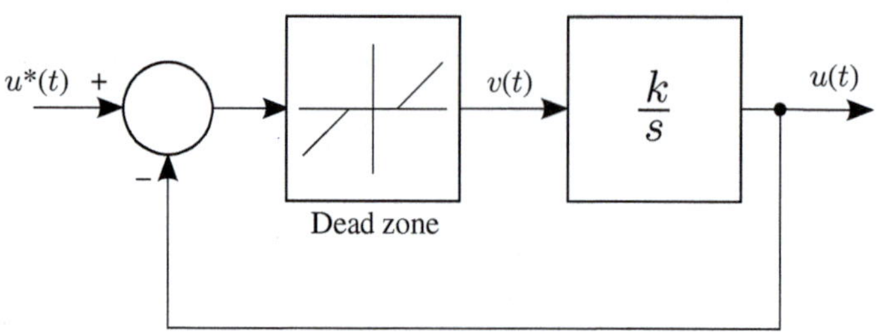

Figure 26.5 Valve with slip-stick friction.

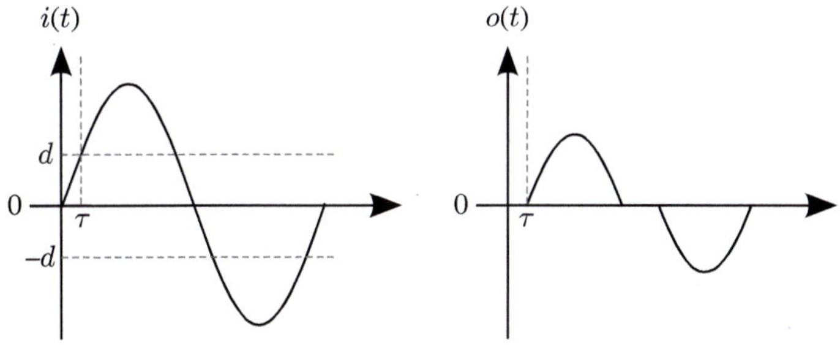

Figure 26.6 The relationship between the input and output for the nonlinear block.

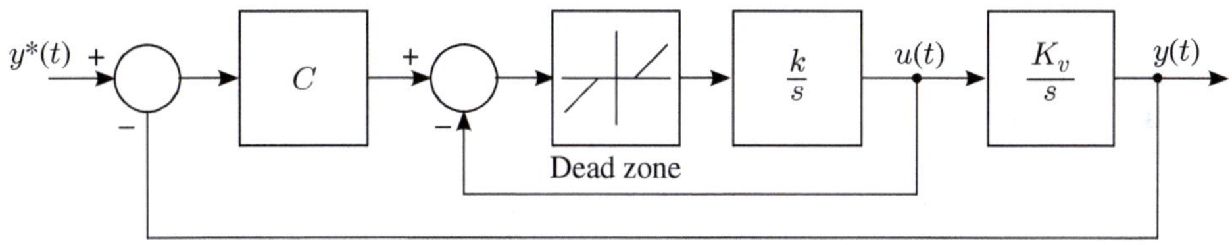

Figure 26.7 Valve with slip stick friction in mould level controller.

The period of the oscillation is first recorded from the experiment of Section 26.3.2. Describing Function methods will then be used to see if slip-stick friction is a probable explanation for the oscillations observed in practice.

As outlined in Section 25.7 of Chapter 25, the essential idea of the Describing Function techniques is to replace the nonlinear element by a 'gain' based on assuming a sine wave input and calculating the first Fourier component of the output of the nonlinearity.

It is now necessary to redraw the system shown in Figure 26.7 so that it takes the form of a single linear transfer function surrounding a nonlinear block. This step is shown in Figure 26.8.

For the case of relevance here, it can be seen that the required single linear transfer function is

$$T(s) = \frac{k}{s} + \frac{k}{s}\frac{K_v}{s}\left[K_p + \frac{K_i}{s}\right].$$

Taking $k = 5$, $K_p = 1.488$, $K_i = 0.298$ and $K_v = 0.28338$ leads to

$$T(s) = \frac{5s^2 + 2.108s + 0.422}{s^3}.$$

Figure 26.8 The schematic of the nonlinear block.

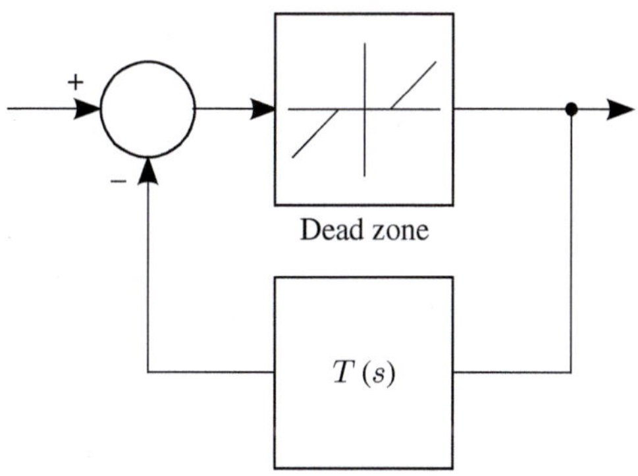

For sustained oscillations the closed-loop-transfer function should have a gain of 1 at a phase angle of 180 degrees ($-\pi$). Therefore, to determine the period of oscillation it is necessary to calculate the value of the frequency, ω, at which the phase angle of $T(s)$ is equal to $-\pi$, i.e. $\angle T(j\omega) = -\pi$.

Setting the phase angle, $\angle T(j\omega)$, to $-\pi$ and solving for ω gives:

$$\omega = \sqrt{\frac{0.422}{5}} = 0.29.$$

Recalling that $\omega = 2\pi f$ and $\tau = \frac{1}{f}$, it follows that the period of the oscillations should be approximately

$$\tau = 21.6 \text{ seconds.}$$

You probably found in the experiment performed in Section 26.3.2 that the period of oscillation was actually about 30 s. The above analysis predicts a period of 21.6 s. This discrepancy can be explained by the fact that Describing Function methods are an approximation to the true nonlinear behaviour.

Nonetheless, the above analysis represents compelling evidence that slip-stick friction is indeed the cause of the observed oscillations.

26.5 A Redesign to Account for Slip-Stick Friction

Now that the likely cause of the oscillation has been confirmed as being slip-stick friction in the SGV, it is now relevant to see how the problem might be alleviated. Removing this oscillation is very difficult in practice. One simple idea that was actually used in the context of continuous casting is to dither the valve. This amounts to rapidly moving the valve back and forth by a small amount thus preventing 'sticking' from occurring. The block diagram in Figure 26.9 shows where the dither signal can be inserted. The core idea behind this design is that the dither keeps the valve moving and thus avoids the slip-stick effect. Of course, the trade-off is that, whilst the output oscillation should reduce, extra wear will be introduced to the valve leading to maintenance considerations. Thus, a careful cost-benefit analysis is required.

26.5.1 Experiment: Testing the Impact of Slip-Stick Friction

The idea described above will be investigated using the virtual laboratory.

Ensure that the system is oscillating. This is done through the valve + actuator block, by ticking the 'Simulation Parameters Machine' and then choosing 'Realistic Valve Characteristics' boxes.

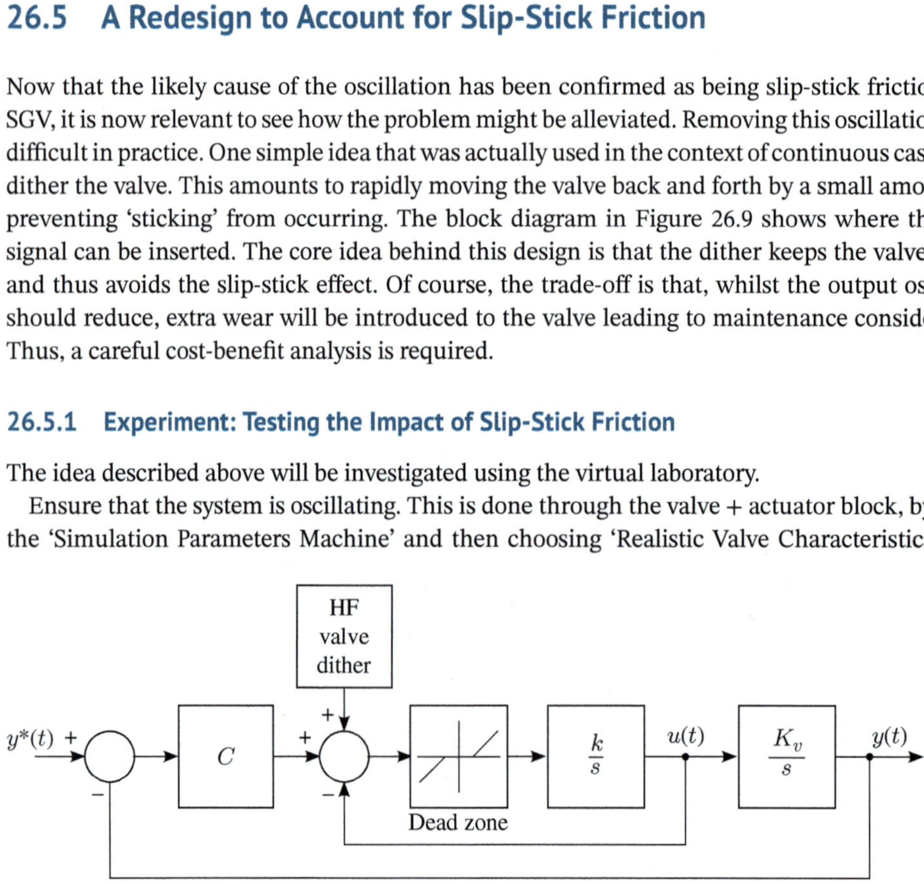

Figure 26.9 Slip stick friction compensated with dither.

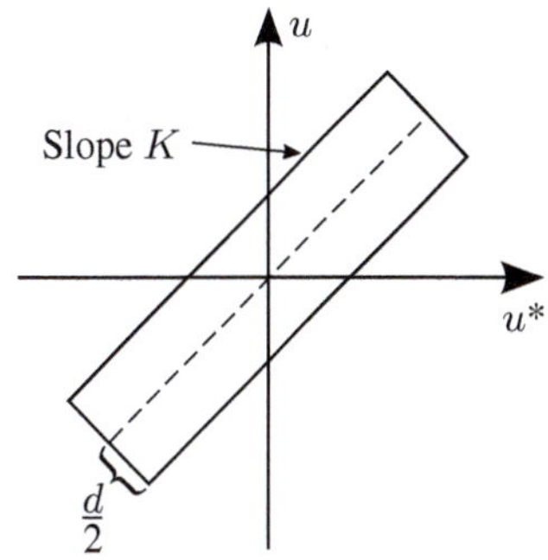

Figure 26.10 Backlash nonlinearity.

Then place dither on the valve by clicking the 'Dither' block and setting the dither to an appropriate value, e.g. a sine wave with offset 0%, amplitude 10% and frequency 10 Hz.

You should observe that dithering the valve does, indeed, lead to much better system performance. The oscillations are significantly reduced.

26.6 Revision Questions

1. Obtain the Describing Function for the backlash nonlinearity shown in Figure 26.10.

Further Reading

The above design essentially mirrors the steps used on a real continuous caster. As expected, dither did resolve the oscillation problem. On the other hand, and as predicted, dither did lead to increased wear on the valve. Hence the valve needed to be replaced more frequently. However, this was more than offset by the improved quality of the cast material. Indeed, several million dollars per year was saved by the final control system. Also, as mentioned previously, this is a very dangerous system. Thus, it was very important that the controller worked the first time! In practice, the controller was extensively tested using a realistic simulation of the system. This was done with two computers so that the controlling computer was connected to a second computer on which a model of the 'system' was implemented. Fortunately, the controller did work first time. It was then in continuous operation until the entire plant was decommissioned almost 10 years later.

Further information on the practical system upon which this virtual laboratory is based may be found in the following references.

1 S.F. Graebe, G.C. Goodwin and G. Elsley, "Control design and implementation in continuous steel casting", *IEEE Control Systems Magazine*, 15(4):64–71, 1995.
2 M.S. Jenkins, B.G. Thomas, W.C. Chen and R.B. Mahapatra, "Investigation of Strand Surface Defects Using Mold Instrumentation and Modelling", *77th Steelmaking Conference*, Chicago, Illinois, USA, Iron and Steel Society, Warrendale, Pennsylvania, USA, pp. 337–345, 1994.

27

Laboratory 13: Cross-Directional Control (Robustness and Impact of Actuator Saturation)

27.1 Introduction

Cross-directional control problems were studied earlier in Chapter 21. In Chapter 21, the fact that the actuators have limited authority was ignored. In practice, the actuators typically have limited authority and thus can easily saturate if 'pushed too hard'. Moreover, when these saturation effects are considered along with the presence of strong multi-variable interactions, then very difficult control issues arise. This is a major issue for this particular application. Moreover, the same issue arises in many other problems where there exists actuator saturation and strong multi-variable iterations. This chapter will take an in-depth look at the impact of limited actuator authority in multivariable problems.

27.2 Effect of Actuator Saturation Without Anti-Windup Protection

The impact of actuator saturation on the cross-directional control problem will be examined using (i) the simple PID structure and using (ii) the decoupled PID structure described in Section 21.7 of Chapter 21.

In the event that actuator saturation prevents the setpoint from being reached, then the integrator in the PID controller will continue to integrate. This phenomenon has been discussed for single-input single-output systems in Section 25.5 of Chapter 25. It would be relatively simple to use the anti-windup methods described in Chapter 25 for the cross-directional control problem. However as will be seen below, these ideas turn out to be inadequate for the problem under study in the current chapter. The reasons will be revealed as the chapter progresses.

27.2.1 Experiment: Impact of Actuator Saturation

Load the SISO controller mode. Ensure that the Random Disturbance profile and the Flat Reference profile are used. Click the System Characteristics block and choose Saturating Actuator. Observe the controller outputs and the actuator profiles.

You will probably find that the controller outputs grow significantly. However, actuator saturation means that the actuator profiles are physically limited to the range $[-1, +1]$. You probably find that the achieved basis weight profile is very poor.

Next, change the Disturbance Profile to Flat and keep the system running.

Practical Control System Design: Real World Designs implemented on Emulated Industrial Systems, First Edition. Adrian Medioli and Graham Goodwin.

Companion website: www.wiley.com/go/medioli/practicalcontrolsystemdesign

You will see that the controller takes a very long time to recover since it needs to wind its way back from the large values reached above. This nicely illustrates the negative consequence of controller windup.

27.2.2 Experiment: Impact of Actuator Saturation with Decoupled PID Design

Next, the effect of actuator saturation will be examined in the context of the decoupled design studied in Section 21.6.

Load the MIMO mode controller. Choose 'No Anti-Windup'. Select the Random Disturbance Profile and the Flat Reference Profile. Set Actuator Saturation to be 'on' by clicking on the machine parameters and choosing saturating actuator. Run the system.

You probably found that decoupling does not mitigate the windup problem.

27.3 PI Decoupled Design with Simple Anti-Windup Protection

Here the single-input single-output anti-windup scheme described in Section 25.5 of Chapter 25.1 will be studied. The core idea is to rearrange the control law so that all of the controller's dynamics are driven by the real (i.e. constrained) actuator signals. (See Section 25.5 of Chapter 25.) This concept is common to many, apparently dissimilar, ideas incorporating anti-windup protection.

The ideas explained earlier in Section 25.5.1 of Chapter 25 are summarised here for completeness in the context of this laboratory. Without loss of generality, it is assumed that the controller transfer function $C(s)$ is biproper. If this is not the case, additional 'fast' zeros can be appended. Then, as in Eq. (25.7) write

$$C(s)^{-1} = \overline{C}_o + D(s), \tag{27.1}$$

where $\overline{C}_o$ is the high frequency gain matrix and $D(s)$ is a strictly proper transfer function. It is also assumed that the controller has a stable inverse, so that $D(s)$ in (27.1) is stable.

The control law $C(s)$ is then implemented as shown in Figure 25.4 of Chapter 25.

When the demanded control signal $\tilde{e}_i$ in Figure 27.1 causes the corresponding plant input to be outside the constraints, then the limiting circuit operates to avoid windup.

Standard PI controllers will be used here. In this case,

$$\begin{aligned} C(s) &= K_p + \frac{K_i}{s} \\ &= K_i \frac{(s\tau + 1)}{s}, \quad \tau = \frac{K_P}{K_i}. \end{aligned} \tag{27.2}$$

Then

$$\begin{aligned} \overline{C}_o &= \frac{1}{K_i \tau}, \\ D(s) &= -\frac{1}{K_i \tau (s\tau + 1)}. \end{aligned} \tag{27.3}$$

In the MIMO case, the design of the limiting circuit in Figure 27.1b presents some subtleties since there are, in general, many different ways of bringing $\mathbf{u}(t)$ back inside the constraint polytope. Here a simple version known as Input Saturation or Actuator Clipping will first be explored as shown in Figure 27.1c. The scalar signals $\hat{u}_i$ are first collected into a vector $\hat{u}$. Then $\hat{u}$ is transformed to give $\overline{u}$. Diagonal saturation is applied to $\overline{u}$ to produce u. Finally $\overline{u}$ is transformed back to $\tilde{u}$. The elements of $\tilde{u}$, i.e. $\overline{u}_i$, are then passed to $D(s)$ as shown in Figure 27.1b.

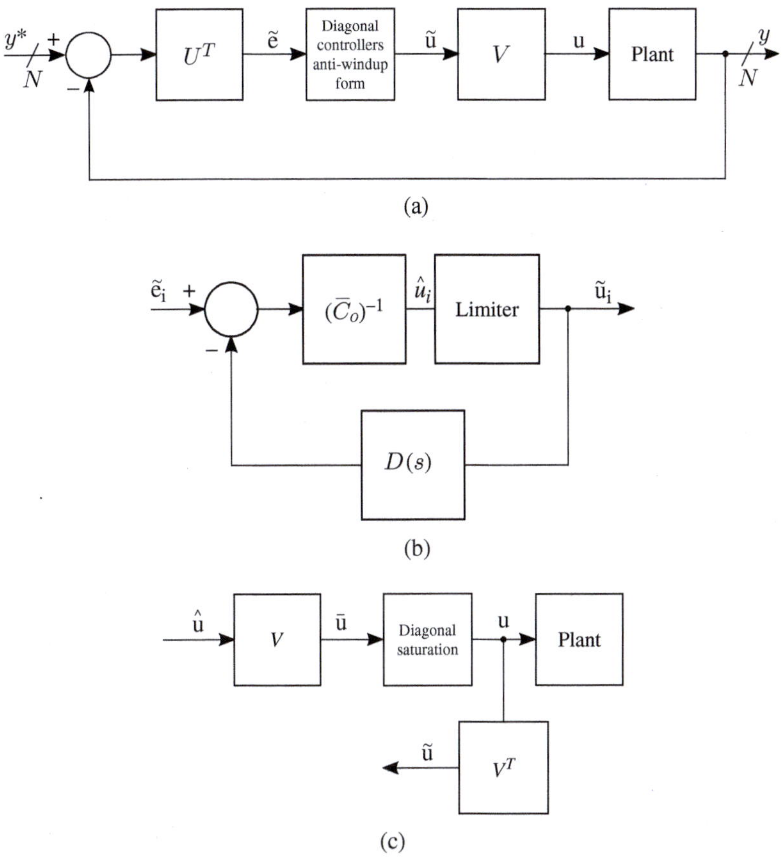

Figure 27.1 Anti-windup controllers in diagonal space. (a) General anti-windup controller implementation in the singular value space (where N is the number of actuators/sensors), (b) Diagonal controllers and (c) Simple saturation.

First, the desired actuator values are calculated ignoring saturation. Then the signals are saturated. Let the saturated vector be denoted as $u_{sat}(t)$. The vector $u^*(t)$ is then passed through V to achieve $u_{sat}(t)$. Thus,

$$u^*(t) = V^T u_{sat}(t). \tag{27.4}$$

Finally, as shown in Figure 27.1, $u^*(t)$ is passed back through $\tilde{C}^*(s)$.

27.3.1 Experiment: Testing the Simple Anti-Windup Scheme

Test the simple anti-windup configuration depicted in Figure 27.1 using the *input-clipping* type limiting circuit. We assume that the process is subject to input constraints of the type

$$|\mathbf{u}(t)_i| \leq 1. \tag{27.5}$$

Set up the system to have a decoupled PI controller by loading the MIMO controller. Select Input Clipping. Set the Disturbance Profile by clicking this block and choosing the Random profile and set the Reference Profile to Flat and keep the actuators Non-Saturating. In the controller, select input clipping and leave the highest control spatial frequency to 'all'. Plot the time response of the controller using the Data Scope Tool.

You should observe that the input signals are indeed constrained to the range ± 1. More importantly, you should observe that due to the input constraints, the controller is not able to fully compensate the applied output disturbance. You probably also found that the overall performance of the closed-loop is actually quite poor. The reasons for this are explained in the sequel.

27.4 Conditioning Problems

Here, problems arising from the conditioning of the interaction matrix will be examined.

The conditioning of any given matrix A is a fundamental property of matrices. The conditioning of A is quantified by the *condition number* defined by:

$$\kappa = \|A\| \, \|A^{-1}\|, \tag{27.6}$$

where $\|A\|$ is a given norm of the matrix A.

When the condition number κ is large, the matrix A is said to be badly (or poorly) conditioned. On the other hand, when the condition number κ is small, the matrix A is said to be well-conditioned.

The condition number, κ, provides an indication of the sensitivity of the inverse A^{-1} to small changes in the entries of A. The reason why this is important is that inversion is a crucial operation in control.[1]

A simple example that illustrates the numerical problems that arise when dealing with poorly conditioned matrices is described below. Here, the solution to a set of linear equations is examined. Thus, consider the following system of linear equations:

$$Ax = b, \tag{27.7}$$

where A is a 2×2 matrix. For example, say that

$$A = \begin{bmatrix} 2.1 & 3.6 \\ 5.4 & 9.257 \end{bmatrix}, \tag{27.8}$$

and

$$b = \begin{bmatrix} 1 \\ 2.571 \end{bmatrix}, \tag{27.9}$$

The solution to the system of linear equations in (27.7) is given by

$$x = A^{-1} b. \tag{27.10}$$

Simple calculations yield:

$$A^{-1} = \frac{-1}{3 \cdot 10^{-4}} \begin{bmatrix} 9.257 & -3.6 \\ -5.4 & 2.1 \end{bmatrix}. \tag{27.11}$$

So that on substituting into (27.10), the following solution is obtained:

$$x = \begin{bmatrix} -4.67 \\ 3 \end{bmatrix}. \tag{27.12}$$

Next, say that b is perturbed slightly to $\begin{bmatrix} 1 \\ 2.572 \end{bmatrix}$. The solution to the system of equations then turns out to be:

$$x = \begin{bmatrix} 7.33 \\ -4 \end{bmatrix}. \tag{27.13}$$

1 Recall that, high-gain feedback loops have the purpose of inverting the plant model in the frequency range of interest.

Comparing (27.13) and (27.12) clearly demonstrates the extremely high sensitivity of the problem to very small variations in the description of the problem. Indeed, a variation of less than 0.1% in the value of the second element of b has resulted in a completely different solution!

The above idea is next examined in the context of the Paper Machine problem.

The poor conditioning of the interaction matrix in the Paper Machine problem, i.e. $\overline{B}$, can be quantified using a Singular Value Decomposition (SVD).

$$\overline{B} = USV^T, \tag{27.14}$$

where $S, U, V \in \mathbb{R}^{m\times m}$. $S = diag\{\sigma_1, \sigma_2, \dots, \sigma_m\}$ is a diagonal matrix with positive singular values arranged in decreasing order.

U and V are orthogonal matrices such that $UU^T = U^TU = I_m$ and $VV^T = V^TV = I_m$. If $\overline{B}$ is symmetric then $U = V$.

If $\overline{B}$ is poorly conditioned then the last singular values on the diagonal of S will be very small compared to the largest singular value. This characteristic implies that the control directions associated with the smallest singular values will be much more difficult to control than those associated with the biggest singular values, in the sense that a larger control effort is required to compensate for disturbances acting in directions associated with small σ_i. Moreover, in the light of the example given above, small changes in the problem specification (including small model errors) are highly likely to lead to dramatically different answers! This constitutes not only a problem in terms of the limited control authority available in the array of actuators in a web-forming application, but it is also an indication of the sensitivity of the entire closed loop to uncertainties in the spatial components of the model.

Recall the model for the system presented in Eq. (21.8) of Chapter 21, i.e.

$$\mathbf{y}(s) = \begin{bmatrix} \dfrac{e^{-s\beta}}{\tau s+1} & & 0 \\ & \ddots & \\ 0 & & \dfrac{e^{-s\beta}}{\tau s+1} \end{bmatrix} \overline{B}\,\mathbf{u}(s). \tag{27.15}$$

In the sequel, an SVD of $\overline{B}$ will be used to design a better anti-windup scheme.

The singular values for the interaction matrix obtained in Chapter 21 have a magnitude that rapidly decreases as the number of the singular value increases.

Indeed,

$$\sigma_{max}(\overline{B}) = 11.82$$
$$\sigma_{min}(\overline{B}) = 0.05. \tag{27.16}$$

Therefore, the condition number, κ, of $\overline{B}$ is

$$\kappa = 236.4. \tag{27.17}$$

This large number is an indication of the strong interaction between actuators. It suggests that any simple design is likely to fail due to these sensitivity issues.

Returning to the anti-windup scheme of Section 27.3, a common disadvantage of this type of anti-windup strategy is that it does not account for the issue of widely different gains encountered in cross-directional control problems. In fact, the condition number of the interaction matrix $\overline{B}$ is usually extremely big, i.e. the difference between the biggest and smallest singular value is considerable. As a result, the system has control directions that are much 'harder' to control than others since the associated gain, i.e. the value of the associated singular value is small.

A frequency analysis framework is adopted to describe the behaviour of the singular values of $\overline{B}$. It can be said that the smallest gains are typically associated with the 'highest' spatial frequencies whilst the biggest gains are associated with the 'lower' spatial frequencies, i.e. it is relatively easy to shape the output as a simple cross-directional shape rather than as an intricate shape. Consequently, the control energy necessary to achieve disturbance compensation for the highest spatial frequencies is much bigger than the control energy required in lower spatial frequencies. If this characteristic of cross-directional control problems is not adequately taken into consideration, the controller risks 'wasting' all the available control energy trying to compensate for disturbances in the highest spatial frequencies, that is, in those directions hardest to control. This leads to a phenomenon known as 'picketing' in which adjacent actuators fight each other and go to alternate saturation limits. You may have observed this already in the experiments. If not carry out the following:

27.4.1 Experiment: Testing Actuator Profile

Closely examine the input (actuator) profile obtained when you retest the simple anti-windup scheme of Section 27.3.

27.5 PI Decoupled Design with Anti-Windup Protection Limited to Low Spatial Frequencies

One way of avoiding the poor performance generated by the 'picket fence' phenomenon described in Section 27.4 is to force the decoupled controller to control only the 'low' spatial frequency components, i.e. those associated with the largest singular values.

In this case, the controller is defined as follows:

$$\mathbf{C}^*(s) = \mathbf{C}_r^*(s) = diag\{c_1^*(s), \dots, c_r^*(s), 0, \dots, 0\}, \tag{27.18}$$

where the control action has been disabled in the 'high' spatial frequency components by setting the corresponding controller gains to zero. The controller is allowed to only compensate the disturbance in the first r spatial components.

27.5.1 Experiment: Limiting Spatial Frequencies Used in the Controller

To test the above idea choose the number of active controllers in the singular value space as $r = 8$.

Implement the previous anti-windup scheme but set the controller gains to zero for $i = 9, \dots, 21$. This is done by clicking in the controller block and from the 'highest spatial controller frequency combo box' select control the first 8 of 21 frequencies. Make sure that the System Characteristics still have 'Actuator Saturating'.

Observe the steady state input (actuator) profile, i.e. the Actuator Profile graph when the system has settled, also observe the disturbance rejection capability by applying the Random disturbance and the Flat reference profile.

You should see that the output disturbance compensation has significantly improved. Of course, complete compensation is now no longer possible since the controller is prevented from acting on the high spatial frequency components. On the other hand, the input profile is now no longer dominated by the 'picket fence' profile.

It is important to note that the approach described here is somehow conservative since it does not 'push the limits', i.e. it does not use, in full, the available control authority. A way to push the limits will be explored in the next section.

27.6 PI Decoupled Design with Adaptive Spatial Frequency Selection

(Note that this section is more advanced and is intended for more advanced students).

A further embellishment to the approach described in Section 27.5 is to allow the selection of the singular value components r, that are used in the controller, to be done adaptively.

This approach will be called the 'Adaptive Spatial Frequency Selection (ASFS)' scheme. The strategy fits into the general structure depicted in Figure 27.1 for a controller incorporating anti-windup protection in the singular value space. In particular, it is to be noted that the singular value decomposition of $\overline{B}$ in (27.14) generates a natural ordering among the plant input directions in terms of the control effort necessary to compensate for a disturbance acting in the corresponding output direction. Indeed, in (27.14) the control directions are ordered starting from the *easiest* to control (associated with the biggest σ_i) and ending with the *hardest* to control (associated with the smallest σ_i). As a result, the ASFS strategy acts in the singular value space by scaling the manipulated variables $\hat{\mathbf{u}}^*(t)$, starting from the highest spatial frequency component and descending towards the lowest spatial frequency components until actuator saturation is avoided. In this way, rather than fixing the number of active directions *a priori* (as would be the case when using a controller structure such as 27.18), the selection is made on-line. There are two key advantages of using the ASFS strategy:

- the choice of active directions is matched to the particular characteristic of the disturbance,
- additional robustness features accrue due to the fact that there is an inherent feedback mechanism involved in making the choice.

The limiting circuit associated with the ASFS strategy defines a diagonal weighting matrix $\mathbf{P}_\gamma(t) \in \mathbb{R}^{N\times N}$ as follows:

$$\mathbf{u}^*(t) = \mathbf{P}_\gamma(t)\,\hat{\mathbf{u}}^*(t)\,, \tag{27.19}$$

where

$$\mathbf{P}_\gamma(t) = \begin{cases} diag\{p_1, \ldots, p_r, p_{r+1}, 0, \ldots, 0\}\,, & \text{when } \|\mathbf{V}\hat{\mathbf{u}}^*(t)\|_\infty \ge \Delta \\ \mathbf{I}_{N\times N}\,, & \text{when } \|\mathbf{V}\hat{\mathbf{u}}^*(t)\|_\infty < \Delta\,. \end{cases} \tag{27.20}$$

$\Delta \in \mathbb{R}$ is the saturation level imposed on every entry of $\mathbf{u}(t)$. The matrix $\mathbf{P}_\gamma(t)$ is parameterised in terms of the variable $\gamma(t) \in \mathbb{R}$ which is allowed to evolve only in the interval $[0\,, N]$. $\gamma(t)$ describes the spatial components that are active at any given time t. If $r \in \mathbb{Z}^+$ is the biggest integer such that $r \le \gamma$, then the entries of $\mathbf{P}_\gamma(t)$ in (27.20) are:

$$p_1 = p_2 = \cdots = p_r = 1, \tag{27.21}$$

$$p_{r+1} = \gamma - r. \tag{27.22}$$

The ASFS strategy computes the value of $\gamma(t)$ in such a way that, if a constraint violation occurs, the control vector $\mathbf{u}(t)$ is brought back on the boundary of the constraint polytope $\mathbb{U}$, that is:

$$\|\mathbf{u}(t)\|_\infty = \|\mathbf{V}\mathbf{P}_\gamma(t)\,\hat{\mathbf{u}}^*(t)\|_\infty = \Delta.$$

Depending on the time evolution of the closed-loop, the ASFS strategy uses a variable set of basis functions that can increase or decrease in number so as to keep the input $\mathbf{u}(t)$ within actuator saturation limits while maintaining good feedback control. The resulting control action achieves zero steady-state error in a subset of the spatial frequency modes of $\overline{B}$.

27.6.1 Experiment: Testing Adaptive Spatial Frequency Selection

Test the scheme by clicking on the Control block and selecting Adaptive Spatial Frequency Anti-Windup type.

Observe the input and output responses to the standard disturbance. Also, observe the steady state input (actuator) and output (basis weight) profiles. Compare your results with those obtained earlier.

You will probably now observe that the input (actuator) profile includes some high-frequency modes but it is not dominated by the 'picket fence' phenomenon. The output (basis weight) profile has improved marginally when compared to that obtained by restricting control to the first eight components only – as was done in Section 27.5.

27.7 Conclusions

The designs outlined in this chapter are all used in various modified forms in several commercial products that implement cross-directional control for various industrial problems. If you have managed to complete this laboratory successfully, then you are well equipped to tackle this class of difficult multi-variable control system design for real systems.

27.8 Revision Questions

1. Consider the problem of controlling the temperature of water in a shower. An approximate model is

$$\begin{bmatrix} y_1 \\ y_2 \end{bmatrix} = e^{-s\tau} \begin{bmatrix} k_1 & -k_1 \\ 1 & 1 \end{bmatrix} \begin{bmatrix} u_1 \\ u_2 \end{bmatrix}, \tag{27.23}$$

   where τ is a time delay; y_1, y_2 are the temperature of the water delivered by the shower and flow rate of water; u_1, u_2 are the flow of hot water and cold water, respectively. Describe the transformation to a different control space that decouples the two control loops.
2. Use your answer to Question 1 to devise a decoupled control architecture.
3. Say that U_1 saturates, i.e. the hot water flow hits a limit. Explain what this might lead to in terms of water temperature delivered by the shower.
4. Transform the constraint from the u_1, u_2 space into your new control space, $\overline{u}$.
5. Explain why restraining $\overline{u}$ back to lie inside the constraint might be a better strategy thean simply saturating u.
6. Why is the delay, τ, significant? What happens if you try to regulate the temperature too quickly?

Further Reading

1 G.H. Golub and C.F. Van Loan, *Matrix computations*, The John Hopkins University Press, Balitmore, Maryland, USA, 3rd Edition, 1996.

2 G.C. Goodwin, S.F. Graebe and M.E. Salgado, *Control system design*, Prentice Hall, Upper Saddle River, New Jersey, USA, 2001.

3 C.D. Myer, *Matrix analysis and applied linear algebra*, SIAM, Philadelphia, Pennsylvania, USA, 2000.

4 O.J. Rojas, G.C. Goodwin and A. Desbiens, "Study of An Adaptive Anti-Windup Strategy for Cross Directional Control System", *Proceedings 41st IEEE Conference on Decision and Control*, Las Vegas, Nevada, USA, 2002.

5 O.J. Rojas, G.C. Goodwin and G.V. Johnston, "Spatial frequency anti-windup strategy for cross directional control problems", *IEE Proceedings: Control Theory and Application*, 149(5):414–422, 2002.

6 G. Strang, *Linear algebra and its applications*, Harcourt Brace and Company, Orlando, Florida, USA, 3rd Edition, 1988.

Part VIII

Modelling and Control of More Complex Nonlinear Systems

This part of the book will be devoted to more complex nonlinear systems. Chapter 28 will develop a model for a rocket operating in three-dimensional space. Chapter 29 will develop a model for a boiler operating under rapidly changing load conditions. Chapter 30 will examine control laws for a boiler.

Practical Control System Design: Real World Designs implemented on Emulated Industrial Systems,
First Edition. Adrian Medioli and Graham Goodwin.

Companion website: www.wiley.com/go/medioli/practicalcontrolsystemdesign

28

Modelling of a Rocket in Three-Dimensional Flight

28.1 Introduction

Chapter 15 dealt with the modelling and control of a rocket in 2D flight.

In this chapter, a model for a rocket in 3D flight will be developed. The derivation of the equations is again based on Newton's Laws. However, here an extra complexity arises, due to the use of different coordinate systems in three dimensions.

Section 28.2 will summarise necessary background. Section 28.3 describes Translational Dynamics. Section 28.4 describes Rotational Dynamics. The associated control problem will not be studied since this would take the reader beyond the scope of this book. However, the details can be found in the references provided in Further Reading.

28.2 Preliminaries

To describe the dynamics of a rocket operating in three dimensions, extra background is necessary to fully appreciate the impact of different coordinate systems. This section gives an overview of the required background.

28.2.1 Coordinate Systems

As has been seen earlier in Chapter 15, when dealing with rockets or aircraft, it is useful to work with different coordinate systems. One useful set of coordinates is fixed in space and termed an 'inertial frame'. Other coordinate systems are defined relative to the body of the aircraft or rocket. Newton's laws still apply but one must account for the motion of the reference frame relative to a fixed reference frame.

Consider the two reference frames shown in Figure 28.1. One is a fixed (or inertial) reference frame with axes $\underline{e}_1, \underline{e}_2, \underline{e}_3$, the other is aligned with the body and has axes $\underline{b}_1, \underline{b}_2, \underline{b}_3$.

Here, and in the sequel, the following notation will be used: $\underline{r}$ denotes a physical vector, $\underline{E}$ denotes a particular set of three (orthogonal) vectors, e.g.

$$\underline{E} = \{\underline{e}_1, \underline{e}_2, \underline{e}_3\}, \tag{28.1}$$

$\mathbf{r}^e$ will denote the coordinates of a vector $\underline{r}$ with respect to the base $\underline{E}$. Similarly, $\mathbf{r}^b$ will denote the coordinate of $\underline{r}$ with respect to the base $\underline{B}$. (Note the use of an upper-case letter for a set of three vectors and a bold letter for coordinates.)

$\underline{E}$ is called an inertial frame if it is stationary (or moving with constant velocity).

Practical Control System Design: Real World Designs implemented on Emulated Industrial Systems,
First Edition. Adrian Medioli and Graham Goodwin.

Companion website: www.wiley.com/go/medioli/practicalcontrolsystemdesign

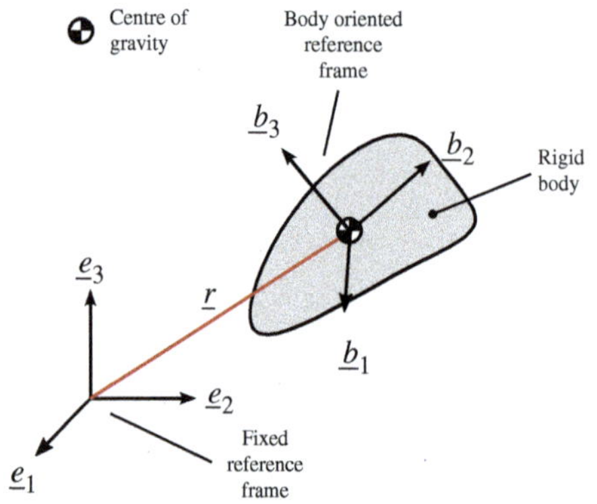

Figure 28.1 Rigid body in free space.

Vectors, such as $\underline{r}$ in Figure 28.1, have a magnitude and a direction, they are described in the fixed reference frame $\underline{E}$ by:

$$\underline{r} = r_1^e \underline{e}_1 + r_2^e \underline{e}_2 + r_3^e \underline{e}_3 = \begin{bmatrix} \underline{e}_1 & \underline{e}_2 & \underline{e}_3 \end{bmatrix} \begin{bmatrix} r_1^e \\ r_2^e \\ r_3^e \end{bmatrix} = \underline{\mathbf{E}}\mathbf{r}^e. \tag{28.2}$$

In (28.2), the superscript e refers to the $\underline{E}$ reference frame. In (28.2), $\underline{e}_i$, for $i = 1, 2$, or 3, represent the basis vectors, which are unit vectors in the specified directions. $\underline{E}$ is called a dextral set, when its vectors are perpendicular to each other. Note that $\underline{\mathbf{E}}$ is a 3×3 matrix formed from the vectors in the set $\underline{E}$.

One can have other frames. Figure 28.1 shows another frame, $B = \{b_1, b_2, b_3\}$, on the body. (Note that $\underline{b}_1$ and $\underline{b}_2$ are aligned with the plane of the body.) The same vector $\underline{r}$ as in (28.2) can be alternatively expressed as

$$\underline{r} = r_1^b \underline{b}_1 + r_2^b \underline{b}_2 + r_3^b \underline{b}_3 = \begin{bmatrix} \underline{b}_1 & \underline{b}_2 & \underline{b}_3 \end{bmatrix} \begin{bmatrix} r_1^b \\ r_2^b \\ r_3^b \end{bmatrix} = \underline{\mathbf{B}}\mathbf{r}^b. \tag{28.3}$$

Here, $\underline{\mathbf{B}} = \begin{bmatrix} \underline{b}_1 & \underline{b}_2 & \underline{b}_3 \end{bmatrix}$ is a 3×3 matrix formed from the vectors in the set $\underline{B}$.

28.2.2 Euler Angles in Three Dimensions

It was shown in Chapter 15 (see Eq. (15.2)) that two sets of basis vectors can be connected by a simple rotation. In the 3D case, it is necessary to deploy multiple (actually 3) successive rotations. These rotations are based on the use of Euler angles. (The term 'Euler angles' acknowledges the contribution of Leonhard Euler who described the orientation of a rigid body with respect to two coordinate systems.)

In the 2D case, it was shown in Chapter 15 that a new coordinate system can be simply obtained by rotating the original coordinates (Figure 15.2 shows a rotation by angle θ about the Z axis pointing out of the page).

It is easily seen that the new basis vectors, $\underline{X}', \underline{Y}'$, can be expressed in terms of the original basis sector $\underline{X}, \underline{Y}$ as follows:

$$\begin{aligned} \begin{bmatrix} \underline{X}', \underline{Y}' \end{bmatrix} &= \begin{bmatrix} \underline{X}, \underline{Y} \end{bmatrix} \begin{bmatrix} \cos\theta & -\sin\theta \\ \sin\theta & \cos\theta \end{bmatrix}, \\ &= \begin{bmatrix} \underline{X}, \underline{Y} \end{bmatrix} \mathbf{R}(\theta). \end{aligned} \tag{28.4}$$

This equation appears as Eq. (15.2) in Chapter 15.

The matrix $\mathbf{R}(\theta)$ is a termed 'rotation matrix' and is given by

$$\mathbf{R}(\theta) = \begin{bmatrix} \cos\theta & -\sin\theta \\ \sin\theta & \cos\theta \end{bmatrix}. \tag{28.5}$$

The situation in three dimensions is a little more complex. In this case, any new set of basis vectors can be expressed in terms of another set by using three elementary rotations performed in sequence.

There are many different conventions. For example, on might first rotate, about the $\underline{z}$-axis, then the new $\underline{y}$-axis and then the new $\underline{x}$-axis or one might choose to first rotate about the $\underline{x}$-axis, then the new $\underline{y}$-axis and then the new $\underline{z}$-axis again.

The set of Euler rotations frequently used in aeronautics in the body axes of an aircraft is as follows:

α	(Yaw)	A counter-clockwise rotation about the b_3 axis
β	(Pitch)	A counter-clockwise rotation about the new b_2' axis
σ	(Roll)	A counter-clockwise rotation about the new b_1'' axis.

The new basis vectors (aligned with the aircraft) can be expressed in terms of the original basis vectors by three rotation matrices. Trigonometry shows that

$$\text{Yaw: } \left[\underline{b}_1' \ \underline{b}_2' \ \underline{b}_3'\right] = \left[\underline{b}_1 \ \underline{b}_2 \ \underline{b}_3\right] \mathbf{R}(\alpha) \text{ where } \mathbf{R}(\alpha) = \begin{bmatrix} \cos\alpha & -\sin\alpha & 0 \\ \sin\alpha & \cos\alpha & 0 \\ 0 & 0 & 1 \end{bmatrix},$$

$$\text{Pitch: } \left[\underline{b}_1'' \ \underline{b}_2'' \ \underline{b}_3''\right] = \left[\underline{b}_1' \ \underline{b}_2' \ \underline{b}_3'\right] \mathbf{R}(\beta) \text{ where } \mathbf{R}(\beta) = \begin{bmatrix} \cos\beta & 0 & -\sin\beta \\ 0 & 1 & 0 \\ \sin\beta & 0 & \cos\beta \end{bmatrix},$$

$$\text{Roll: } \left[\underline{b}_1''' \ \underline{b}_2''' \ \underline{b}_3'''\right] = \left[\underline{b}_1'' \ \underline{b}_2'' \ \underline{b}_3''\right] \mathbf{R}(\sigma) \text{ where } \mathbf{R}(\sigma) = \begin{bmatrix} 1 & 0 & 0 \\ 0 & \cos\sigma & -\sin\sigma \\ 0 & \sin\sigma & \cos\sigma \end{bmatrix}.$$

The combined set of rotations is given by

$$\left[\underline{b}_1''' \ \underline{b}_2''' \ \underline{b}_3'''\right] = \left[\underline{b}_1 \ \underline{b}_2 \ \underline{b}_3\right] \mathbf{R}(\alpha, \beta, \sigma), \tag{28.6}$$

where $\mathbf{R}(\alpha, \beta, \sigma) = \mathbf{R}(\alpha)\mathbf{R}(\beta)\mathbf{R}(\sigma)$

The matrices $\mathbf{R}(\alpha), \mathbf{R}(\beta), \mathbf{R}(\sigma), \mathbf{R}(\alpha, \beta, \sigma)$ satisfy:

$$\mathbf{R}(\alpha)^{-1} = \mathbf{R}(\alpha)^T, \tag{28.7}$$

$$\mathbf{R}(\beta)^{-1} = \mathbf{R}(\beta)^T, \tag{28.8}$$

$$\mathbf{R}(\sigma)^{-1} = \mathbf{R}(\sigma)^T. \tag{28.9}$$

It is also easy to express the coordinates of a vector in the new basis system in terms of the coordinates in the original basis system. For example, consider a vector $\underline{r}$ having the following representation with respect to $\underline{b}_1, \underline{b}_2, \underline{b}_3$:

$$\underline{r} = \left[\underline{b}_1, \underline{b}_2, \underline{b}_3\right] \mathbf{r}^b. \tag{28.10}$$

Here $\mathbf{r}^b$ is the vector of coordinates in the $[\underline{b}_1\ \underline{b}_2\ \underline{b}_3]$ system. Say that a new set of basis vectors satisfies

$$[\underline{b}_1'''\ \underline{b}_2'''\ \underline{b}_3'''] = [\underline{b}_1\ \underline{b}_2\ \underline{b}_3]\,\mathbf{R}(\alpha, \beta, \sigma), \tag{28.11}$$

then

$$\begin{aligned}\underline{r} &= [\underline{b}_1, \underline{b}_2, \underline{b}_3]\,\mathbf{r}^b, \\ &= [\underline{b}_1''', \underline{b}_2''', \underline{b}_3''']\,(\mathbf{r}^b)''', \\ &= [\underline{b}_1''', \underline{b}_2''', \underline{b}_3''']\,\mathbf{R}(\alpha, \beta, \sigma)^T\mathbf{r}^b,\end{aligned} \tag{28.12}$$

where (28.6) has been used.

Hence, the coordinates are related by

$$(\mathbf{r}^b)''' = \mathbf{R}(\alpha, \beta, \sigma)^T\mathbf{r}^b. \tag{28.13}$$

28.2.3 Time Derivative of Rotation Matrices

In the subsequent development, it will sometimes be necessary to compute the time derivative of rotation matrices.

Here, the time derivative of rotation matrices such as $\mathbf{R}(\theta)$ and $\mathbf{R}(\alpha, \beta, \sigma)$ will be computed. First recall, that, using the chain rule, $\frac{d}{dt}\sin\theta = \frac{d\theta}{dt}\cos\theta$ and $\frac{d}{dt}\cos\theta = -\frac{d\theta}{dt}\sin\theta$.

For a single rotation in the $\underline{b}_1, \underline{b}_2$ plane, i.e. about the $\underline{b}_3$ axis, it is then easily seen that

$$\begin{aligned}\mathbf{R}^T(\alpha)\frac{d}{dt}\mathbf{R}(\alpha) &= \begin{bmatrix}\cos\alpha & \sin\alpha & 0\\ -\sin\alpha & \cos\alpha & 0\\ 0 & 0 & 1\end{bmatrix}\begin{bmatrix}-\left(\frac{d}{dt}\alpha\right)\sin\alpha & -\left(\frac{d}{dt}\alpha\right)\cos\alpha & 0\\ \left(\frac{d}{dt}\alpha\right)\cos\alpha & -\left(\frac{d}{dt}\alpha\right)\sin\alpha & 0\\ 0 & 0 & 1\end{bmatrix}, \\ &= \begin{bmatrix}0 & -\left(\frac{d}{dt}\alpha\right) & 0\\ \left(\frac{d}{dt}\alpha\right) & 0 & 0\\ 0 & 0 & 1\end{bmatrix}, \\ &= \begin{bmatrix}0 & -\omega_3^b & 0\\ \omega_3^b & 0 & 0\\ 0 & 0 & 1\end{bmatrix},\end{aligned} \tag{28.14}$$

where ω_3^b denotes the angular velocity about the $\underline{b}_3$ axis.

In a similar fashion, it can be shown that for $\mathbf{R}(\alpha, \beta, \delta)$

$$\mathbf{R}^T\left(\frac{d}{dt}\mathbf{R}\right) = \begin{bmatrix}0 & -\omega_3^b & \omega_2^b\\ \omega_3^b & 0 & -\omega_1^b\\ -\omega_2^b & \omega_1^b & 0\end{bmatrix}. \tag{28.15}$$

A simplified notation, $\tilde{\omega}^b$, will be used for the matrix on the right-hand side of (28.15), i.e.

$$\tilde{\omega}^b = \begin{bmatrix}0 & -\omega_3^b & \omega_2^b\\ \omega_3^b & 0 & -\omega_1^b\\ -\omega_2^b & \omega_1^b & 0\end{bmatrix}. \tag{28.16}$$

Hence $\left(\frac{d\mathbf{R}}{dt}\right) = \mathbf{R}\tilde{\omega}$, and

$$\frac{d\underline{\mathbf{B}}}{dt} = \underline{\mathbf{E}}\frac{d\mathbf{R}}{dt}, \tag{28.17}$$

$$= \underline{\mathbf{E}}\mathbf{R}\tilde{\omega}^{\mathbf{b}}, \tag{28.18}$$

$$= \underline{\mathbf{B}}\tilde{\omega}^{b}. \tag{28.19}$$

28.2.4 Angular Velocities

The following angles are defined:

$$\theta_1^b = \alpha, \tag{28.20}$$

$$\theta_2^b = \beta, \tag{28.21}$$

$$\theta_3^b = \sigma, \tag{28.22}$$

These angles can be expressed as a vector which, by convention, is perpendicular to the plane in which the rotation lies. Hence in a reference frame $\underline{B}$, one can define the angle vector, $\underline{\theta}$ as

$$\underline{\theta} = \begin{bmatrix} \underline{b}_1 & \underline{b}_2 & \underline{b}_3 \end{bmatrix} \begin{bmatrix} \theta_1^b \\ \theta_2^b \\ \theta_3^b \end{bmatrix} = \underline{\mathbf{B}}\theta^b. \tag{28.23}$$

Next, consider the angular velocities $\omega_1^b, \omega_2^b, \omega_3^b$ defined as follows:

$$\omega_1^b = \frac{d}{dt}\alpha, \tag{28.24}$$

$$\omega_2^b = \frac{d}{dt}\beta, \tag{28.25}$$

$$\omega_3^b = \frac{d}{dt}\gamma. \tag{28.26}$$

When computing the angular velocity, one needs to account for the time derivative of the reference frame. Hence,

$$\underline{\omega} = \frac{d}{dt}\underline{\theta}, \tag{28.27}$$

$$= \frac{d\underline{\mathbf{B}}}{dt}\theta^b + \underline{\mathbf{B}}\frac{d}{dt}\theta^b. \tag{28.28}$$

Then using (28.19),

$$\underline{\omega} = \underline{\mathbf{B}}\left[\tilde{\omega}^b\theta^b + \frac{d}{dt}\theta^b\right], \tag{28.29}$$

$$= \underline{\mathbf{B}}\left[\tilde{\omega}^b\theta^b + \omega^b\right]. \tag{28.30}$$

28.2.5 Angular Acceleration

Next the rate of change of angular velocity, i.e. angular acceleration, will be considered.

To obtain the acceleration in the body frame, it is necessary to differentiate (28.30) this leads to

$$\frac{d}{dt}\underline{\omega} = \frac{d\underline{\mathbf{B}}}{dt}\left[\tilde{\omega}^b\theta^b + \boldsymbol{\omega}^b\right] + \underline{\mathbf{B}}\left[\left(\frac{d}{dt}\tilde{\omega}^b\right)\theta^b + \tilde{\omega}^b\left(\frac{d}{dt}\theta^b\right) + \frac{d}{dt}\boldsymbol{\omega}^b\right]. \tag{28.31}$$

Recall from (28.19)

$$\frac{d\underline{\mathbf{B}}}{dt} = \underline{\mathbf{B}}\tilde{\omega}^b. \tag{28.32}$$

Substituting (28.32) into (28.31) yields

$$\frac{d}{dt}\underline{\omega} = \underline{\mathbf{B}}\left[\left(\tilde{\omega}^b\right)^2\boldsymbol{\theta}^b + \tilde{\omega}^b\boldsymbol{\omega}^b\right] + \underline{\mathbf{B}}\left[\left(\frac{d}{dt}\tilde{\omega}^b\right)\boldsymbol{\theta}^b + \tilde{\omega}^b\boldsymbol{\omega}^b + \frac{d}{dt}\boldsymbol{\omega}^b\right]. \tag{28.33}$$

An alternative expression for $(\tilde{\omega}^b)^2$ can be obtained as follows:

$$\left(\tilde{\omega}^b\right)^2 = \begin{bmatrix} 0 & -\omega_3^b & \omega_2^b \\ \omega_3^b & 0 & -\omega_1^b \\ -\omega_2^b & \omega_1^b & 0 \end{bmatrix}\begin{bmatrix} 0 & -\omega_3^b & \omega_2^b \\ \omega_3^b & 0 & -\omega_1^b \\ -\omega_2^b & \omega_1^b & 0 \end{bmatrix}, \tag{28.34}$$

or

$$\left(\tilde{\omega}^b\right)^2 = \begin{bmatrix} -\left[\left(\omega_2^b\right)^2 + \left(\omega_3^b\right)^2\right] & \left(\omega_1^b\right)\left(\omega_2^b\right) & \left(\omega_1^b\right)\left(\omega_3^b\right) \\ \left(\omega_1^b\right)\left(\omega_2^b\right) & -\left[\left(\omega_1^b\right)^2 + \left(\omega_3^b\right)^2\right] & \left(\omega_2^b\right)\left(\omega_3^b\right) \\ \left(\omega_1^b\right)\left(\omega_3^b\right) & \left(\omega_2^b\right)\left(\omega_3^b\right) & -\left[\left(\omega_1^b\right)^2 + \left(\omega_2^b\right)^2\right] \end{bmatrix}, \tag{28.35}$$

$$= \left(\boldsymbol{\omega}^b\right)\left(\boldsymbol{\omega}^b\right)^T - \left(\boldsymbol{\omega}^b\right)^T\left(\boldsymbol{\omega}^b\right) I. \tag{28.36}$$

Substituting (28.36) into (28.33) yields

$$\frac{d\underline{\omega}}{dt} = \underline{\mathbf{B}}\left\{\left[\left(\boldsymbol{\omega}^b\right)\left(\boldsymbol{\omega}^b\right)^T - \left(\boldsymbol{\omega}^b\right)^T\left(\boldsymbol{\omega}^b\right) I\right]\boldsymbol{\theta}^b + 2\tilde{\omega}^b\boldsymbol{\omega}^b + \left(\frac{d}{dt}\tilde{\omega}^b\right)\boldsymbol{\theta}^b + \frac{d}{dt}\boldsymbol{\omega}^b\right\}. \tag{28.37}$$

where I is a 3×3 identity matrix. Equation (28.37) can be expressed in many alternative forms. For example, analogously to (28.16), a matrix $\tilde{\theta}^b$ can be defined by

$$\tilde{\theta}^b = \begin{bmatrix} 0 & -\theta_3^b & \theta_2^b \\ \theta_3^b & 0 & -\theta_1^b \\ -\theta_2^b & \theta_1^b & 0 \end{bmatrix}. \tag{28.38}$$

The second last term in (28.37) contains the expression

$$\frac{d}{dt}\left(\tilde{\omega}^b\right)\theta^b = \frac{d}{dt}\begin{bmatrix} 0 & -\omega_3^b & \omega_2^b \\ \omega_3^b & 0 & -\omega_1^b \\ -\omega_2^b & \omega_1^b & 0 \end{bmatrix}\begin{bmatrix} \theta_1^b \\ \theta_2^b \\ \theta_3^b \end{bmatrix}. \tag{28.39}$$

Hence by rearranging the terms and using (28.38):

$$\frac{d}{dt}\left(\tilde{\omega}^b\right)\theta^b = -\tilde{\theta}^b\begin{bmatrix} \frac{d}{dt}\omega_1^b \\ \frac{d}{dt}\omega_2^b \\ \frac{d}{dt}\omega_3^b \end{bmatrix}, \tag{28.40}$$

$$= -\tilde{\theta}^b\frac{d}{dt}\omega^b.$$

Substituting (28.40) into (28.37) yields the following alternative form for the angular acceleration

$$\frac{d}{dt}\underline{\omega} = \underline{\mathbf{B}}\left\{\left[\left(\boldsymbol{\omega}^b\right)\left(\boldsymbol{\omega}^b\right)^T - \left(\boldsymbol{\omega}^b\right)^T\left(\boldsymbol{\omega}^b\right) I\right]\boldsymbol{\theta}^b + 2\tilde{\omega}\boldsymbol{\omega}^b - \tilde{\theta}^b\left(\frac{d}{dt}\boldsymbol{\omega}^b\right) + \frac{d}{dt}\boldsymbol{\omega}^b\right\}. \tag{28.41}$$

28.2.6 Cross-Products

In the sequel, cross-product notation will be used to simplify expressions such as (28.33) or (28.41).

The cross-product (denoted $\underline{x} \times \underline{y}$) between two vectors $\underline{x}$ and $\underline{y}$ is defined as a 3D vector that is perpendicular to the plane containing $\underline{x}, \underline{y}$ and having magnitude $|\underline{x}||\underline{y}| \sin\theta$ where $|\underline{x}|, |\underline{y}|$ are the magnitudes of $\underline{x}, \underline{y}$ and θ is the angle between them. A right-hand convention applies, i.e. the direction is given by the thumb if one curls the fingers of the right hand from $\underline{x}$ to $\underline{y}$.

A cross-product can also be expressed in matrix notation by utilising the notation $\tilde{x}^b$ introduced in (28.38). Specifically, if

$$\underline{x} = \underline{\boldsymbol{B}}\boldsymbol{x}^b, \tag{28.42}$$

$$\underline{y} = \underline{\boldsymbol{B}}\boldsymbol{y}^b. \tag{28.43}$$

Then

$$\begin{aligned} \underline{x} \times \underline{y} &= \underline{\boldsymbol{B}}\tilde{x}^b\boldsymbol{y}^b \\ &= \underline{\boldsymbol{B}}\begin{bmatrix} 0 & -x_3^b & x_2^b \\ x_3^b & 0 & -x_1^b \\ -x_2^b & x_1^b & 0 \end{bmatrix}\begin{bmatrix} y_1^b \\ y_2^b \\ y_3^b \end{bmatrix}, \\ &= \underline{\boldsymbol{B}}\boldsymbol{x}^b \times \boldsymbol{y}^b. \end{aligned} \tag{28.44}$$

Using this notation, a torque vector, t, about a centre of gravity g, can be described using the cross-product notation as

$$\underline{t} = \underline{\mu}_{gt} \times \underline{f}_t. \tag{28.45}$$

where μ_{gt} is the vector from the point of application of the force to the point g and f_t denotes the force vector.

28.3 Translational Dynamics

It is now possible to describe the translational dynamics.

28.3.1 Forces

As in the 2D case, three forces act on the rocket, namely, Gravity, Thrust and Drag. These forces are described in 3D space as follows:

Gravity acts downwards and is thus easily described in the $\underline{\mathbf{E}}$ frame as

$$\text{Gravity} : \underline{f}_g = \underline{\mathbf{E}}\begin{bmatrix} 0 \\ -mg \\ 0 \end{bmatrix} = \underline{\mathbf{E}}\mathbf{f}_g^e. \tag{28.46}$$

where m is the mass and g is the gravitational constant. Thrust is considered to be orientated at an angle u to the body of the rocket. Hence,

$$\text{Thrust} : \underline{f}_t = \underline{\mathbf{B}}\begin{bmatrix} f_t^{b1} \\ f_t^{b2} \\ 0 \end{bmatrix} = \underline{\mathbf{B}}\mathbf{f}_t^b = \underline{\mathbf{E}}\ \mathbf{R}\mathbf{f}_t^b, \tag{28.47}$$

where $f_t^{b1} = (\cos u)T, f_t^{b2} = (\sin u)T$ and T is the magnitude of the thrust.

The drag force is typically modelled as a function of the velocity with magnitude proportional to the speed of the rocket and direction opposing the velocity. Hence, in the $\underline{\mathbf{E}}$ coordinates, the drag force is

$$\underset{-}{f}_d = \underline{\mathbf{E}} \begin{bmatrix} -v_1 \\ -v_2 \\ -v_3 \end{bmatrix} \left(\frac{1}{2}\sigma\right) \left(v_1^2 + v_2^2 + v_3^2\right)^{\frac{1}{2}},$$
$$= \underline{\mathbf{E}}\mathbf{f}_d^e, \tag{28.48}$$

where v_1^2, v_2^2, v_3^2, denote the velocity and direction e_1, e_2, e_3.

28.3.2 Model for Translational Dynamics

The translational dynamics are conveniently described in the fixed frame. Let m denote the mass of the rocket.[1] Also, the location of the rocket's centre of gravity in the fixed frame is described by

$$\underset{-}{r} = \underline{\mathbf{E}}\mathbf{r}^e. \tag{28.49}$$

The equation describing translational motion is then

$$m\frac{d^2\underset{-}{r}}{dt^2} = \underset{-}{f}, \tag{28.50}$$

where $\underset{-}{r}$ is the vector from the origin of the fixed frame to the centre of gravity of the body, $\underset{-}{f}$ is the sum of forces acting on the vehicle. The forces are described in equations (28.46), (28.47). Hence (28.50) is equivalent to:

$$m\frac{d^2\underset{-}{r}}{dt^2} = m\underline{\mathbf{E}}\frac{d^2}{dt^2}\mathbf{r}^e = \underline{\mathbf{E}}\mathbf{f}_g^e + \mathbf{E}\ \mathbf{R}\mathbf{f}_t^b + \mathbf{E}\ \mathbf{f}_d^e. \tag{28.51}$$

28.4 Rotational Dynamics

Rotation about the centre of gravity will be considered. The core modelling equation is Newton's Law (see Eq. (15.18) of Chapter 15).

$$J\frac{d\omega}{dt} = \text{torque}, \tag{28.52}$$

where J is the inertia matrix.

28.4.1 Torque

Torque is produced by $\underset{-t}{f}$ (the thrust acting on the gimbal), $\underset{-d}{f}$ (the force due to drag) and $\underset{-g}{f}$ (the force due to gravity). Note that $\underset{-g}{f}$ acts through the centre of gravity. Thus, the torque resulting from f_g can be ignored since rotation about the centre of gravity is being considered. The torques resulting from $\underset{-t}{f}$ and $\underset{-d}{f}$ can be found as follows:

Let $\underset{-gp}{\mu}$ be the vector from the centre of gravity to the centre of pressure. Let $\underset{-gt}{\mu}$ be the vector from the centre of gravity to the centre of the pivot of rotation of the engine gimbal.

1 The mass will be taken as constant but, in practice, the mass will change as fuel is consumed.

These two vectors can be expressed in the body reference frame as

$$\underline{\mu}_{gp} = \underline{\mathbf{B}} \begin{bmatrix} \gamma_{gp}^{b1} \\ \gamma_{gp}^{b2} \\ \gamma_{gp}^{b3} \end{bmatrix} = \underline{\mathbf{B}} \begin{bmatrix} d_{gp} \\ 0 \\ 0 \end{bmatrix}, \tag{28.53}$$

$$\underline{\mu}_{gt} = \underline{\mathbf{B}} \begin{bmatrix} \gamma_{gt}^{b1} \\ \gamma_{gt}^{b2} \\ \gamma_{gt}^{b3} \end{bmatrix} = \underline{\mathbf{B}} \begin{bmatrix} -d_{gt} \\ 0 \\ 0 \end{bmatrix}, \tag{28.54}$$

where d_{gp} is the distance from the centre of gravity to the centre of pressure and d_{gt} is the distance from the centre of gravity to the gimbal point. (If the gimbal point is behind the centre of gravity, then there will be a *–ve* sign as shown in equation (28.54).) The torques *about the centre of gravity* due to $\underline{f}_t$ and $\underline{f}_d$ are, respectively, given by

$$\underline{t}^t = \underline{\mu}_{gt} \times \underline{f}_t, \tag{28.55}$$

$$\underline{t}^d = \underline{\mu}_{gp} \times \underline{f}_d. \tag{28.56}$$

As usual, define

$$\tilde{\mu}_{gp}^b = \begin{bmatrix} 0 & -\gamma_{gp}^{b3} & \gamma_{gp}^{b2} \\ \gamma_{gp}^{b3} & 0 & -\gamma_{gp}^{b1} \\ -\gamma_{gp}^{b2} & \gamma_{gp}^{b1} & 0 \end{bmatrix}, \tag{28.57}$$

and similarly for $\tilde{\mu}_{gt}^b$.

Then, the cross-products in (28.55), (28.56) can alternatively be expressed as

$$\underline{\mu}_{gt} \times \underline{f}_t = \underline{\mathbf{B}}\tilde{\mu}_{gt}^b \mathbf{f}_t^b, \tag{28.58}$$

$$\underline{\mu}_{gp} \times \underline{f}_d = \underline{\mathbf{B}}\tilde{\mu}_{gp}^b \mathbf{f}_d^b, \tag{28.59}$$

where $\underline{f}_t = \underline{\mathbf{B}}\mathbf{f}_t^b, \underline{f}_d = \underline{\mathbf{B}}\mathbf{f}_d^b$. The net torque is expressed as $\underline{t}^{nt} = \underline{\mathbf{B}}\mathbf{t}^{nt}$.

Then using (28.58), (28.59)

$$\mathbf{t}^{nt} = \tilde{\mu}_t^b \mathbf{f}_t^b + \tilde{\mu}_{gp}^b \mathbf{f}_d^b. \tag{28.60}$$

28.4.2 Model for Rotational Dynamics

The model for rotational dynamics is simply obtained by substituting (28.41), (28.60) into (28.52).

28.5 Stable or Unstable Rocket

The full model for motion in 3D has been described in Sections 28.1–28.4. An interesting question relates to whether a rocket is stable or unstable in open-loop. This depends on the relative location of the centre of gravity and the centre of pressure. If the centre of gravity is above the centre of pressure, then the rocket is said to be stable. However, when the centre of pressure is above the centre of gravity, then the rocket is said to be unstable. See also Section 15.5 in Chapter 5.

28.6 Revision Questions

1. An inverted pendulum has some characteristics that resemble those of a rocket in flight. The model for an inverted pendulum was presented in Section 3.4.1 of Chapter 3. The linearised model was presented in Eqs. (3.53), (3.54), (3.55).
   What is the transfer function from u to $\tilde{x}_c$? (Ignore the inertia of the rod and friction.)
2. What are poles and zeros?
3. Do you feel that you could design a robust control system using $\tilde{x}$ as the measured output and u as the input? Justify your answer.
4. Would it help your design if $\tilde{x}, \dot{\tilde{x}}, \tilde{\theta}$ and $\dot{\tilde{\theta}}$ were all measured?
5. Would it help the design if one used an observer to estimate $\dot{\tilde{x}}$, $\tilde{\theta}$ and $\dot{\tilde{\theta}}$ when only $\tilde{x}$ is measured?

Further Reading

This chapter has given a very brief introduction to the dynamics of a rocket. For further information on rocket control and rocket aerodynamics the reader can consult the following:

1 In 1957 soviet rocket scientists surprised the world with the launch of a rocket, carrying the world's first artificial satellite Sputnik 1 *https://en.wikipedia.org/wiki/sputnik_1*.

2 Ambitious targets were announced to equal and surpass the Russian achievements, and the Space Race began (see Tom Wolf's book the Right Stuff (Bantam Press, 1980) for a readable account of this period).

3 A.W. Babister, *Aircraft dynamic stability and response*, Pergamon Press, 1980.

4 M.J. Abzug and E.G. Larrabee, *Airplane stability and control*, Cambridge Aerospace Series, 2008.

5 Euler angles -see L. Euler, "Novi commentari academiae scientiarum", *Petropolotanae*, 20:189–207, 1776.

6 European Space Agency, https://www.esa.int/.

7 Modern rocketry has its roots in the German V1 and V2 rocket program during the second world war, see http://www.v2rocket.com/.

8 NASA, https://www.nasa.gov/missionpages/station/main/index.html.

9 R.F. Stengel, *Flight dynamics*, Princeton University Press, 2004.

10 A. Tewari, *Advanced control of aircraft, spacecraft and rockets*, John Wiley & Sons, 2011.

11 The following movie is also worth watching, October Sky, Universal Pictures, 1999.

12 Wikipedia, https://en.wikipedia.org/wiki/Rocket.

13 R. Langton, *Stability and control of aircraft systems – introduction to classical feedback control*, John Wiley & Sons, 2006.

29

Modelling of a Steam-Generating Boiler

29.1 Introduction

Steam boilers play a central role in modern society. Their history can be traced back thousands of years. During the industrial revolution, they played a pivotal role in providing energy to machines. Of course, transportation, especially steam trains, as a very visible and important part of those developments. Also, important aspects of classical control can be traced back to these early steam engines. For example, Watt's Flyball Governor (used to regulate the speed of steam engines) was a pivotal development in this era.

Chapters 29 and 30 grew out of a three-year project[1] conducted in collaboration with Wilmar Sugar at their Proserpine Sugar Mill in Northern Australia. Sugar Mills burn Bagasse (the sugar cane residual after the sugar has been extracted) to produce steam in a Boiler. The steam, so generated, is used throughout the factory and as a source of co-generated electrical energy.

Unlike boilers used in conventional gas- or coal-fired power stations, the boilers in sugar mills are subject to very large and rapid load changes, e.g. when cane crushers are started or stopped. Also, the fuel (Bagasse) has variable calorific value due to different types of cane or residual moisture content. Thus, boilers in sugar mills present unique challenges which are usually not encountered in power plant installations.

Before reading on, it is suggested that the reader review basic thermodynamic principles including the concepts of internal energy and enthalpy. If the reader has not met these ideas previously, then it is possible to skip this chapter and proceed to Chapter 30 which addresses the associated control problems, since its contents will not be used in the sequel. On the other hand, if you are familiar with internal energy and enthalpy, then the modelling exercise outlined below will provide a beautiful example of how a relatively complex process can be modelled using basic physical principles.

The core idea used when building a complex model is actually relatively simple, i.e. one first constructs a list of relevant variables and then one makes sure that the number of independent equations (including algebraic and differential equations) is equal to the number of variables. As the modelling proceeds, it is usual that extra variables will be introduced. Of course, these require extra equations and so on.

1 The authors acknowledge Diego Carrasco who carried out this project jointly with the second author of this book. Diego's work underlies much of the development in this chapter.

Practical Control System Design: Real World Designs implemented on Emulated Industrial Systems,
First Edition. Adrian Medioli and Graham Goodwin.

Companion website: www.wiley.com/go/medioli/practicalcontrolsystemdesign

29.2 Physical Principles

29.2.1 Internal Energy and Enthalpy

In the sequel, the concepts of internal energy and enthalpy will be used.

Internal energy (which is denoted by U) is the energy associated with molecular activity. On the other hand, enthalpy (which is denoted by H) is a measure of the thermodynamic potential of a system. It accounts for both the internal energy as well as the work done on a system. These variables are related as follows:

$$H = U + PV, \tag{29.1}$$

where H, U, P, V denote the enthalpy, internal energy, pressure and volume, respectively.

29.2.2 Ideal Gases

An 'Ideal Gas' is one in which simple relationships exist between pressure, temperature and internal energy.

An ideal gas has the property that the internal energy U is only a function of temperature T, which is denoted as $U(T)$. Another well-known relationship for an ideal gas is Boyles's Law which states that

$$PV = RT, \tag{29.2}$$

where R is a constant.

Hence, the enthalpy of an ideal gas can be expressed as

$$\begin{aligned} H &= U + PV \\ &= U(T) + RT. \end{aligned}$$

In other words, the enthalpy of an ideal gas is only a function of temperature.

It can also be shown that, for an ideal gas, the ratio of enthalpy to internal energy is constant (it is actually 1.4). Most real-world gases are not 'Ideal' but can be approximated as 'Ideal' under certain conditions.

29.2.3 Steam

Steam is certainly not an ideal gas. For example, one needs two quantities, in general, to define the 'state' of steam. For example, one could use temperature and pressure. In that case, other variables, e.g. enthalpy, can be expressed as a function of the given two variables.

Since steam is used so frequently in industrial processes, the properties of steam have been well documented in the literature. The data appears in the form of 'Steam Tables'. These tables relate two independent variables to one dependent variable.

One exception to the claim that two independent variables are needed to describe the state of steam is when the steam is saturated (i.e. in contact with its liquid phase). In this case, the boiling temperature changes with pressure alone and hence only one independent variable is needed to describe saturated steam.

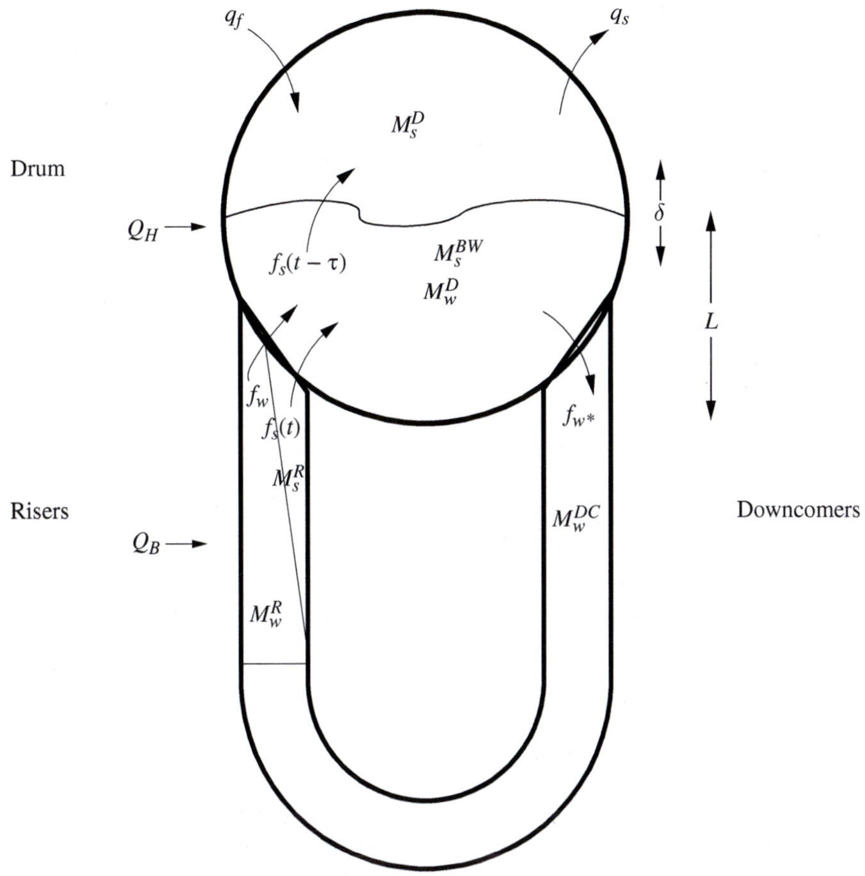

Figure 29.1 Cross-section of boiler.

29.3 Physical Principles Used in Boiler Modelling

In the sequel, the following basic physical principles will be utilised:

- Conservation of mass
- Conservation of energy
- Conservation of momentum
- Constant volume of a closed system.

A schematic diagram of a boiler is shown in Figure 29.1. For convenience some of the key variables of relevance to the model are summarised in Table 29.1.

All key assumptions will be explicitly stated as the modelling proceeds.

29.4 Mass Balances

The development of the model begins by stating the mass balance relationships.

The mass of steam and water in the drum and risers satisfies conservation equations. In particular, the time rate of change of mass contained in an open system is equal to the difference between mass inflow and outflow. This leads to:

Table 29.1 List of variables.

Symbol	Description	See Eq.
α^i	Steam quality in riser section i	(29.48)
δ	Drum water level deviations	(29.38)
f_ℓ	Mass flow of water converted into steam	(29.29)
f_s	Steam mass flow from risers into drum	(29.50) and (29.43d)
f_w	Water mass flow from risers into drum	(29.52) and (29.43c)
f_{w^*}	Water mass flow from drum into risers	
h_s	Enthalpy of steam	Steam tables
h_w	Enthalpy of water	Steam tables
h_s	Enthalpy of feedwater	Assumed known
M_s^D	Mass of steam in the drum (above water line)	(29.3)
M_w^D	Mass of water in the drum	(29.4)
M_s^R	Mass of steam in the risers	(29.5)
M_w^R	Mass of water in the risers	(29.6)
M_s^{BW}	Mass of steam below the water line	(29.35)
P	Drum pressure	(29.34)
Q^T	Total heat flow	Control variable
Q^H	Heat flow needed to raise water temperature	(29.25)
Q^B	Heat flow used to turn water into steam	(29.31)
q_f	Mass flow of feedwater	Control variable
q_s	Mass flow of steam exiting the drum	Assumed known
ρ_s	Steam density	Steam tables
ρ_w	Water density	Steam tables
V^D	Total volume of the drum	Assumed known
V_s^D	Volume of steam in the drum	(29.7)
V_w^D	Volume of water in the drum	(29.7)
V_s^{BW}	Volume of steam below the water line	(29.37)
V^R	Total volume of the risers	Assumed known
V_s^R	Volume of steam in the risers	(29.8)
V_w^R	Volume of water in the risers	(29.8)

Mass of steam in drum:

$$\dot{M}_s^D = f_s - q_s. \tag{29.3}$$

Mass of water in drum:

$$\dot{M}_w^D = q_f + f_w - f_w^*. \tag{29.4}$$

Mass of steam in risers:

$$\dot{M}_s^R = f_\ell - f_s. \tag{29.5}$$

Mass of water in risers:

$$\dot{M}_w^R = f_w^* - f_w - f_\ell, \tag{29.6}$$

where M_s^D, M_w^D, M_s^R, M_w^R denote the mass of steam in the drum, the mass of water in the drum, the mass of steam in the risers, and the mass of water in the risers, respectively, and where f_s, q_s, q_f, f_w, f_w^*, f_ℓ denote the mass flow of steam from the risers into the drum, the mass flow of steam out of the drum, the mass flow of feedwater into the drum, the mass flow of water from the risers into the drum, the mass flow of water from the drum into the downcomers and the mass flow of water converted into steam in the risers, respectively.

Remark 29.1 Note that M_s^D is defined as the total mass of steam in the drum, i.e. it includes both the mass of steam above and below the water level.

Remark 29.2 Equations (29.3), (29.4), (29.5), (29.6) give four equations for M_s^D, M_w^D, M_s^R, M_w^R. Note that the right hand sides of these equations depend on f_s, q_s, q_f, f_w, f_w^*, f_ℓ.

To proceed further equations are introduced (and further variables).

Remark 29.3 The above equations consider the risers as one section. That is sufficient for the moment. In Section 29.9, a multiple-section model for the risers will be described.

29.5 Constant Volume of Drum, Risers and Downcomers

Three different sections of the boiler, i.e. drum, risers and downcomers, will be considered. Each of these volumes is fixed. This leads to the following conservative equations:

$$V^D = V_w^D + V_s^D, \tag{29.7}$$

$$V^R = V_w^R + V_s^R, \tag{29.8}$$

$$V^{DC} = V_w^{DC}, \tag{29.9}$$

where V^D, V^R, V^{DC} denote the volume of the drum, the volume of the risers and the volume of the downcomers, respectively. Here, V_w^D, V_s^D, V_w^R, V_s^R, V_w^{DC} denote the volume of water in the drum, the volume of steam in the drum (both above and below the water line), the volume of water in the risers, the volume of steam in the risers and the volume of water in the downcomers, respectively. Note that it is assumed that there is no steam in the downcomers.

Remark 29.4 Note that the total volume of the boiler, V^T, is given by $V^T = V^D + V^R + V^{DC}$.

Assumption 29.1 It is assumed that boiling of water only occurs in the risers.

Assumption 29.2 It is assumed that all steam in the boiler is saturated. Hence, the variables P, ρ, h, T, u for steam are all algebraically related to one variable. (see Section 29.2.3).

29.5.1 Consequence of Constant Volume of the Drum

Since the volume of the drum is constant, using (29.7) and taking its derivative, leads to:

$$
\begin{aligned}
0 =& \frac{d}{dt}\left\{V_w^D + V_s^D\right\}\\
=& \frac{d}{dt}\left\{\frac{M_w^D}{\rho_w} + \frac{M_s^D}{\rho_s}\right\}.
\end{aligned}
$$

Here, the fact that $V = M/\rho$, where $M,\ \rho$ denote mass and density, has been used. Expanding the derivative, leads to:

$$
0 = \frac{\dot{M}_w^D}{\rho_w} - \frac{M_w^D}{\rho_w^2}\cdot\dot{\rho}_w + \frac{\dot{M}_s^D}{\rho_s} - \frac{M_s^D}{\rho_s^2}\cdot\dot{\rho}_s. \tag{29.10}
$$

Since, the water is assumed to be at its saturation point, density will be a function of pressure only. It follows that:

$$
\dot{\rho} = \frac{d\rho}{dt} = \frac{\partial\rho}{\partial P}\cdot\frac{dP}{dt} = \frac{\partial\rho}{\partial P}\cdot\dot{P}, \tag{29.11}
$$

where P denotes the pressure of the boiler. Using (29.11), and substituting the mass balance results (Eqs. (29.4) and (29.3)) into (29.10), leads to:

$$
0 = \frac{q_f + f_w - f_w^*}{\rho_w} + \frac{f_s - q_s}{\rho_s} - \left(\frac{M_w^D}{\rho_w^2}\frac{\partial\rho_w}{\partial P} + \frac{M_s^D}{\rho_s^2}\frac{\partial\rho_s}{\partial P}\right)\dot{P}. \tag{29.12}
$$

To simplify the algebra, the following variable is introduced:

$$
C_1 \triangleq \frac{M_w^D}{\rho_w^2}\frac{\partial\rho_w}{\partial P} + \frac{M_s^D}{\rho_s^2}\frac{\partial\rho_s}{\partial P}. \tag{29.13}
$$

Then Eq. (29.12) becomes:

$$
0 = \frac{q_f + f_w - f_w^*}{\rho_w} + \frac{f_s - q_s}{\rho_s} - C_1\cdot\dot{P}.
$$

Rearranging the above equation leads to:

$$
\dot{P} = C_1^{-1}\left(\frac{q_f - f_w^*}{\rho_w} - \frac{q_s}{\rho_s} + \frac{f_w}{\rho_w} + \frac{f_s}{\rho_s}\right). \tag{29.14}
$$

This equation will be used later.

29.5.2 Consequence of Constant Volume of the Risers

In a similar fashion, since the volume of the risers is constant, then using (29.8) and taking its derivative leads to:

$$
\begin{aligned}
0 =& \frac{d}{dt}\left\{V_w^R + V_s^R\right\},\\
=& \frac{d}{dt}\left\{\frac{M_w^R}{\rho_w} + \frac{M_s^R}{\rho_s}\right\},\\
=& \frac{\dot{M}_w^R}{\rho_w} - \frac{M_w^R}{\rho_w^2}\cdot\dot{\rho}_w + \frac{\dot{M}_s^R}{\rho_s} - \frac{M_s^R}{\rho_s^2}\cdot\dot{\rho}_s,\\
=& \frac{f_w^* - f_w - f_\ell}{\rho_w} + \frac{f_\ell - f_s}{\rho_s} - \left(\frac{M_w^R}{\rho_w^2}\frac{\partial\rho_w}{\partial P} + \frac{M_s^R}{\rho_s^2}\frac{\partial\rho_s}{\partial P}\right)\cdot\dot{P}.
\end{aligned}
$$

To simplify the algebra, the following variable is introduced:

$$C_2 \triangleq \frac{M_w^R}{\rho_w^2}\frac{\partial \rho_w}{\partial P} + \frac{M_s^R}{\rho_s^2}\frac{\partial \rho_s}{\partial P}, \tag{29.15}$$

then

$$0 = \frac{f_w^* - f_w - f_\ell}{\rho_w} + \frac{f_\ell - f_s}{\rho_s} - C_2 \cdot \dot{P}.$$

Rearranging the above equation leads to:

$$\frac{f_w}{\rho_w} + \frac{f_s}{\rho_s} = \frac{f_w^* - f_\ell}{\rho_w} + \frac{f_\ell}{\rho_s} - C_2 \cdot \dot{P}. \tag{29.16}$$

This equation will also be used in Section 29.7 to obtain a model for boiler pressure.

Remark 29.5 Equations (29.14) and (29.16) give two simultaneous equations for $\dot{P}$ in terms of the variables introduced so far.

29.6 Energy Balances

Next the fact that the time rate of change of energy contained in an open system is equal to the difference between energy inflow and energy outflow of the system will be used. The energy contained in the drum is thus given by the energy contained in the masses of water and steam in the drum, plus the energy contained in the metal walls of the drum. The energy inflows and outflows are given by the energy carried by the masses flowing into and out of the drum, plus a component of heat from the furnace. The energy balance for the risers follows the above principle. The following energy balance equations result:

Drum:

$$\frac{d}{dt}\left\{M_s^D u_s + M_w^D u_w + M_m^D C_p T_m\right\} = \left(f_s - q_s\right) h_s + \left(f_w - f_w^*\right) h_w + q_f h_w^f + Q^H. \tag{29.17}$$

Risers:

$$\frac{d}{dt}\left\{M_s^R u_s + M_w^R u_w + M_m^R C_p T_m\right\} = -f_s h_s + \left(f_w^* - f_w\right) h_w + Q^B, \tag{29.18}$$

where u_s , u_w, M_m^D, M_m^R, C_p, T_m denote the internal energy of steam, internal energy of water, mass of metal in the drum, mass of metal in the risers, heat capacity of metal temperature of metal. Also, h_s, h_w, h_w^f, Q^H, Q^B denote the enthalpy of steam, enthalpy of water, enthalpy of feedwater, heat flow used for heating the water to boiling temperature and heat flow used for boiling water in the risers, respectively.

Remark 29.6 Note that the total heat flow generated by the furnace, Q^T, is given by

$$Q^T = Q^H + Q^B. \tag{29.19}$$

Assumption 29.3 It is assumed that the flow rate and properties of the feed water are known. Hence q_f and h_w^f are given quantities and do not need extra equations.

29.6.1 Consequence of Drum Energy Balance

From Section 29.2.1, it follows that $h = u + P/\rho$. Hence $u = h - P/\rho$. Substituting into the left-hand side of (29.17), leads to:

$$\frac{d}{dt}\left\{M_s^D h_s + M_w^D h_w - \left(\frac{M_s^D}{\rho_s} + \frac{M_w^D}{\rho_w}\right)P + M_m^D C_p T_m\right\} = \left(f_s - q_s\right)h_s + \left(f_w - f_w^*\right)h_w + q_f h_w^f + Q^H.$$

Noting that

$$\frac{M_s^D}{\rho_s} + \frac{M_w^D}{\rho_w} = V_s^D + V_w^D = V^D,$$

then

$$\frac{d}{dt}\left\{M_s^D h_s + M_w^D h_w - V^D P + M_m^D C_p T_m\right\} = \left(f_s - q_s\right)h_s + \left(f_w - f_w^*\right)h_w + q_f h_w^f + Q^H.$$

Expanding the LHS of the above equation leads to:

$$\dot{M}_s^D h_s + M_s^D \dot{h}_s + \dot{M}_w^D h_w + M_w^D \dot{h}_w - V^D \dot{P} + M_m^D C_p \dot{T}_m = \left(f_s - q_s\right)h_s + \left(f_w - f_w^*\right)h_w + q_f h_w^f + Q^H. \tag{29.20}$$

Since water is considered to be at its saturation point, then its enthalpy is a function of pressure alone. Hence:

$$\dot{h} = \frac{dh}{dt} = \frac{\partial h}{\partial P}\cdot\frac{dP}{dt} = \frac{\partial h}{\partial P}\cdot\dot{P}. \tag{29.21}$$

Assumption 29.4 It is assumed that T_m is the saturation temperature of the steam, T_s.

It follows that:

$$\dot{T}_m = \frac{dT_s}{dt} = \frac{\partial T_s}{\partial P}\cdot\frac{dP}{dt} = \frac{\partial T_s}{\partial P}\cdot\dot{P}. \tag{29.22}$$

Substituting (29.21) and (29.22), as well as the mass balance Eqs. (29.3) and (29.4), into (29.20), leads to:

$$\left(f_s - q_s\right)h_s + \left(q_f + f_w - f_w^*\right)h_w + \left(M_s^D\frac{\partial h_s}{\partial P} + M_w^D\frac{\partial h_w}{\partial P} - V^D + M_m^D C_p\frac{\partial T_s}{\partial P}\right)\dot{P}$$
$$= \left(f_s - q_s\right)h_s + \left(f_w - f_w^*\right)h_w + q_f h_w^f + Q^H.$$

Cancelling the common terms on both sides leads to:

$$q_f h_w + \left(M_s^D\frac{\partial h_s}{\partial P} + M_w^D\frac{\partial h_w}{\partial P} - V^D + M_m^D C_p\frac{\partial T_s}{\partial P}\right)\dot{P} = q_f h_w^f + Q^H.$$

To simplify the algebra, the following variable is introduced:

$$K_1 \triangleq M_s^D\frac{\partial h_s}{\partial P} + M_w^D\frac{\partial h_w}{\partial P} - V^D + M_m^D C_p\frac{\partial T_s}{\partial P}, \tag{29.23}$$

which leads to:

$$K_1\dot{P} = q_f\left(h_w^f - h_w\right) + Q^H, \tag{29.24}$$

Rearranging the above equation leads to:

$$Q^H = K_1\dot{P} - q_f\left(h_w^f - h_w\right). \tag{29.25}$$

Remark 29.7 Equation (29.25) relates Q^H to the variables introduced so far, including $\dot{P}$.

29.6.2 Consequences of Energy Balance in the Risers

In a similar fashion, substituting $u = h - P/\rho$ into the left-hand side of (29.18), leads to:

$$\frac{d}{dt}\left\{M_s^R h_s + M_w^R h_w - \left(\frac{M_s^R}{\rho_s} + \frac{M_w^R}{\rho_w}\right)P + M_m^R C_p T_m\right\} = -f_s h_s + (f_w^* - f_w)\,h_w + Q^B.$$

Again, noting that

$$\frac{M_s^R}{\rho_s} + \frac{M_w^R}{\rho_w} = V_s^R + V_w^R = V^R,$$

yields

$$\frac{d}{dt}\{M_s^R h_s + M_w^R h_w - V^R P + M_m^R C_p T_m\} = -f_s h_s + (f_w^* - f_w)\,h_w + Q^B.$$

Expanding the LHS of the above equation leads to:

$$\dot{M}_s^R h_s + M_s^R \dot{h}_s + \dot{M}_w^R h_w + M_w^R \dot{h}_w - V^R \dot{P} + M_m^R C_p \dot{T}_m = -f_s h_s + (f_w^* - f_w)\,h_w + Q^B. \qquad (29.26)$$

Since it is assumed that the water is at its saturation point, then equations (29.21) and (29.22) follow. In addition, substituting the mass balance Eqs., (29.5) and (29.6), into (29.26),leads to:

$$(f_\ell - f_s)\,h_s + (f_w^* - f_w - f_\ell)\,h_w + \left(M_s^R\frac{\partial h_s}{\partial P} + M_w^R\frac{\partial h_w}{\partial P} - V^R + M_m^R C_p\frac{\partial T_s}{\partial P}\right)\dot{P}$$
$$= -f_s h_s + (f_w^* - f_w)\,h_w + Q^B.$$

Cancelling the common terms on both sides leads to:

$$f_\ell h_s - f_\ell h_w + \left(M_s^R\frac{\partial h_s}{\partial P} + M_w^R\frac{\partial h_w}{\partial P} - V^R + M_m^R C_p\frac{\partial T_s}{\partial P}\right)\dot{P} = Q^B.$$

To simplify the algebra, the following variable is introduced:

$$K_2 \triangleq M_s^R\frac{\partial h_s}{\partial P} + M_w^R\frac{\partial h_w}{\partial P} - V^R + M_m^R C_p\frac{\partial T_s}{\partial P}, \qquad (29.27)$$

which leads to

$$(h_s - h_w)\,f_\ell + K_2\dot{P} = Q^B. \qquad (29.28)$$

Rearranging the above equation leads to:

$$f_\ell = (h_s - h_w)^{-1}\,(-K_2\dot{P} + Q^B)\,. \qquad (29.29)$$

Remark 29.8 Equation (29.29) relates f_ℓ to the variables introduced so far.

29.7 A Model for Boiler Pressure

So far, the model for the boiler dynamics comprises: (i) the four conservation of mass equations, (29.3)–(29.6), (ii) the two equations derived from the constant volume of the drum and risers, (29.14) and (29.16), and (iii) the two equations derived from energy balance in the drum and in the risers, (29.25) and (29.29). Next Eqs. (29.25) and (29.29) will be used to derive an expression for boiler pressure, P. Substituting (29.16) into (29.14), leads to:

$$\dot{P} = C_1^{-1}\left(\frac{q_f - f_w^*}{\rho_w} - \frac{q_s}{\rho_s} + \frac{f_w^* - f_\ell}{\rho_w} + \frac{f_\ell}{\rho_s} - C_2\cdot\dot{P}\right).$$

Solving for $\dot{P}$:

$$(C_1 + C_2)\,\dot{P} = \left(\frac{q_f}{\rho_w} - \frac{q_s}{\rho_s} - \left(\frac{1}{\rho_w} - \frac{1}{\rho_s}\right) f_\ell\right). \tag{29.30}$$

On the other hand,

$$Q^B = Q^T - Q^H, \tag{29.31}$$

where Q^T is the total heat flow input. Next substituting (29.25) into (29.29) leads to:

$$\begin{aligned} f_\ell =& (h_s - h_w)^{-1}\left(-K_2\dot{P} + Q^T - \left(K_1\dot{P} - q_f\left(h_w^f - h_w\right)\right)\right) \\ =& (h_s - h_w)^{-1}\left(-\left(K_1 + K_2\right)\dot{P} + Q^T + q_f\left(h_w^f - h_w\right)\right). \end{aligned} \tag{29.32}$$

Substituting (29.32) into (29.30) leads to:

$$(C_1 + C_2)\,\dot{P} = \left(\frac{q_f}{\rho_w} - \frac{q_s}{\rho_s} - \left(\frac{1}{\rho_w} - \frac{1}{\rho_s}\right)(h_s - h_w)^{-1}\left(-\left(K_1 + K_2\right)\dot{P} + Q^T + q_f\left(h_w^f - h_w\right)\right)\right).$$

Again to simplify the algebra, the following variable is introduced.

$$C_3 \triangleq \left(\frac{1}{\rho_w} - \frac{1}{\rho_s}\right)(h_s - h_w)^{-1}. \tag{29.33}$$

Then, rearranging the above equation leads to:

$$\left(C_1 + C_2 - C_3\left(K_1 + K_2\right)\right)\dot{P} = \left(\frac{q_f}{\rho_w} - \frac{q_s}{\rho_s} - C_3\left(Q^T + q_f\left(h_w^f - h_w\right)\right)\right).$$

Finally, solving for $\dot{P}$ gives:

$$\dot{P} = \left(C_1 + C_2 - C_3\left(K_1 + K_2\right)\right)^{-1}\left(\frac{q_f}{\rho_w} - \frac{q_s}{\rho_s} - C_3\left(Q^T + q_f\left(h_w^f - h_w\right)\right)\right). \tag{29.34}$$

Remark 29.9 Equation (29.34) is a differential equation for P, Eqs. (29.29) and (29.25) are algebraic equations for f_ℓ and Q^H, respectively, and Eqs. (29.3)–(29.6) are differential equations for the steam and water masses in the boiler. Note, however, that no new equations have been introduced since all that has been done is to manipulate the equations obtained so far.

Remark 29.10 Note that, to evaluate (29.34), it is necessary to know f_e. However, it is not necessary to know the values of f_s, f_w and f_w^*. Indeed, C_3 is a function of only the current pressure, $(K_1 + K_2)$ and $(C_1 + C_2)$ depend on pressure and the *total* mass of water and steam in the boiler. Next recall that $\dot{M}_w^D + \dot{M}_w^R = q_f - f_\ell$ and $\dot{M}_s^D + \dot{M}_s^R = f_\ell - q_s$. Thus, f_s, f_w and f_w^* are not needed to obtain pressure. This is reasonable since pressure is a global property of the boiler, and therefore it stands to reason that it should not depend on internal properties of the boiler. However, unlike pressure, drum water level is not a global property of the boiler. Hence, as shown below, in order to obtain a model for drum water level, it is necessary to know internal quantities such as f_s, f_w and f_w^*.

29.8 A Model for Drum Water Level

A model for drum water level is next developed.

Consider a section of the boiler drum as shown in Figure 29.1, where L is the steady state (nominal) height of the water, and δ denotes variations around L. Note that, to determine the height of the water one must know, not only the mass of water in the drum, M_w^D, but also the mass of steam below the water line, M_s^{BW}. An expression for M_s^{BW} can be obtained by applying the conservation of mass principle, i.e.:

$$\dot{M}_s^{BW}(t) = f_s(t) - f_s(t-a), \tag{29.35}$$

where $f_s(t)$, $f_s(t-a)$ denote the mass flow of steam out of the risers and a delayed version of $f_s(t)$, respectively. The time delay a can be estimated by noting that it is the time taken for a given mass to cover a certain distance, i.e.:

$$a = \frac{\text{distance}}{\text{speed}} = \frac{L+\delta}{f_s(t)/\left(\rho_s A^R\right)}, \tag{29.36}$$

where A^R denotes the total area of the risers.

In steady-state, it is assumed that the drum is half full of water, and that the corresponding height of water is L. Therefore,

$$V_w^D + V_s^{BW} = \overline{V}_w^D + \overline{V}_s^{BW} + \tilde{V}_w^D + \tilde{V}_s^{BW} = \frac{V^D}{2} + A^D\delta, \tag{29.37}$$

where A^D denotes the area at the centre line of the drum. Using the fact that $V = M/\rho$, then Eq. (29.37) can be rewritten as:

$$\frac{M_w^D}{\rho_w} + \frac{M_s^{BW}}{\rho_s} = \frac{V^D}{2} + A^D\delta.$$

Finally, the following expression for the drum water level deviation, δ is obtained:

$$\delta = \left(A^D\right)^{-1}\left(\frac{M_w^D}{\rho_w} + \frac{M_s^{BW}}{\rho_s} - \frac{V^D}{2}\right). \tag{29.38}$$

Remark 29.11 Equations (29.38), (29.35) and (29.4) show that to determine the incremental drum water level, δ, it is necessary to be able to calculate f_s, f_w, f_w^*.

Remark 29.12 Note that (29.14) and (29.16) are linearly dependent when viewed as a function of f_s and f_w. Therefore, with these two equations alone it is not possible to obtain independent expressions for f_s and f_w. This problem will be resolved in Section 29.9.

Remark 29.13 An expression for f_w^* will be obtained in Section 29.10 through a conservation of momentum analysis in the downcomer-riser system.

29.9 Spatial Discretisation and Homogeneous Mixing in the Risers

In this section, the model of the risers is embellished to account for the spatial distribution of the boiling process along the length of the risers. An additional assumption is made which enables separate expressions for f_s and f_w to be obtained.

29.9.1 Spatial Discretisation

The volume of the risers is uniformly divided into n sections each of fixed volume. For each section, there are three core equations, resulting from mass balance, constant volume and energy balance. Hence for each section:

$$\dot{M}_s^{R_i} = f_\ell^i + f_s^{i-1} - f_s^i \quad i = 1, \dots, n, \tag{29.39}$$

$$\dot{M}_w^{R_i} = f_w^{i-1} - f_w^i - f_\ell^i \quad i = 1, \dots, n, \tag{29.40}$$

$$V^{R_i} = V_w^{R_i} + V_s^{R_i} \quad i = 1, \dots, n, \tag{29.41}$$

$$\frac{d}{dt}\left\{M_s^{R_i} u_s + M_w^{R_i} u_w + M_m^{R_i} C_p T_m\right\} = -f_s^i h_s + \left(f_w^{i-1} - f_w^i\right) h_w + Q^{B_i}; \quad i = 1, \dots, n, \tag{29.42}$$

where the superscript R_i denotes the ith section of the risers, f_ℓ^i denotes the flow of water converted into steam in section i, and where f_s^i, f_w^i denote the mass flow of steam and water leaving section i, respectively. Similarly, f_s^{i-1}, f_w^{i-1} denote the mass flow of steam and water entering section i, respectively. Q^{B_i} denotes the heat flow directly affecting section i, for $i = 1, \dots, n$.

Assumption 29.5 It is assumed that $Q^{B_i} = Q^B/n, \ \forall i$.

Assumption 29.6 It is assumed that pure water enters the rises.

Hence,

$$f_w^0 = f_w^*, \tag{29.43a}$$

$$f_s^0 = 0. \tag{29.43b}$$

Note that f_w and f_s have been used to denote the flow of water and steam from the top of the risers. Hence,

$$f_w^n = f_w, \tag{29.43c}$$

$$f_s^n = f_s, \tag{29.43d}$$

$$\sum_{i=1}^{n} f_\ell^i = f_\ell, \tag{29.43e}$$

$$\sum_{i=1}^{n} M_w^{R_i} = M_w^R, \tag{29.43f}$$

$$\sum_{i=1}^{n} M_s^{R_i} = M_s^R. \tag{29.43g}$$

It also follows that the equations derived from the constant volume of the risers and the energy balance of the risers obtained in Sections 29.5.2 and 29.6.2 can also be rederived. We next define:

$$C_2^i \triangleq \frac{M_w^{R_i}}{\rho_w^2}\frac{\partial \rho_w}{\partial P} + \frac{M_s^{R_i}}{\rho_s^2}\frac{\partial \rho_s}{\partial P}, \tag{29.44}$$

$$K_2^i \triangleq M_s^{R_i}\frac{\partial h_s}{\partial P} + M_w^{R_i}\frac{\partial h_w}{\partial P} - V^{R_i} + M_m^{R_i} C_p \frac{\partial T_s}{\partial P}, \tag{29.45}$$

then, from equations (29.41) and (29.42):

$$\frac{f_w^i}{\rho_w} + \frac{f_s^i}{\rho_s} = \frac{f_w^{i-1} - f_\ell^i}{\rho_w} + \frac{f_\ell^i + f_s^{i-1}}{\rho_s} - C_2^i \cdot \dot{P}, \tag{29.46}$$

$$f_\ell^i = \left(h_s - h_w\right)^{-1}\left(-K_2^i \dot{P} + Q^{B_i}\right). \tag{29.47}$$

Equations (29.46) and (29.47), together with (29.39) and (29.40), give a complete account of the dynamics of the ith section of the risers. Alas, one cannot independently describe f_s^i and f_w^i. Hence, further information is needed so as to independently describe these variables. In Section 29.9.2 an additional assumption is introduced that allows one to distinguish between f_s^i from f_w^i.

29.9.2 Homogeneous Mixing in a Section of the Risers

Consider a section of the risers. Then, over an infinitesimal period of time Δ, the mass of steam and water leaving that section are given by $f_s^i\Delta$ and $f_w^i\Delta$, respectively. Define the steam quality of the section as follow:

$$\alpha^i = \frac{M_s^{R_i}}{M_s^{R_i} + M_w^{R_i}}. \tag{29.48}$$

A key simplifying assumption is now made, namely:

Assumption 29.7 Perfect mixing of steam and water occurs in each section of the risers.

A consequence of this assumption is that the mass of steam and water leaving the section over a period of time Δ has the same ratio α^i. Therefore,

$$\alpha^i = \frac{M_s^{R_i}}{M_s^{R_i} + M_w^{R_i}} = \frac{f_s^i\Delta}{f_s^i\Delta + f_w^i\Delta}. \tag{29.49}$$

Solving for f_s^i, leads to:

$$f_s^i = \frac{M_s^{R_i}}{M_w^{R_i}} f_w^i. \tag{29.50}$$

Equation (29.50) is the missing piece of the puzzle since it provides and algebraic relationship between $f_s^i, f_w^i, M_s^R i$ and $M_w^R i$.

Substituting (29.50) into (29.46), leads to:

$$\frac{f_w^i}{\rho_w} + \frac{1}{\rho_s}\frac{M_s^{R_i}}{M_w^{R_i}} f_w^i = \frac{f_w^{i-1} - f_\ell^i}{\rho_w} + \frac{f_\ell^i + f_s^{i-1}}{\rho_s} - C_2^i \cdot \dot{P}. \tag{29.51}$$

Solving for f_w^i, yields:

$$f_w^i = \left(\frac{1}{\rho_w} + \frac{1}{\rho_s}\frac{M_s^{R_i}}{M_w^{R_i}}\right)^{-1}\left(\frac{f_w^{i-1} - f_\ell^i}{\rho_w} + \frac{f_\ell^i + f_s^{i-1}}{\rho_s} - C_2^i \cdot \dot{P}\right). \tag{29.52}$$

29.10 Water Flow in the Downcomers

Mass flow of water in the downcomers has been denoted, f_w^*. In this section, momentum balance is used to obtain a model for f_w^*.

It is known that the differential pressure at the bottom of a column of liquid satisfies

$$P = \rho g h, \tag{29.53}$$

where h is the height of liquid, ρ is density and g is the gravitational constant.

Hence the differential force acting at the bottom of the column of water in the downcomers, F^{DC} is

$$F^{DC} = \rho_w g h^{DC} A, \tag{29.54}$$

where h^{DC} is the height of the downcomers, and A is the area.

The force at the same level in the risers is more difficult to compute. However, it is known that the differential pressure at the bottom of a column of height h_i and density ρ_i is $\delta P = \rho_i g h_i$.

Recall that the water plus steam mixture in the risers has been modelled as an 'amalgam' in which steam and water are locked together. (Recall the homogeneous mixing assumption.)

Hence, the sum of the differential pressures from top to bottom of the risers gives the pressure of the bottom. Equations have already been derived that describe the mass of water M_w^i and mass of steam, M_s^i in the ith section of the risers. Also, V^R has been used to denote the total volume of the risers. Hence, the volume of each section is

$$V_R^i = \frac{V^R}{n}. \tag{29.55}$$

Then the density of the 'amalgam' mixture of steam plus water in the ith section is that quality, ρ_A^i which satisfies

$$\rho_A^i \frac{V^R}{n} = M_s^i + M_w^i, \tag{29.56}$$

or

$$\rho_A^i = \frac{[M_s^i + M_w^i]n}{V^R}. \tag{29.57}$$

Finally, the differential force applied to the liquid at the bottom of the risers is

$$F^R = \sum_{i=1}^{n} \rho_A^i g \frac{h^{DC}}{n} A. \tag{29.58}$$

Remark 29.14 Note that the total boiler pressure is very large and thus the small increase in pressure can be ignored as one moves down the risers when describing the other boiler dynamics. However, this change in pressure drives the flow of water in the downcomers by convection.

The mass of water is:

$$\rho_w A L_e, \tag{29.59}$$

where L_e the equivalent length of pipe in this region. Using Newton's first law a force is necessary to accelerate the above mass of water. Moreover, the available force is the difference between F^{DC} and F^R minus friction forces.

Assumption 29.8 Frictional forces are assumed to be proportional to the square of velocity.

Hence

$$[\rho_w A L_e] \frac{d}{dt}\left\{\frac{f_w^*}{\rho_w A}\right\} = F^{DC} - F^R - c\left(f_w^*\right)^2, \tag{29.60}$$

where c is a constant.

Finally, the following differential equation for f_w^* is obtained.

$$L_e \frac{d}{dt} f_w^* = F^{DC} - F^R - c\left(f_w^*\right)^2. \tag{29.61}$$

Remark 29.15 Note that, in steady-state,

$$f_w^* = \sqrt{\frac{F^{DC} - F^R}{c}}. \tag{29.62}$$

Equation (29.62) allows one to evaluate the constant c from design data for the steady-state downcomer water velocity. Also, note that the steady-state relationship in (29.62) contains a square root. Such a square root is commonly used in steady-state modelling of flow in pipes.

29.11 Superheaters

Many boilers have superheaters.

Actually, boiler number 3 at the Wilmar Sugar Factory did not have a superheater. Hence, the model for this component is not relevant to the Laboratory described in Chapter 30. A brief description follows.

A superheater is a heat exchanger used to convert saturated steam generated in a boiler into superheated steam by adding heat, thus drying the steam. Superheated steam is used in steam turbines to generate electricity. The main difference between saturated and superheated steam is that when considering saturated steam, it suffices to use one state variable. (Actually, the pressure of the water/steam mixture was used above.) This allowed the density, enthalpy, temperature and other state variables, for both liquid and vapour phases to be evaluated. However, to describe the state of superheated steam it is necessary to consider two independent state variables. For example, the state variables pressure and enthalpy can be used.

The reader is asked to construct a suitable model for a superheater in question 2 in the Revision Questions.

29.12 Steam Receiver

A steam receiver is a metal vessel that is frequently used to collect the steam generated by one or more boilers, and from where steam is drawn when it is needed.

Modelling a steam receiver follows the same principles used to model a superheater, i.e. one needs to account for the use of superheated steam. The differences lie in that (i) the steam receiver can have many different steam flow inputs (in the case of multiple boilers), and (ii) there is no heat input.

Let the steam receiver have volume V^E. Then, mass balance, energy balance and constant volume equations can be developed as follows:

29.12.1 Mass Balance

$$\dot{M}_s^E = \sum_i q_s^i - q_s^E, \tag{29.63}$$

where M_s^E, q_s^i, q_s^E denote the mass of steam in the steam receiver, the steam mass flow out of the boiler and into the steam receiver, and the steam mass flow out of the steam receiver (load). The notation $\sum_i q_s^i$ is used to denote

$$\sum_i q_s^i = (q_s)^1 + (q_s^{SH})^3 + (q_s^{SH})^4, \tag{29.64}$$

where $(q_s)^i$ refers to the steam mass flow from each Boiler.

29.12.2 Energy Balance

$$\frac{d}{dt}\{M_s^E u_s^E\} = \sum_i q_s^i h_s^i - q_s^E h_s^E, \tag{29.65}$$

where h_s^i denotes the enthalpy of q_s^i from each boiler/superheater. By definition $u = h - P/\rho$, therefore:

$$\frac{d}{dt}\left\{M_s^E h_s^E - \frac{M_s^E}{\rho_s^E} P^E\right\} = \sum_i q_s^i h_s^i - q_s^E h_s^E.$$

Noting that $M_s^E/\rho_s^E = V^E$, and expanding the derivative,leads to

$$\dot{M}_s^E h_s^E + M_s^E \dot{h}_s^E - V^E \dot{P}^E = \sum_i q_s^i h_s^i - q_s^E h_s^E.$$

Finally, using Eq. (29.63) and cancelling the common terms, leads to

$$M_s^E \dot{h}_s^E - V^E \dot{P}^E = \sum_i q_s^i \left(h_s^i - h_s^E\right). \tag{29.66}$$

29.12.3 Constant Volume of the Steam Receiver

Since the volume of the steam receiver is constant, then

$$\frac{d}{dt}\{V^E\} = 0. \tag{29.67}$$

By definition, $V = M/\rho_S$. Hence,

$$\frac{d}{dt}\{V^E\} = \frac{d}{dt}\left\{\frac{M_s^E}{\rho_s^E}\right\} = \frac{\dot{M}_s^E}{\rho_s^E} - \frac{M_s^E}{\left(\rho_s^E\right)^2}\dot{\rho}_s^E. \tag{29.68}$$

However, as mentioned earlier, the density of superheated steam is no longer a function of pressure only, but is a function of two independent variables say pressure and enthalpy. Hence,

$$\dot{\rho}_s^E = \frac{\partial \rho_s^E}{\partial h_s^E}\dot{h}_s^E + \frac{\partial \rho_s^E}{\partial P^E}\dot{P}^E. \tag{29.69}$$

Substituting (29.68) and (29.69) into (29.67), and noting that $M/\rho = V$, leads to:

$$\frac{\dot{M}_s^E}{\rho_s^E} - \frac{V^E}{\rho_s^E}\left(\frac{\partial \rho_s^E}{\partial h_s^E}\dot{h}_s^E + \frac{\partial \rho_s^E}{\partial P^E}\dot{P}^E\right) = 0.$$

Using (29.63) and reordering terms, yields:

$$V^E\left(\frac{\partial \rho_s^E}{\partial h_s^E}\dot{h}_s^E + \frac{\partial \rho_s^E}{\partial P^E}\dot{P}^E\right) = \sum_i q_s^i - q_s^E. \tag{29.70}$$

29.12.4 Summary of the Model for the Steam Receiver

Equations (29.63), (29.66) and (29.70) determine a complete model describing the dynamics of a steam receiver. To implement the model it is helpful to decouple equations (29.66) and (29.70). From (29.70):

$$V^E\dot{P}^E = \left(\frac{\partial \rho_s^E}{\partial P^E}\right)^{-1}\left(\sum_i q_s^i - q_s^E - V^E\frac{\partial \rho_s^E}{\partial h_s^E}\dot{h}_s^E\right).$$

Substituting into (29.66), leads to:

$$M_s^E\dot{h}_s^E + \left(\frac{\partial \rho_s^E}{\partial P^E}\right)^{-1}V^E\frac{\partial \rho_s^E}{\partial h_s^E}\dot{h}_s^E = \sum_i q_s^i\left(h_s^i - h_s^E\right) + \left(\frac{\partial \rho_s^E}{\partial P^E}\right)^{-1}\left(\sum_i q_s^i - q_s^E\right).$$

Solving for $\dot{h}_s^E$:

$$\dot{h}_s^E = \left(M_s^E + \left(\frac{\partial \rho_s^E}{\partial P^E}\right)^{-1}V^E\frac{\partial \rho_s^E}{\partial h_s^E}\right)^{-1}\left(\sum_i q_s^i\left(h_s^i - h_s^E\right) + \left(\frac{\partial \rho_s^E}{\partial P^E}\right)^{-1}\left(\sum_i q_s^i - q_s^E\right)\right).$$

Then, from (29.66):

$$\dot{P}^E = \left(V^E\right)^{-1}\left(M_s^E\dot{h}_s^E - \sum_i q_s^i\left(h_s^i - h_s^E\right)\right).$$

In summary, the model for a steam receiver is given by the following equations:

$$\dot{M}_s^E = \sum_i q_s^i - q_s^E, \tag{29.71}$$

$$\dot{h}_s^E = \left(M_s^E + \left(\frac{\partial \rho_s^E}{\partial P^E}\right)^{-1}V^E\frac{\partial \rho_s^E}{\partial h_s^E}\right)^{-1}\left(\sum_i q_s^i\left(h_s^i - h_s^E\right) + \left(\frac{\partial \rho_s^E}{\partial P^E}\right)^{-1}\left(\sum_i q_s^i - q_s^E\right)\right), \tag{29.72}$$

$$\dot{P}^E = \left(V^E\right)^{-1}\left(M_s^E\dot{h}_s^E - \sum_i q_s^i\left(h_s^i - h_s^E\right)\right). \tag{29.73}$$

29.13 Other Model Components

29.13.1 Mass Flow out of Drum

In the model as presented above, it has been assumed that q_s was a controllable (i.e. independent) variable. This is not true in general, since the flow of steam out of a boiler is a function of the throttle valve opening, the piping resistance between the boiler and the steam receiver, and the pressure difference between drum and steam receiver.

Assumption 29.9 The following simple relationships are assumed to apply to steam flow.

$$q_s = K_F u_T\left(P - P^E\right) + K^*, \tag{29.74}$$

where K_F, K^* are constants specific to each boiler, u_T denotes the throttle valve opening and $P - P^E$ is the pressure difference between the boiler and the steam receiver.

29.13.2 Feedwater Mass Flow

It has been assumed in the above development that the mass flow rate of feedwater, q_f, is a control variable. This flow actually results from operating a valve. Moreover, the feedwater valve has its own dynamics and is constrained both in amplitude and slew rate. Those extra details need to be included in the model if it is to truly represent the system.

29.13.3 Total Heat

Whilst the model described above has used the total heat Q^T as an input, it is not a directly manipulated variable. Instead, when total heat is needed to, say, increase, then in reality a chute is opened to feed more fuel into the furnace. Also, fans are required to increase airflow to burn the extra fuel. Additionally, hard constraints usually apply to the maximum and minimum fuel and airflow. Again, these factors are relevant to the development of a comprehensive model.

29.13.4 Disturbances

The steam flow q_s^E out of the steam receiver is the key process disturbance. This flow is demanded by external operational requirements. In some cases, these flows can be highly unpredictable and can cause up to 50% variations in the total load.

29.13.5 A Preliminary Simulation

To complete the model and to provide a proof of concept, a simple simulation is included showing a step change in q_s for a particular Boiler from 12.5 (kg/s) to 18.75 (kg/s). To maintain variables in a sensible range, two simple PID controllers are used, one for feedwater and one for heat flow. Figure 29.2 shows q_s, Q^T, q_f, P and δ, respectively, for the step in q_s applied at time $t = 200$ [s]. An interesting phenomenon in boilers is that pressure changes directly effect the drum water level by compressing the steam bubbles under water. This phenomenon has been captured in the model derived in this chapter. It is commonly known as the 'Shrink-Swell' effect. It often means that the drum water level initially goes in the opposite direction to the one which would be expected from simple mass flow considerations. The effect of the shrink-swell effect is clearly seen in the δ plot given in Figure 29.2.

29.14 Revision Questions

1. Continuous Stirred Tank Reactors (CSTRs) form a commonly used processing unit in chemical and biotechnology plants. The modelling principle involved in describing such processes include
   i) Conservation of mass
   ii) Conservation of momentum
   iii) Conservation of energy.

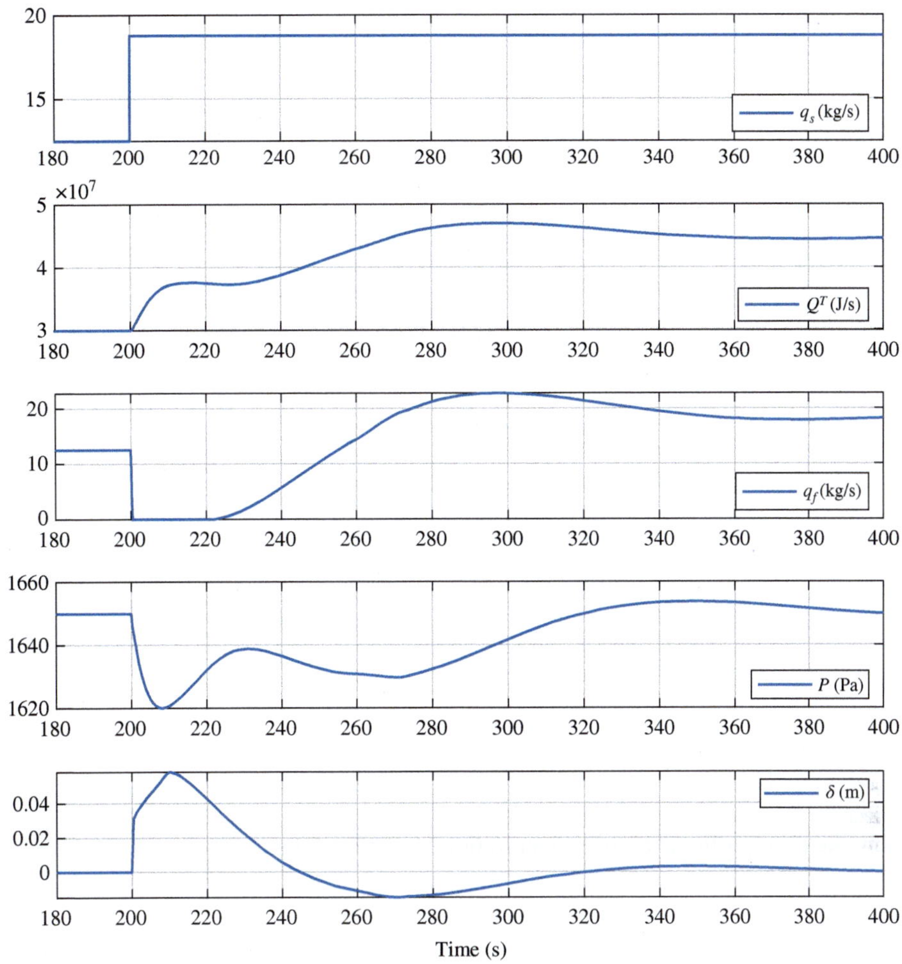

Figure 29.2 Simulation results.

A CSTR is schematically shown in Figure 29.3. It is suggested that you make several simplifying assumptions, e.g.

i) Adiabatic operation (i.e. constant temperature and no heat generated or used by the conversion process).
ii) One species entering, namely, A having entry concentration C_A and flow rate $F_i(t)$.
iii) Conversion in the reactor occurs to a new species, namely, B.
iv) The content of the reactor is well stirred so that the concentrations of material at the exit is the same as the concentration in the vessel.
v) All materials have the same density.

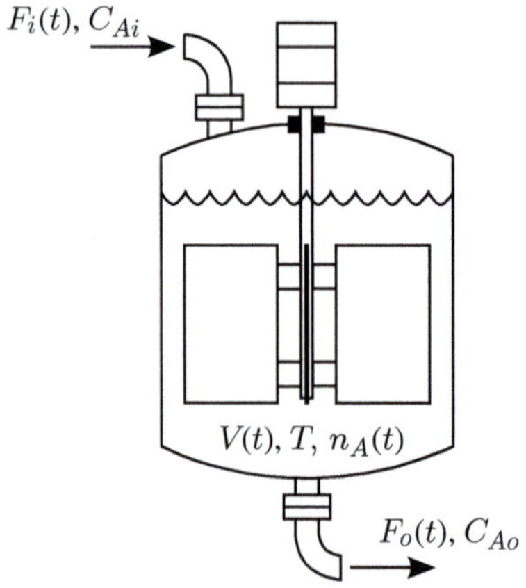

Figure 29.3 Continuously stirred tank reactor schematic.

The following notation can be used

$V(t)$	Volume of liquid in the reactor (m^3)
$\rho(t)$	Density of material (kg/m^3)
$F_i(t)$	Flow rate of material into the vessel, m^3/s
$F_o(t)$	Flow rate of material out of the vessel, m^3/s
$C_{Ai}(t)$	Concentration of A at inlet (mmol/m^3)
$C_{Ao}(t)$	Concentration of A in vessel, (mmol/m^3)
$C_{Bo}(t)$	Concentration of B in vessel, (mmol/m^3)
T	Temperature of reactor.

The reaction rate is proportional to the concentration of species A, i.e,

$$r = kC_{Ao} \quad mmol/m^3, \tag{29.75}$$

where the constant k typically takes the form: $k = k_0 \exp\left(-\frac{E}{RT}\right)$ where k_0 is the frequency factor, E is the activation energy, R is the gas constant and T is the temperature. Obtain a model for this process.

2. Consider a superheater having volume V^{SH} and a heat flow input Q^{SH}. Obtain the mass balance, energy balance and constant volume equations for the superheater.

Further Reading

1 K.J. Åström and R.D. Bell, "Drum boiler dynamics", *Automatica*, 36:363–378, 2000.

2 D.S. Carrasco, G.C. Goodwin and R.D. Peirce, "Novel modelling for a steam boiler under fast load dynamics with implications to control", *Automatica*, 156, 2023, https://doi.org/10.1016/j.automatica.2023.111184.

30

Laboratory 14: Control of a Steam Boiler

30.1 Introduction

This chapter will be concerned with the design and implementation of a control system for a boiler. This is actually a demanding problem because of the presence of core nonlinearities and strong multivariable interactions. The reader may feel this suggests that a full nonlinear multivariable design methodology is necessary. However, this approach can hide underlying fundamental design trade-offs. Thus, a more direct approach will be taken in which the problem will be divided into a set of small sub problems.

The starting point will be the nonlinear model developed in Chapter 29. The first step will be to obtain a simplified linear model to provide insight into the control design problem. Then a separate design will be carried out for

(i) drum water level regulation
(ii) boiler pressure regulation
(iii) steam receiver pressure regulation.

The control laws will then be implemented on the emulated system and tested. The system studied here is typical of complex multivariable control problems. Fortunately, these are the exceptions rather than the rule in industry. A host of complex trade-offs will be revealed in this system. These are typical of such systems. This problem has been included to conclude the book since it will take the reader out of the 'nice' realm of simple feedback control into the realm of real-world complex design trade-offs in control system design.

30.2 Extracting an Approximate Linear Model

30.2.1 Introduction

Chapter 29 described the physics behind the operation of the boiler. This model gives important physical insights into the operation of boilers. However, to better understand control system design, it is usually desirable, at least initially, to work with an approximate linear model.

There are several ways that one could extract a linear model from the nonlinear model. For example, one could use a First-Order Taylor's series expansion at an operating point.

The sequel will describe an alternative method based on sine wave testing of the nonlinear model in closed loop. The advantage of this procedure is that it can, in principle, be conducted on a real plant.

Practical Control System Design: Real World Designs implemented on Emulated Industrial Systems,
First Edition. Adrian Medioli and Graham Goodwin.

Companion website: www.wiley.com/go/medioli/practicalcontrolsystemdesign

30.2.2 Sine Wave Testing in Closed-Loop (Scalar Case)

The core idea is first described in the scalar case. The key closed-loop transfer functions are

$$\frac{y}{d} = \frac{H}{1+GC}, \tag{30.1}$$

$$\frac{u}{d} = \frac{-CH}{1+GC}, \tag{30.2}$$

$$\frac{y}{r} = \frac{GC}{1+GC}, \tag{30.3}$$

$$\frac{u}{r} = \frac{C}{1+GC}. \tag{30.4}$$

where C, G, H denote the transfer functions of the controller; the plant; and from the disturbance, d, to the output, y, respectively. Hence, if a sinusoidal signal is applied at the reference input and one measures the gain and phase of the two ratios: $\frac{y}{r}$ and $\frac{u}{r}$, then

$$\frac{y}{u} = \frac{y/r}{u/r} = G. \tag{30.5}$$

This seems very simple. Indeed, this method was used in Section 5.3.6 of Chapter 5 to obtain the model for the door-opening system.

However, it is important that the same sine wave does *not* appear in d. For example, if the sine wave actually arose from d but not in r, then

$$\frac{y}{u} = \frac{y/d}{u/d} = -\frac{1}{C}. \tag{30.6}$$

In other words, in this case, the ratio $\frac{y}{u}$ is an estimate of the inverse of the control law not the plant. This is a well-known problem with closed-loop identification.

This idea can be extended to cover the case when the same sine wave appears simultaneously on r and d. Then neither (30.5) nor (30.6) will strictly hold. One needs to account for the coupling. To illustrate, say that the connection between d and r is of the form:

$$d = \lambda r, \tag{30.7}$$

where λ is a complex number representing the amplitude and phase relationship between r and d.

Then the system described by (30.1) to (30.2) is

$$\begin{aligned} y &= \left[\frac{GC}{1+GC}\right] r + \left[\frac{H}{1+GC}\right] d, \\ &= \left[\frac{GC+H\lambda}{1+GC}\right] r, \end{aligned} \tag{30.8}$$

and

$$u = \left[\frac{C-\lambda CH}{1+GC}\right] r. \tag{30.9}$$

Hence

$$\begin{aligned} \frac{y}{u} &= \left[\frac{GC+\lambda H}{C-\lambda CH}\right] \\ &= \beta G + (1-\beta)\left[\frac{1}{C}\right], \end{aligned} \tag{30.10}$$

where $\beta = 1/(1-\lambda H)$.

In other words, the ratio $\frac{y}{u}$ will actually be a linear combination of the plant transfer function, G, and the negative inverse of the controller $(-1/C)$.

One way out of this dilemma is, if possible, to organise an experiment in which two tests are carried out at the same frequency. One then obtains values of u and d which, though coupled, are not linearly related.

Say, that the two experiments yield values of $u^{(1)}, y^{(1)}, d^{(1)}$ and $u^{(2)}, y^{(2)}, d^{(2)}$ at the same frequency. Then one has the following relationships:

$$y^{(1)} = Gu^{(1)} + Hd^{(1)} \tag{30.11}$$

$$y^{(2)} = Gu^{(2)} + Hd^{(2)}. \tag{30.12}$$

Thus, at the frequency used in the two experiments:

$$\left[y^{(1)}, y^{(2)}\right] = [G, H] \begin{bmatrix} u^{(1)} & u^{(2)} \\ d^{(2)} & d^{(2)} \end{bmatrix}. \tag{30.13}$$

The values of G and H can then be determined by solving the above two simultaneous equations. This leads to

$$[G, H] = \left[y^{(1)}, y^{(2)}\right] \begin{bmatrix} u^{(1)} & u^{(2)} \\ d^{(2)} & d^{(2)} \end{bmatrix}^{-1}, \tag{30.14}$$

where it is assumed that $u^{(1)}, u^{(2)}, d^{(1)}, d^{(2)}$ are all measured.

30.2.3 Application to the Boiler Model

The boiler model has multiple inputs. Nonetheless, the same principle as described in Section 30.2.2 can be applied, i.e. one needs to ensure that multiple experiments at the same frequency lead to non-linearly dependent multiple inputs and disturbances at each frequency.

The boiler has two inputs $[u_1, u_2]$ = [feedwater, heat], two outputs $[y_1, y_2]$ = [drum water level, pressure] and one disturbance d = [steam flow]. Here, if three independent sets of measurements are taken at each frequency then at each frequency

$$Y = \begin{bmatrix} y_1^{(1)} & y_1^{(2)} & y_1^{(3)} \\ y_2^{(1)} & y_2^{(2)} & y_2^{(3)} \end{bmatrix} = \begin{bmatrix} G_{11} & G_{12} & H_1 \\ G_{21} & G_{22} & H_2 \end{bmatrix} \begin{bmatrix} u_1^{(1)} & u_1^{(2)} & u_1^{(3)} \\ u_2^{(1)} & u_2^{(2)} & u_2^{(3)} \\ d^{(1)} & d^{(2)} & d^{(3)} \end{bmatrix}. \tag{30.15}$$

Hence,

$$\begin{bmatrix} G_{11} & G_{12} & H_1 \\ G_{21} & G_{22} & H_2 \end{bmatrix} = \begin{bmatrix} y_1^{(1)} & y_1^{(2)} & y_1^{(3)} \\ y_2^{(1)} & y_2^{(2)} & y_2^{(3)} \end{bmatrix} \begin{bmatrix} u_1^{(1)} & u_1^{(2)} & u_1^{(3)} \\ u_2^{(1)} & u_2^{(2)} & u_2^{(3)} \\ d^{(1)} & d^{(2)} & d^{(3)} \end{bmatrix}^{-1}. \tag{30.16}$$

This can be repeated at multiple frequencies, $w_i; i = 1, \dots, M$, giving the frequency response of $\{G_{11}, G_{12}, H_1, G_{21}, G_{22}, H_2\}$ at those frequencies. One can then use curve-fitting ideas to fit a rational transfer function to each frequency response. When this was done for the boiler model described in Chapter 29, then the resulting transfer functions were found to be as follows:

$$G_{11} = \frac{0.153(-s+1)}{s(s/4+1)^2}, \tag{30.17}$$

$$G_{12} = \frac{-0.0254(-s/0.2+1)}{s(s/4+1)^2(s/0.05+1)}, \tag{30.18}$$

$$G_{21} = \frac{-0.571}{s}, \tag{30.19}$$

$$G_{22} = \frac{0.03664}{s(s/0.05+1)}, \tag{30.20}$$

$$H_1 = \frac{0.05(s/0.006+1)}{s(s/4+1)^2}, \tag{30.21}$$

$$H_2 = \frac{-2.36}{s}. \tag{30.22}$$

Some observations from this linear model are as follows:

(i) The two transfer functions, G_{11}, G_{12} (linking feedwater and heat to drum water level), are both non-minimum phases since they both contain zeros in the right-half-complex plane.
(ii) All transfer functions contain an integrator. This is easily understood from the physics since, for example, increasing feedwater flow rate (u_1) will, in steady state, lead to a ramp upwards in the drum water level.
(iii) At first sight, it is surprising that H_1 (the transfer function from steam flow to drum water level) has a positive gain. However, the explanation is that when steam flow increases then the boiler pressure drops (see H_2). Moreover, when the pressure drops, then the steam bubbles in the boiler risers will expand and hence the drum water level will rise. This phenomenon is generally called the 'shrink-swell' effect.

30.2.4 The Steam Receiver

The model described above has treated q_s (the steam flow out of the boiler) as an exogenous variable. However, in reality, this flow is a result of the pressure drop from the boiler to the steam receiver and the opening of the throttle valve see Section 29.13.1.

An approximate model is

$$q_s = Ku_{tv}\left(P_B - P_{SR}\right), \tag{30.23}$$

where

u_{tv} is the value opening.
P_B is the boiler pressure.
P_{SR} is the pressure in the steam receiver.

A more refined model would use a square root of the pressure difference in (30.23). However, for the purpose of this laboratory, the simple linear relationship in (30.23) will suffice.

For the boiler used as an illustrative example, under nominal conditions, the pressure drop is 100 kPa and a valve opening of 70% gives a flow of 12.5 kg/s. This information can be used to give an estimate of K as

$$K = \frac{12.5}{70 \times 100} = 0.00178. \tag{30.24}$$

The nonlinearity described in (30.23) is fundamental to the boiler control problem and thus will be retained in the model.

The steam receiver acts as a simple pressure vessel. The pressure can be modelled approximately by

$$\frac{d}{dt}P_{SR} = k\left[q_s - q_L\right], \tag{30.25}$$

where q_L is the steam flow to the load (which is assumed to be an exogenous disturbance). (A more general model was developed for the steam receiver in Chapter 29.)

Since the steam receiver in the application under study is quite small, then k in (30.25) is quite large. A value of 200 gives responses that approximate those found in practice.

30.3 The Control Architecture

The basic control architecture comprises four separate loops, namely

(i) a PID controller regulating the steam flow out of the boiler using the throttle valve u_{tv} in (30.23).
(ii) a PID controller regulating boiler pressure by adjusting fuel flow.
(iii) a PID controller regulating boiler drum water level by adjusting the feedwater flow.
(iv) a PID controller regulating P_{SR} by changing the *desired* steam flow out of the boiler.

The sequel will describe each of the above control laws:

30.4 Regulating Steam Flow from the Boiler

Since the 'plant' is basically a nonlinear gain, then a simple integral controller will be used. The following value is suggested:

$$K_I = 5. \tag{30.26}$$

This gives a relatively fast response for the closed loop. Note, however, that the input, u, is constrained to lie in the range [0, 100]. Thus, it is important that anti-windup be implemented in the PI block.

The saturation limit on u_{tv} is particularly important when $P_{SR} - P_B$ drops since this will cause the valve to quickly go into saturation.

A typical scenario is as follows:

q_L	goes down
P_{SR}	rises (see equation (30.25))
$P_B - P_{SR}$	falls

Then it is necessary to open u_{tv} to maintain the steam flow. However, the valve will quickly hit the 100% saturation limit. Hence, q_s will actually drop below q_s^*.

30.5 Boiler Pressure Controller

It is suggested that the reader begins with a PID control law of the form:

$$PID = K_P + \frac{K_I}{s} + \frac{K_D s}{(\gamma s + 1)}, \tag{30.27}$$

where $K_P = 1.892, K_I = 0.048, K_D = 17.503, \gamma = 1.$

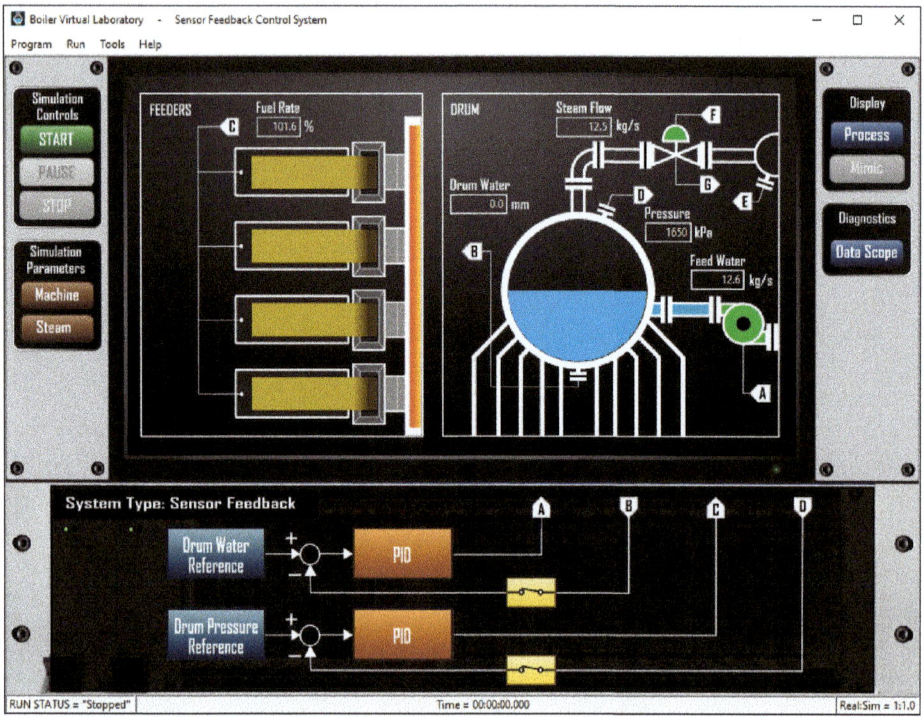

Figure 30.1 Boiler laboratory drum water and pressure controller.

It is important that anti-windup be used since the fuel flow is limited to a minimum value and maximum value.

30.6 Drum Water Level Controller

The drum water level controller is implemented as in Figure 30.1.

Obviously, the feedwater flow is constrained to lie between 0 and some maximum value (which for the boiler being used for the purposes of illustration is about 22.5 kg/s). Hence, it is very important that the PID block have anti-windup. Also, even at maximum feedwater flow, the drum water level rises very slowly. (The reader might care to think of filling a swimming pool (the boiler drum) with a garden hose.) Indeed, a quick check of the physical model suggests that it is impossible to raise (or lower) the water level in the drum by more than about 1 mm/s. Since, typical disturbances are of the order of 100 mm over a few seconds, then it can take several minutes to restore the drum water level using the feedwater pump. Hence, the PID controller for drum water level is really useful only under semi-steady state conditions.

30.6.1 Experiment: Implementing Drum Water Level Control Law

It is suggested the reader try a PI controller with $K_P = 2.5, K_I = 0.125$.

30.7 Steam Receiver Controller

The control architecture used here comprises two cascaded loops. The outer loop compares the pressure in the steam receiver with the set point for that quantity. The error is passed to a PID controller which outputs a steam flow demand for the inner loop. The inner loop compares the steam flow demand with the actual steam flow. The error is passed to a PID controller whose output regulates the boiler steam flow via the steam flow relationship given in (30.23).

The steam flow controller is used to maintain P_{SR} near set point. (Actually, the value of 1550 kPa applies to the illustrative boiler.) The basic cascaded control architecture is shown in Figure 30.2. To get started, it is suggested that the outer loop controller be chosen with the following controller gains: $K_P = 0.008$ and $K_I = 0.0008$. Also, choose the inner loop control law parameters as $K_I = 5.0$.

Note that there is a fundamental design trade-off related to the design of the associated controllers. In particular:

(i) If the steam receiver control loop is 'fast', then q_s will be rapidly regulated to achieve $q_L \simeq q_s$. This means that P_{SR} will not change by much when q_L changes. However, changing q_s rapidly will mean that the boiler pressure changes rapidly (see the transfer function H_2 in Eq. (30.22)). This, in turn, will cause P_B to change rapidly which will cause large variations in the drum water level. Moreover, as noted in Section 30.6, these changes in drum water level can only be slowly remedied via the feedwater control loop.

(ii) On the other hand, if the loop is 'slow' then q_s will change slowly when P_{SR} changes. This means that the shock to the drum pressure and water level will be relatively small but the price paid will be a significant deterioration in the regulation of P_{SR}.

This is a fundamental design trade-off.

30.7.1 Experiment: Testing Steam Receiver Control Law

It is suggested that the reader try three settings for PID as follows:

Very Fast:	$K_P = 0.036, K_I = 0.012$
Fast:	$K_P = 0.008, K_I = 0.0008$
Slow:	$K_P = 0.0027, K_I = 0.0001$

Note that all controllers need to be implemented in anti-windup form.

30.8 Experiments

30.8.1 Set Up

1. Choose the inner loop controller for the steam receiver pressure cascaded controller as in Section 30.4.
2. Choose the outer loop controller for the steam receiver pressure controller as the 'Fast' option in Section 30.7.

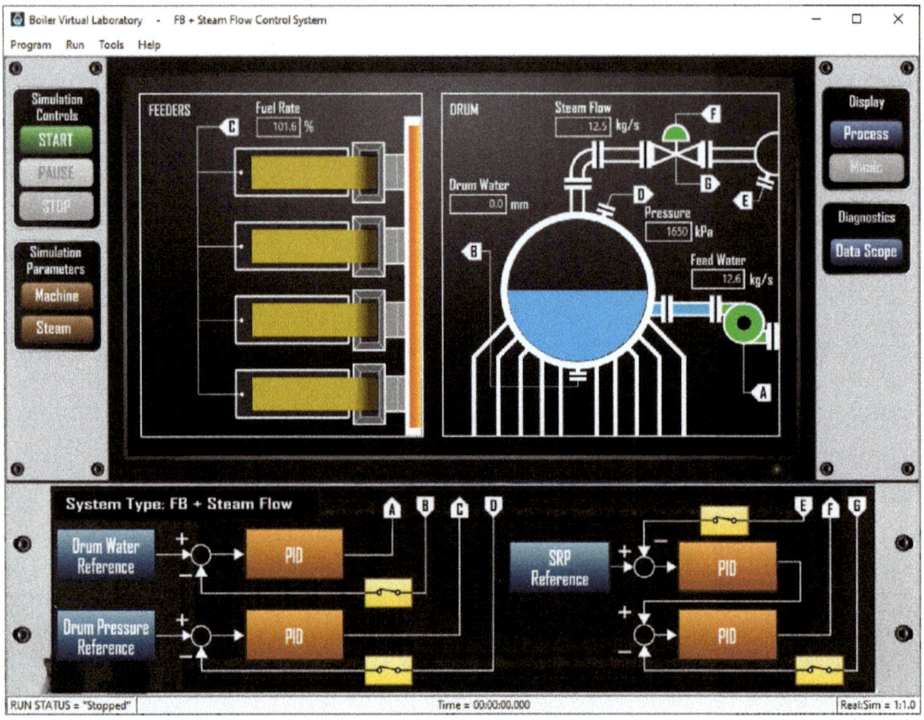

Figure 30.2 Boiler laboratory drum water, drum pressure and steam receiver controllers.

3. Choose the drum water level controller as in Section 30.6.
4. Choose the boiler drum pressure controller as in Section 30.5.
5. Set the steam load (using the simulation parameters) as value = 12.5.
6. Run the system and check that it settles.

30.8.2 Small Load Change

Make sure the only constraint is on the valve opening in the inner steam receiver loop.

Change the steam flow to 10.5 kg/s.

Comment on the response of all variables.

30.8.3 Faster Outer Loop

Change the Outer Loop Controller Parameters to the 'Very Fast' settings given in Section 30.7.

Repeat the experiment in Section 30.8.2.

Comment on the transient responses of every variable.

30.8.4 Slower Outer Loop

Change the Outer Loop Controller Parameters to the 'Slow' settings in Section 30.7.

30.8.5 Large Decrease in Load

Return the Outer Loop Controller Parameters to the 'Very Fast' settings given in Section 30.7.

Do a very large load change from 12.5 to 5 kg/s. (This size change can actually occur frequently in a Sugar Mill.)

30.8.6 Constraints

Enter the Limits on the Drum Water Level PID: Max 22.2, Min 0, Fuel PID Max 141.6, Min 20

Inner Loop Max 100, Min 0.

Repeat Section 30.8.5.

30.8.7 Large Load Change with 'Fast' Outer Loop

Repeat Section 30.8.6. However, change the Outer Loop Controller on the Steam Receiver to the 'Fast' setting.

30.8.8 Large Increase in Load

Repeat Section 30.8.7. However, use a load change of 4 to 12.5 kg/s.

The reader is warned of the presence of strong multivariable interactions, nonlinear phenomena and multiple hard constraints. Thus, the experiments outlined above may not be quite as simple as one might expect from simple heuristic arguments.

30.9 Summary

This laboratory is typical of a complex industrial control problem. The reader should gain an appreciation of the difficulties of real-world control. The designs suggested in this laboratory were implemented on a real-world boiler in the sugar industry. The final results were excellent. For example, the standard deviation in the drum water level excursions was reduced by approximately 50% compared to that achieved with the existing control laws.

30.10 Revision Questions

1. Consider a swimming pool of depth ranging from 1 m, at the shallow end, to 2 m, at the deep end. The pool is rectangular in shape (2 m wide by 4 m long)[1] .
   (a) How much water (in m^3) does the swimming pool hold?
   (b) Roughly estimate the weight of water in the swimming pool.
   (c) Say that a garden hose is able to fill a 15-litre bucket in one minute. If the garden hose is used to top-up the swimming pool, how long will it take for the level to rise 1mm?
   (d) How long would it take to fill the pool from empty?

1 This question is a much-simplified version of the drum water level control problem.

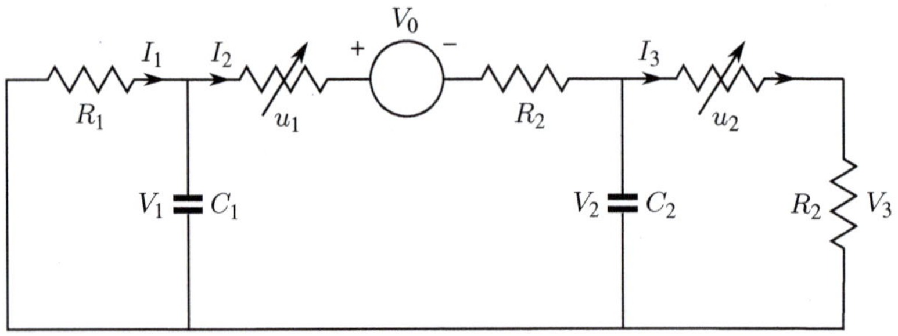

Figure 30.3 Electrical analogue of a furnace.

2. An electrical analogue of a furnace system is shown in Figure 30.3. Here, the variable resistor u_1 and u_2 represent dampers in the ducts, V_0 denotes a constant velocity pump, V_1, V_2, V_3 are the electrical analogues of pressures, I_1, I_2, I_3 are electrical analogues of air flow.
   Derive a model for this system.
3. It has been suggested that the two dampers (u_1, u_2) can be used to control V_2 and I_3. Does this make sense?
4. Would the situation described in Question 2 change if one wanted to control I_3 and V_3?
5. Consider the Steam Receiver described in Question 2. What could be the advantage in terms of drum water level control of increasing the size of the steam receiver?

Further Reading

1 K.J. Åström and R.D. Bell, "Drum boiler dynamics", *Automatica*, 36:363–378, 2000.
2 D.S. Carrasco, G.C. Goodwin, and R.D. Peirce, "Novel modelling for a steam boiler under fast load dynamics with implications to control", *Automatica*, 156, 2023. https://doi.org/10.1016/j.automatica.2023.111184.

Index

Practical Control System Design: Real World Designs implemented on Emulated Industrial Systems,
First Edition. Adrian Medioli and Graham Goodwin.

Companion website: www.wiley.com/go/medioli/practicalcontrolsystemdesign